ROOT DRUG PLANTS OF ETHNIC INDIA

ROOT DRUG PLANTS OF ETHNIC INDIA

S.K. SOOD
SUDERSHNA

2008

SATISH SERIAL PUBLISHING HOUSE
403, Express Tower, Commercial Complex, Azadpur, Delhi-110033 (India)
Phone : 011-27672469, 27672852, Fax : 91-11-27672046
E-mail :info@satishserial.com, hkjain1975@yahoo.com
Website : www.satishserial.com

Published by :

SATISH SERIAL PUBLISHING HOUSE
403, Express Tower, Commercial Complex, Azadpur, Delhi - 110033 (INDIA)
Phone : 011-27672852 Fax : 91-11-27672046
E-mail : info@satishserial.com, hkjain1975@yahoo.com

ISBN 81-89304-57-7

Printed at
SALASAR IMAGING SYSTEMS, DELHI-35

Preface

India has a pioneer role from time immemorial in utilizing plants in the preparation of drugs of indigenous systems of medicine. As many as 2500 plant species are used as medicinal plants in India, and the supply base of 90% of herbal raw drugs (medicinal plants) used in the manufacture of Ayurveda, Unani and Homeopathic systems of medicine is largely met from the wild natural source. Recognition of traditional systems of medicine owing to their low cost, easy availability, time-tested efficacy and with no side effects, warrants their development and promotion. According to a WHO estimate more than 80 % of the world's population relies on traditional medicinal practices for primary health care needs. From the Indian perspective, considerable work has been done in the past to strengthen the medicinal plants resource base in the context of primary health care and for meeting the increasing demand of the pharmaceutical industry. In spite of a plethora of literature available on the subject, the need for a systematic compilation of data on root drug plants of ethnic India in particular has been long-felt by almost all researchers engaged in phytomedicine, pharmacology, ethnobotany, economic botany, chemistry etc. as the data available are widely scattered in various books and journals and require enormous efforts to be pooled at one place. An earnest attempt has, therefore, been made in this direction and it is hoped that the compilation will narrow the gap between the followers of traditional medicines and scientific researchers of phytomedicine, as well may help to find out many new leads for alleviating human suffering.

This venture could not have been possible without the valuable advice and incessant encouragement of Dr Anil Wilson, Vice-Chancellor, Himachal Pradesh University, Shimla, for which we express our sincere gratitude to him. Well deserved appreciation is due to Prof. T.N. Lakhanpal, Prof. D.C. Kalia , Prof. V.K. Mattu and other colleagues of Department of Biosciences for their co-operation and well wishes; to Mr Suresh, Ms Anjna and Mr Munish for their ungrudging help in critically reading the various drafts of this manuscript, and to Mr Bal Krishan for his help in various ways. Thanks are also due to authors of books, papers and figures listed in the compendium. Finally, we would like to thank the members of our families for moral support and unflinching patience during the course of the work.

S.K. SOOD
SUDERSHNA
Department of Biosciences
H.P. University
Shimla 171 005

Abbreviations

OAC.	Acetoxy
Alc.	Alcoholic
α	Alpha
&	And
A.P.	Andhra Pradesh
Auct. Non	Auctorum non (Not of author (s))
β	Beta
B.P.	Blood Pressure
Ca^{2+}	Calcium ion
cm	Centimeter
C.	Central
CNS	Central Nervous System
°C	Degree Celsius
Δ	Delta
δ	Delta
E.	East/Eastern
Eng.	English
et	And
et al.	*et alii* (and others)
etc.	*et cetera* (so on)
EtOH	Ethanol
Extr.	Extract
Extrs.	Extracts
γ	Gamma
g	Gram
H.P.	Himachal Pradesh
hrs.	Hours
ssp.	Subspecies
Syn.	Synonym
U.P.	Uttar Pradesh
var.	Variety
I.U.	International Unit
in vitro	In glass
in vivo	In living tissue
J. & K.	Jammu and Kashmir
i.e.	*id est* (that is)
Kg	Kilogram
Me	Methyl
M.P.	Madhya Pradesh
Mg	Magnesium
m	Meter
ìm	Micrometer
mg	Milligram
ml	Millilitre
N	Nitrogen
N.	North/Northern
N.E.	North-East
N.W.	North-West
O	Oxygen
OH	Hydroxy
%	Per cent
Pl. (s)	Plate (s)
ψ	Psi
Sans.	Sanskrit
S.	South/Southern
S.E.	South-East
S.W.	South-West
sp.	Species
spp.	Species
vit.	Vitamin
Viz.	*Vide licet* (namely)
Vol. (s)	Volume (s)
W.	West/Western

Contents

1

Plants Based Medicines

Knowledge of treating dis1eases using plants is traced back to ancient times. Nearly all cultures from ancient times to the present day have used plants as a source of medicines. India's use of plants for medicinal treatment dates back over 5,000 years and has become codified in the Ayurveda, which contains over 8,000 herbal remedies. As indicated by Rubin (1960) "by necessity man has undoubtedly always been concerned with the question of health and survival and has sought within the framework of his knowledge, solutions to problems to illness." It is estimated that around 70, 000 species of higher plants are used as medicines in various cultures of the world. Also, it is a fact undeniable that indigenous communities living in biodiversity rich areas possess a wealth of knowledge on the utilization and conservation of plants. There are an estimated 300 million tribal people all over the world. In India alone, there are over 53.8 million tribal people under 427 communities inhabiting more than 5,000 villages in different geographical and climatic zones, and overall accounts for almost 7 per cent of its total population (Ragupathy & Mahadevan, 1991). There is a great wealth of knowledge on tribal medicines available in India, as over 7,500 plants are used in traditional, folk and herbal medicine, representing about 75 per cent of the medical needs of The Third World (Huxley, 1984). In India, about 1.5 million practitioners of Indian Systems of Medicines use around 25,000 effective plant based formulations (Mehrotra & Mehrotra, 2005). Even today 2/3rd of the world population relies on plant-derived drugs as these are relatively inexpensive, safe to use, and are usually readily accepted by the local populace (Prescott-Allen, 1982). This is why Alma-Ata Conference in 1978 recommended that governments should give high priority to the utilization of traditional medicine and the incorporation of proven traditional remedies into national drug policies and regulations. All over the world, the tribal traditions are fast disappearing due to urbanization, rapid industrialization and changes in sustenance economy.

It is clear that plants can be useful in their crude or advanced forms as drugs having curative properties attributed to the presence of various active principles, which are found as secondary plant metabolites and grouped according to their composition as alkaloids, glycosides, corticosteroids, essential oils, etc. (Waller & Nowacki, 1978). Only about 15 per cent or so of all angiospermous species on earth have been examined chemically and pharmacologically, in detail or superficially, for their medical potential (Farnsworth & Soejarto, 1991). With in India out of 7, 500 species of medicinal plants used by the indigenous tribal communities, about 500 species are used in the Indian market yielding good revenues. There are some 0.4 million registered practitioners in traditional medicine supported by over 5,000 licensed manufacturing units, producing herbal drugs worth Rs. 3,400 million every year (Gupta, 1986). Conservative estimates put the economic value of medicinal plant-related trade in India to be of the order of Rs. 1,000 crores per year, and world trade in medicinal plants to be over US $ 60 billion (Principe, 1991). Available statistical data (Farnsworth & Soejarto, 1991) have revealed that the production and sharing of Indian medicinal plants and their products in the international market is negligible. However, the importance of ethnobotanical inquiry as a cost-effective means of locating new and useful tropical plant compounds cannot be emphasized. Most of the secondary plant compounds employed in modern medicine were first "discovered" through ethnobotanical investigation. There are some 119 pure chemical substances extracted from higher plants that are used in medicine throughout the world and 74 per cent of these compounds have the same or related use as the plants from which they were derived (Farnsworth, 1998). Evidently, the plant drugs serve as chemical blueprints and starting materials for the commercial production of related synthetic substitutes for which now many pharmaceutical companies all over the world are eagerly looking for (Sinha & Sinha, 2001).

❑❑❑

2

Present Status of Work on Phyto-Medicines in India-An Overview

India has a long history of plant based medicines. The earliest credit to document this knowledge goes to Jones (1799) who recorded their use in his book entitled '*Botanical Observations on Select Plants*' followed by other valuable contributions, viz., *Catalogue of Medicinal Plants* (Fleming, 1810), *Materia Medica of Hindoostan* (Ainslie 1813, 1826), *Flora Indica* (Roxburgh, 1832), *An Essay on the Antiquity of Hindu Medicine* (Royle, 1837) and *The Bengal Dispensatory* (the first ever book dealing exclusively with the properties and use of medicinal plants-O'Shaugnessy, 1841). Thereafter, the other valuable works are that of Dutt (1877)-*Materia Medica of the Hindus*, Dymock (1883)-*Vegetable Materia Medica of Western India*, Khory (1887)-*The Bombay Materia Medica and Their Therapeutics*, Dey (1896)-*Indigenous Drugs of India* while *Pharmacographia Indica* (Dymock *et al.*, 1889-1893) and *Dictionary of Economic Products of India* (Watt, 1889-1896) are the most comprehensive contributions of the 19th century. Besides, the other works which added to the wealth of literature on medicinal plants pertain to compilations, like *Indian Materia Medica* (Nadkarni, 1926), *Medicinal and Economic Aspects of Some Indian Medicinal Plants* (Chopra, 1932), *Pharmacognosy of Indian Root and Rhizome Drugs* (Datta, 1949), *Pharmacognosy of Indian Leaf Drugs* (Datta & Mukerji, 1952), *Review of Work on Indian Medicinal Plants* (Chopra & Chopra, 1955), *Medicinal Plants of India* (Satyavati *et al.*, 1976), *Herbal Drugs in Indian Pharmaceutical Industry* (Kapoor, 1979), *Marketing Directory of Medicinal Plants* (Kumar, 1979), *Cultivation and Utilisation of Medicinal Plants* (Atal, 1982), *Drug Plants of Indian:*

Root Drugs (Agarwal, 1985), *Bibliography on Pharmacognosy of Medicinal Plants* (Mitra, 1985), *Medicinal Chemistry Research in India* (Singh, 1985), *Current Research on Medicinal Plants* (Dhawan, 1986), *Drug Plant Sources of Central Indian* (Srivastava, 1989), *Medicinal Plants of India* (Jain & De Filipps, 1991), *Recent Advances in Medicinal, Aromatic and Spice Crops* (Raychaudhuri, 1991), *Glossary of Indian Medicinal Plants with Active Principles* (Asolkar *et al.*, 1992), *History of Indian Medicine* (Mukhopadhyaya, 1992), *Production Technology of Medicinal and Aromatic Crops* (Farooqi, 1993), *Tropical Indian Medicinal Plants Propagation Methods* (Ved, 2000), and *Cultivation Practices of Some Commercially Important Medicinal Plants* (Rawat, 2002). Not only this, a large number of medicinal flora of different States of the country, namely, Arid zone (Chopra, 1960), A.P. (Raghunathan, 1976; Anonymous, 1999), Garhwal Himalayas (Uniyal, 1989), Gwalior (Singh, 1984), H.P. (Chauhan, 1999), Kachchh-Gujarat (Goyal, 1999), Kerala (Anonymous, 1981; Nambiar, 1985), M.P. (Tiwari, 1998), Nilgiris (Krishnamurthy, 1953), North East India (Kumar, 2002), Patalkot (Saxena, 1971), Saharanpur (Ahuja, 1965), Sikkim Himalayas (Biswas & Chopra, 1982; Rai, 1994), S. India (Sundararaj, 1959), Uttarankhand Himalaya (Uniyal, 1997) and research publications including ethnobotanical ones (*see also* Jain *et al.* 1984; Jain 2002) have also enriched our knowledge on the subject. Despite these efforts, no systematic compilation on root drug plants of ethnic India has appeared so far. The present work is an attempt in this direction, and it is hoped that this attempt though belated, will fill the gap adequately.

❑❑❑

3

Enumeration of Root Drug Plants

Collection of Data

The present comprehensive manual on root drug plants of India was largely compiled and collated from scrutiny of outdated classical literature, viz., *Vegetable Materia Medica of Western India* (Dymock, 1883), *A Dictionary of Economic Products* (Watt, 1889-1896), *Indigenous Drugs of India* (Dey, 1896), *Materia Medica of Indian and their Therapeutics* (Katark, 1903), *Vegetable Drugs of India* (Sanyal, 1924), *Materia Indica* (Ainslie, 1926), *Indian Materia Medica* (Nadkarni, 1926), *Commercial Drugs of India* (Dutt, 1928), *Vegetable Drugs of India* (Robert, 1932), *Pharmacopoeia Indica* (Bose, 1932), *Indigenous Drugs of India* (Chopra, 1933), *Indian Medicinal Plants* (Kirtikar & Basu, 1935), *The Wealth of India: A Dictionary of Indian Raw Materials and Industrial Products* (Anonymous, 1948-1976), *Pharmacognosy of Indian Root and Rhizome Drugs* (Datta, 1949), *Glossary of Indian Medicinal Plants* (Chopra *et al.*, 1956), *Medicinal Plants of the Arid Zone* (Chopra, 1960), *Supplement to Glossary of Indian Medicinal Plants* (Chopra *et al.*, 1969), *Dictionary of Economic Plants of India* (Maheshwari & Singh, 1965), *Medicinal Plants of India and Pakistan* (Dastur, 1970); scarcely available standard works (Atkinson, 1982; Singh *et al.*, 1983; Dutta, 1985; Agarwal, 1986; Ambasta, 1986; Singh 1996; Uphof, 1986; Nayar *et al.*, 1989; Thakur *et al.*, 1989; Jain, 1991; Jain & De Filipps, 1991; Jain *et al.*, 1991; Chatterjee & Pakrashi, 1991-1995; Ambasta *et al.* 1992; Hussain, 1992; Saklani & Jain, 1994; Warrier *et al.*, 1994-1996; Pandey, 1995; Maheshwari, 1996; Rao & Henry, 1996; Varghese, 1996; Kaul, 1997, Sharma

& Sood, 1997, Vedavathy *et al.*, 1997; Rawat & Chowdhury, 1998; Singh & Pandey, 1998; Chauhan, 1999; Kumar, 1999; Joshi, 2000; Maheshwari, 2000; Bhattacharjee, 2001; Sharma *et al.*, 2001; Sood *et al.*, 2001; Kumar, 2002; Caius, 2003, Sood & Thakur, 2004; Sood *et al.*, 2005 a-c, Prajapati *et al.*, 2006); and a large number of research papers published in various journals on economic botany and ethnobotany. The methodology as outlined by Schultes (1960), Jain (1987, 1989) and Jain & Mudgal (1999) was adopted for the collection of ethnobotanical literature. However, the information on English, Hindi and Sanskrit names, distribution, active constituents and biological activity was largely based upon Hooker (1872-1897), Rastogi & Mehrotra (1998, 1999a, b, 2001, 2002) and Asolkar *et al.* (1992). Wherever possible data gathered from own field ethnobotanical investigations was also included.

Presentation and Arrangement of Data

Altogether systematic enumeration of 917 species of root drug plants has been presented alphabetically with updated botanical names (nomenclature based on Bennet, 1986) and synonyms (wherever necessary), followed by name of the family, English, Hindi and Sanskrit names, distribution, part(s) used, utilization, active constituents, biological activity and relevant references. 36 photographs, 3 histograms and 3 tables have been also included to explain the textual matter. Appendices include information on total number of genera and species of root drug plants under various divisions of plant kingdom (Appendix-I), families of root drug plants (Appendix-II), predominant Indian genera of root drug plants (Appendix-III), predominant Indian root drug families with three or more species (Appendix-IV), index to Botanical names (Appendix-V), index to English names (Appendix-VI), index to Hindi names (Appendix-VII) and index to Sanskrit names (Appendix-VIII).

❑❑❑

4

Root Drug Plants of Ethnic India

Abelmoschus crinitus Wall.

Syn: *A. racemosus* Wall.; *Bamia cancellata* Wall.; *B. fusiformis* Wall.; *Hibiscus cancellatus* Roxb.; *H. crinitus* (Wall.) G. Don
Family: Malvaceae.

Distribution : N.W. Himalaya (Kumaon, Garhwal). Sub-Himalayan Tract from Kashmir Eastwards.

Part Used : Root.

Utilization : Root paste heals fresh cuts. Aq. extr. of root (2 teaspoons, 2 times a day for 2 days) given for the treatment of fever accompanying child birth. Decoction given to cure diarrhoea and dysentery.

References : Ambasta (1986), Singh (1996), Singh & Prakash (1994), Sood & Prakash (2007), Sood & Thakur (2004).

Abelmoschus esculentus W. & A.

Syn: *Hibiscus esculentus* L.; *H. longifolius* Roxb.
Family: Malvaceae.

English, Hindi & Sanskrit Names:

Eng. : Gumbo, Lady's Finger, Okra;
Hindi : Bhindi, Katavandai, Ramturai;
Sans. : Bhenda, Darivka, Dindish, Gandhamula, Tindisa.

Distribution : Cultivated as a Garden Crop or Mixed Field Crop in India Upto 1,200 m & Elsewhere in the Tropics & Sub-Tropics.

Part Used : Root.

Utilization : About 5-10 g powdered root given with water or cow milk twice a day for a month to cure tuberculosis and burning sensation during urination. Paste of the root applied on wound for quick healing. Also, root powder taken with sugar orally to check undesirable discharge of sperm with urine and to increase sexual potentiality. Its decoction recommended for inflamed mucous membrane of the urethra and vagina.

Active Constituents : Roots rich in okra-mucilage R.

References : Ambasta (1986), Hooker (1872), Jadhav (2006), Khanna (2002a, b), Kurian (1995), Parrotta (2001), Rastogi & Mehrotra (2002), Saini (1996a), Singh & Pandey (1996, 1998).

Abelmoschus manihot (L.) Medic.

Syn: *Hibiscus manihot* L.
Family: Malvaceae.

Distribution : Often Naturalized in W. Bengal & S.W. India from Gujarat to Kerala.

Part Used : Root.

Utilization : Root paste applied for healing cuts and wounds. Its juice with a cup of warm water given in pneumonia.

Active Constituents : Roots contain 'Abelmoschus-mucilage M' which comprised of 82 % polysaccharide and 17% protein.

References : Asolkar *et al.* (1992), Saini (1996b), Sharma & Singh (1999), Sinha & Sinha (2001), Parrotta (2001).

Abelmoschus moschatus Medic. (Pl. 1A)

Syn: *Hibiscus abelmoschus* L.; *H. chinensis* Wall.; *H. flavescens* Cav.; *H. sagittifolius* Kurz.; *H. spathaceus* Wall.; *H. ricinifolius* Wall.

Family: Malvaceae.

English, Hindi & Sanskrit Names:

Eng. : Ambrette Plant, Musk Mallow, Muskseed;
Hindi : Kasturidana, Mask Dana;
Sans. : Latakasturika, Latakasturikam.

Distribution : Hilly Regions Throughout Peninsular India & Himalayan Foot Hills; Cultivated in Maharashtra, U.P & H.P Upto 2,000 m.

Parts Used : Root. Root bark.

Utilization : Roots considered useful in gonorrhoea, pain and swelling of the testicles and venereal diseases. Roots along with those of *Byttneria pillosa* Roxb. are made into paste to apply externally for treating chronic ulcers and rheumatism. One teaspoonful root paste taken with a cup of warm milk twice a day for two weeks in haematuria, gout and rheumatism. Extr. of root and root bark given orally to the patient suffering from dysentery, diarrhoea and abdominal pain.

References : Chatterjee & Pakrashi (1997), Lalramanghinglova & Jha (1997), Parrotta (2001), Reddi *et al.* (2005), Singh & Pandey (1998), Sood & Prakash (2007), Tirkey (2004), Varghese (1996).

Abroma augusta Linn. (Pl. 1B)

Syn: *A. angulata* Lam.; *A. fastuosum* Gaertn .f.

Family: Sterculiaceae.

English, Hindi & Sanskrit Names:

Eng. : Devil's Cotton, Perennial Indian Hemp;
Hindi : Ulatkambal;
Sans. : Peevaree, Ultkambal, Yoshinee.

Distribution : Wild & Cultivated in U. P. & E. Parts of India.

Parts Used : Root. Root bark.

Utilization : One glassful decoction of root taken twice a day for a month to treat tuberculosis and menstrual disorders. Crushed roots (30 g) boiled with seed oil of *Madhuca longifolia* are massaged on affected organs twice daily for 7-8 days to treat pain in backbone. Roots considered diuretic and uterine tonic and used in dysmenorrhoea and amenorrhoea. Paste of the root also used internally and externally to cure abscess. The root bark has long been used in Bengal as an emmenagogue for congestive and nervous dysmenorrhoea and also, for sores in Lakhimpur (Assam).

Active Constituents : These are cholin, betaine, β-sitosterol, stigmasterol friedeline, abromine, abromasterol A (root), maslinic acid α-amyrin, proto catechuic, vanillic and affeic acids, rhamnose, arabinose, xylose, mannose, galactose, glucose, galactouronic acid and glucouronic acid (root bark).

Biological Activity : Roots exhibit anti-implantation, abortifacient, depressant and spasmogenic actions.

References : Ambasta (1986), Anonymous (1948, 1968), Caius (2003), Chatterjee & Pakrashi (1997), Chopra *et al.* (1956, 1969), Dastur (1970), Dey (1973), Dwivedi (1993), Hooker (1872), Kayal *et al.* (2003), Khare (2004), Kirtikar & Basu (1935), Rastogi & Mehrotra (1998, 1999b), Rawat & Chowdhury (1998), Thakur *et al.* (1989), Tirkey (2004), Tiwari & Tiwari (1996), Watt (1889).

Abrus fruticulosus Wall.

Syn: *A. laevigatus* E.Mey.; *A. pulchellus* Wall.
Family: Fabaceae.

Distribution : Throughout India; Tropical Region.

Part Used : Root.

Utilization : Root juice applied on the teeth to treat dental caries, tooth decay and toothache.

References : Ambasta (1986), Asolkar *et al.* (1992), Hooker (1879), Saikia & Nath (2003).

Abrus precatorius L. (Pl. 1C)

Syn: *Glycine abrus* L.

Family: Fabaceae.

English, Hindi & Sanskrit Names:

Eng. : Abrus, Crab's Eye, Indian Licorice, Rosary Pea;

Hindi : Chirmiti, Ganchi, Ghungchi, Gumchi, Guncha, Gunj, Karjini, Pondkawat, Rati;

Sans. : Gunja, Kakananti, Kakanantika, Krishnala, Raktika, Tulabija.

Distribution : Common Upto 1,000 m in the Semi-Arid Tropical Dry Deciduous Areas of the Country.

Parts Used : Root. Root bark.

Utilization : Roots considered emetic, diuretic, tonic, alexiteric and used in the preparation for night blindness, leucorrhoea, hyperdiuresis, gonorrhoea, jaundice, haemoglobinuric bile, painful swellings, wounds, ulcer in cattle and as a sweat demulcent. Dry powder of the root used as contraceptive, and its decoction taken to increase sexual vigour in man. Extr. of the root given on empty stomach in cases of chronic cough, sore throat, rheumatism (for a week), dysentery, diarrhoea, retention of urine and for the treatment of syphilis (2 teaspoons, once a day for 3-4 days). Also the root extr. together with the root extr. of *Asparagus racemosus* and plant of *Cuscuta reflexa* given (1 teaspoon, thrice a day) after menstruation to check conception. Paste of the root applied to cure alti (reddish, small boil-like structure). On mixing with cane sugar (*Saccharum officinarum*) and water it is administered orally for peptic ulcer by 'Kolams'-2-3 spoonfuls till cure. Decoction of the root also given in cough, cold, colic pain and urinary disorders twice a day for 7 days. Decoction prepared from fresh roots (50g, three times daily) useful against abortion. Oil prepared with root paste good for paralysis. Root bark considered a good uterine tonic and found useful in neuralgic dysmenorrhoea.

Active Constituents : Roots contain oleanic acid, glycyrrhizin, precol, abrol, abrasine, precasine, iso-flavoquinones I, II and III, galactose, arabinose and xylose.

Biological Activity : Alc. extr. of the roots caused a transient fall in blood pressure of anaesthetized rats and dogs. Root extr. showed antifertility activity in rats.

References : Agarwal (2003), Ambasta (1986), Anonymous (1948, 1968), Asolkar *et al.* (1992), Chatterjee & Pakrashi (1997), Chopra *et al.* (1956), Dastur (1970), Deshmukh *et al.* (2000), Dey (1973), Drury (1978), Dwivedi (1993), Gogoi & Das (2003), Islam & Jha (2003), Jadeja (1999), Kumar & Chauhan (2005), Maheshwari *et al.* (1997), Meena *et al.* (2003), Parrota (2001), Rao & Henry (1996), Rastogi & Mehrotra (1999b, 2001), Ravisankar & Henry (1992), Reddy *et al.* (2003), Sarkar *et al.* (2000), Sen & Behera (2003), Sharma (2003), Singh (1996, 2002), Singh & Pandey (1998), Singh & Prakash (1994, 1996), Thakur *et al.* (1989), Varghese (1996), Watt (1889).

Abutilon indicum (L.) Sweet

Syn: *Abutilon asiaticum* (L.) Sweet; *A. guineense* (Schum.) Baker.f. & Exell; *Sida guineensis* Schum.

Family: Malvaceae.

English, Hindi & Sanskrit Names:

Eng. : Chinese Bell-Flower, Country Mallow, Flowering Maples;

Hindi : Jhampi, Kandhi, Kanghi, Kangli, Potari, Tepari, Tutri;

Sans. : Atibala, Balika, Balya, Chanta Rishiprokta, Shila, Shitapurshpa, Vikankala.

Distribution : Widely Distributed Throughout the Tropics & Sub-Tropics; Cultivated and also Occurring Naturally as a Common Weed of Roadsides & Degraded Lands Upto 1,200 m.

Part Used : Root.

Utilization : Root given for neurological disorders like hemiplegia, facial paralysis, sciatica, debility, piles, leucoderma, stones in bladder, eye diseases, painful child birth, paralysis, cough and as demulcent and diuretic. Its decoction alleviates all types of dysuria. Root paste applied externally in leprosy and skin diseases. An infusion of the roots considered useful for relieving fever and also given in strangury and haematuria. About 10 g root powder given with water twice a day for 3 months to cure tuberculosis, cough and epilepsis. Decoction of about 10 g root crushed with 2-3 black pepper given with 200 ml milk in case of dog bite. A decoction (200 ml) prepared by boiling the roots (25 g) in water mixed with sugar candy (10 g) and cow milk (100 ml) or curd is prescribed once a day in burning sensation of urinary tract and bacterial dysentery. Also, 150 g fresh crushed root boiled with 200 ml 'sarson' (Mustard oil) and massaged on an affected part (twice a day for 4-5 days) in case of gout.

Biological Activity : Fixed oil from roots showed significant analgesic activity.

References : Agarwal (2003), Alagesaboopathi *et al.* (2000), Anonymous (1948), Asolkar (1992), Caius (2003), Chatterjee & Pakrashi (1997), Chopra *et al.* (1956), Dastur (1970), Drury (1978), Dutta & Nath (2000), Dwivedi (1996), Girach & Aminuddin (1995), Khanna *et al.* (1996), Khare (2004), Painuli & Maheshwari (1996), Parrotta (2001), Rastogi & Mehrotra (2002), Saini (1996a), Tirkey (2004), Varghese (1996).

Acacia catechu (Roxb.) Willd. (Pl. 1D)

Syn: *A. catechuoides* Wall.; *A. polycantha* Willd.; *A. sundra* Bedd., *A. wallichiana* DC.; *Mimosa catechu* Roxb.; *M. catechuoides* Roxb.

Family: Fabaceae.

English, Hindi & Sanskrit Names:

Eng. : Black Catechu, Cutch Tree;

Hindi : Alagjaie, Katha, Khair, Khair-Babul, Khaivar, Khayer, Sandramand;

Sans. : Balapatra, Dantadhavan, Gayatri, Khadira, Khadirah.

Distribution : Open Grassland & Teak Forests in M.P. & Dry Open Thorn Forests in Maharashtra & Gujarat.

Part Used : Root.

Utilization : Root paste applied for 7 days on affected parts in rheumatism.

References : Hooker (1879), Parrotta (2001), Samvatsar & Diwanji (2000), Watt (1889).

Acacia chundra (Roxb. *ex* Rottl.) Willd.

Syn : *Acacia catechu* auct. non Willd.; *A. catechu* Willd. var. *sundra* Prain.; *A. sundra* (Roxb.) DC.; *Mimosa chundra* Roxb. *ex* Rottl.; *M. sundra* Roxb.

Family : Fabaceae.

English, Hindi & Sanskrit Names:

Eng. : Red Kutch;

Hindi : Kath;

Sans. : Khadira.

Distribution : Scrub Forests of W. & S. India in the States of Rajasthan, Gujarat, Maharashatra, Karnataka, A.P. & Tamil Nadu on dry rocky soils.

Part Used : Root.

Utilization : Root paste (50 g) administered orally as an antiseptic after delivery. Its paste applied externally to cure headache and rheumatism.

References : Dey (1973), Hosagoudar & Henry (1996a), Islam & Jha (2003), Parrotta (2001).

Acacia pennata (L.) Willd.

Syn : *A. megaladena* Desr.; *A. paludosa* Miq.; *A. polycephala* Graham; *A. pluricapitata* Steud.; *A. tomentella* Zippel *ex* Spangoghe; *Mimosa ferruginea* Rottl.; *M. pennata* L.; *M. torta* Roxb.

Family: Fabaceae.

English, Hindi & Sanskrit Names:

Eng. : Rusty Mimosa;
Hindi : Agla-Bel, Aila, Biswul, Shembi;
Sans. : Ari, Vallikhadira.

Distribution : Dry & Moist Deciduous Areas Upto 1,500 m in the C. & E. Himalaya & E. Slopes of the W. Ghats.

Parts Used : Root. Root bark.

Utilization : Root considered useful in menstrual disorders and digestive complaints. 15 g of its powder mixed with honey given orally thrice a day to cure rickets. Root bark extr. along with jaggery (*Saccharum officinarum*) administered (3 spoonfuls, twice a day for 3 days) for fever. Paste of the root bark with turmeric (*Curcuma longa*) applied on ulcerous wounds.

References : Agarwal (2003), Hooker (1879), Jha & Verma (1996), Parrotta (2001), Rao & Henary (1996), Reddi *et al.* (2005), Varghese (1996).

Acalypha fruticosa Forssk.

Syn: *A. amentaceae* Roxb.; *A. betulina* Thw.; *A. capitata* Wall.
Family: Euphorbiaceae.

English & Hindi Names:

Eng. : Birch Leaved Acalypha;
Hindi : Chinni, Chinnikajhar.

Distribution : Weed in Gardens & Cultivated Fields of S. India from Orissa to Tamil Nadu, Karnataka & Kerala.

Part Used : Root.

Utilization : Reportedly used to treat gonorrhoea.

References : Asolkar *et al.* (1992), Hooker (1890), Parrotta (2001).

Acalypha indica Linn.

Syn: *A. canescens* Wall.; *A. ciliata* Wall.; *A. spicata* Forssk.
Family: Euphorbiaceae.

English, Hindi & Sanskrit Names:

Eng. : Chenilie Plant, Indian Acalypha, Red Hot Cat-Tail;

Hindi : Khokali, Kupikhokli, Kuppi, Kuppu;
Sans. : Harita-Manjri, Muktavarchas, Rudra.

Distribution : Throughout the Plains of India as an ornamental plant.

Part Used : Root.

Utilization : Considered emetic, cathartic, expectorant and nauseant. Its powder administered along with cow milk or water at bed time to get rid of constipation. Extr. of root and black pepper (*Piper nigrum*) given orally as an antiemetic.

Active Constituents : Root comprises of stigmasterol.

Biological Activity : Alc. extr. of roots showed antimicrobial activity.

References : Agarwal (2003), Ambasta (1986), Anonymous (1968), Beg *et al.* (2005), Chatterjee & Pakrashi (1997), Chopra *et al.* (1956), Dastur (1970), Dey (1973), Drury (1978), Hooker (1890), Jain (1968), Khare (2004), Rao & Henry (1996), Rastogi & Mehrotra (2002), Singh & Pandey (1998), Watt (1889).

Acanthospermum hispidum DC.

Syn: *A. humile* auct. non DC.
Family: Asteraceae.

Distribution : Moist, Disturbed Habitats Throughout Most of the Country.

Part Used : Root.

Utilization : Fresh root juice useful against jaundice.

References : Parrotta (2001), Reddi *et al.* (2005).

Acanthus ilicifolius L.

Family: Acanthaceae

English, Hindi & Sanskrit Names:

Eng. : Sea Holly;
Hindi : Hargoza, Harikusha;
Sans. : Harkusa.

Distribution : Common Along the Ditches, River Banks & Sea Coast in Different Parts of India.

Part Used : Root.

Utilization : Useful in cough, asthma, leucorrhoea, general debility and paralysis.

Active Constituents : Roots contain acanthicine, benzoxazoline -2-one, trigonelline, octacosyl alcohol,[α–L-arabinofuranosyl -(1→4)-β–D-glucuronopysansosyl (1→3)]-3 β-hydroxy-lup-20(29)–ene, stigmasterol and its â–D-glucopyroanoside, lupeol, oleanolic acid, quercetin, and octacosanol.

References : Agarwal (2003), Ambasta (1986), Chatterjee & Pakrashi (1997), Rastogi & Mehrotra (2002).

Achyranthes aspera L. (Pl. 2A)

Family: Amarantaceae.

English, Hindi & Sanskrit Names:

Eng. : Prickly Chaff Flower;

Hindi : Apang, Apmarga, Chichra, Chirchira, Chirchita, Latjira, Puthkanda;

Sans. : Aghata, Apamargah, Apangaka, Kharamari, Mauuraka, Shaikharika.

Distribution : Pastures, Wastelands & Roadsides Throughout India Upto 900 m.

Parts Used : Root. Root bark.

Utilization : Roots dug out on Sunday and tied on the head of the women to facilitate delivery. If inserted into vagina, it is equally good for expelling the dead child from the womb. To treat goitre, 10-12 pieces of roots tied to a string are hung around the neck. Ash of root with honey given for treating cough and external application on piles. Root paste given for cuts, wounds, scorpion sting, leucoderma, stomachache, cholera, hardening gums, cleansing of teeth, tooth caries, decay, pyorrhoea and bleeding after abortion. Dried root powder (6g) given with water in epileptic condition to regain consciousness. Root powder also used against tumours, leucorrhoea, tuberculosis, gonorrhoea, fever, wounds, urinary infections and swelling of eyes. Its decoction given to ladies after conception with cow's milk in cases of sterility, stomach troubles and for stones in the bladder whereas extr. considered good for treating menstrual disorder, dysentery and pneumonia. Root infusion regarded as an astringent. Root juice used as an eye drops for cataract. Powdered root bark along with sugar given for dysentery.

Active Constituents : Root contains ecdysteron, ecdysone and inokosteron.

Biological Activity : Roots antidepressant and hypotensive.

References : Agarwal (2003), Aminuddin & Girach (1991), Anonymous (1968), Asolkar *et al.* (1992), Bhatt *et al.* (2002), Bhattacharjee (2001), Chandrasekar & Srivastava (2003), Chatterjee & Pakrashi (1997), Chauhan (1999), Chopra *et al.* (1956), Dastur (1970), Dwivedi (1993), Girach *et al.* (2000), Hembrom (1996), Hosagoudar & Henry (1996a), Jain & Patole (2001), Katewa *et al.* (2001), Khan *et al.* (2003), Khare (2004), Kumar & Chauhan (2005), Kurian (1995), Maheshwari *et al.* (1997), Meena *et al.* (2003), Pandey & Rout (2006), Parrota (2001), Patil & Patil (2002), Pawar & Patil (2004), Rastogi & Mehrotra (1999b), Reddi *et al.* (2005), Saklani & Jain (1994), Sharma (2004), Singh *et al.* (1997, 2003), Singh & Kumar (2000a), Singh & Maheshwari (1992), Sinha & Sinha (2001), Sood & Thakur (2004), Sur & Halder (2002), Thakur *et al.* (1989), Tiwari *et al.* (2001), Varghese (1996), Verma & Chauhan (2006).

Achyranthes bidentata Blume

Family: Amarantaceae.

Distribution : Temperate & Sub-Tropical Himalaya (Kishtwar to Sikkim): Khasia Hills: 1,300-2,000 m, Bihar, Konkan, Nilgiris & Travancore Hills.

Part Used : Root.

Utilization : Root paste applied in hemicrania and on the infected gum in dental caries.

Active Constituents : These are ecdysterone, inokosterone, rubrosterone, emodin, physcion, arginine, glycine, serin, aspartic acid, glutamic acid, threonine, praline, tyrosine, tryptophan, valine, phenylalanine and leucine.

Biological Activity : Root extr. exhibits antimicrobial, antitumourous and anticancerous (inhibits growth of sarcoma-180 in ICR mice) properties.

References : Asolkar *et al.* (1992), Chopra *et al.* (1956), Jain *et al.* (1991), Molla & Roy (1996), Rastogi & Mehrotra (1998), Sood *et al.* (2005c).

Achyranthes coynei Santapau

Family : Amarantaceae.

Part Used : Root.

Utilization : Powder given internally to kill intestinal worms.

Reference : Bhogaonkar & Devarkar (2002).

Achyranthes porphyristachya Wall.

Syn : *A. aspera* var. *porphyristachya* (Wall.) Hk.f.

Family : Amarantaceae.

English Name:

Eng. : Prickly Chaff Flower.

Distribution : Sikkim Himalaya: 600-1,600 m. Chittagong.

Part Used : Root.

Utilization : Useful in the treatment of cancer, stomach disorder and stones in the bladder. Its decoction mixed with *Piper nigrum* given as an antidote to snake-bite.

References : Hooker (1885), Jain *et al.* (1991), Kumar (2002).

Aconitum atrox (Bruh.) Mukerjee

Family : Ranunculaceae.

Distribution : Garhwal to Nepal: 3,000-4,500 m.

Part Used : Root.

Utilization : Paste applied on snake bite. For relieving cough and cold, a little amount of root powder mixed with tobacco is taken. Roots fried in butter oil (ghee) and rubbed on the skin for the treatment of rheumatism.

Biological Activity : Roots are diuretic, diaphoretic and antiperiodic.

References : Bhattacharjee (2001), Negi *et al.* (1999).

Aconitum balfouri Stapf.

Family : Ranunculaceae.

Hindi Name:

Hindi : Gobriya.

Distribution : Kumaon & Garhwal in the Shady Slopes & Forest Edges: 3,000-4,200 m.

Part Used : Root.

Utilization : Beneficial for curing sciatica, gastritis, swelling, wounds, rheumatism, tonsil and neuralgia.

Active Constituents : Pseudaconitine.

References : Agarwal (2003), Ambasta (1986), Anonymous (1948), Bhattacharjee (2001), Kala & Rawat (2001), Kirtikar & Basu (1935), Samant *et al.* (2001), Sharma (2003).

Aconitum chasmanthum Stapf. *ex* Holms

Syn : *A. dissectum* Watt.; *A. hians* Wall.

Family : Ranunculaceae.

English & Hindi Names:

Eng.: Indian Napellus;
Hindi: Mohri, Vatsnabhi.

Distribution : Sub-Alpine & Alpine Zone of the W. Himalaya: 1,300-4,000 m.

Part Used : Root.

Utilization : Tuberous roots used externally for rheumatism and internally to relieve pain and fever.

Active Constituents : Roots contain chasmanine, indaconine, chasmaconitine, chasmanthinine and homochasmanine.

References : Agarwal (2003), Ambasta (1986), Asolkar *et al.* (1992), Bhattacharjee (2001), Kirtikar & Basu (1935), Rastogi & Mehrotra (1999b), Sharma & Rana (2000).

Aconitum denorrhizum Stapf.

Family : Ranunculaceae.

Hindi Name:

Hindi : Safed Vish.

Distribution : Kumaon to Nepal.

Part Used : Root.

Utilization : Root smoked in toothache and body pains. Also, it is used against rheumatism, fever, acute headache, leprosy, cholera, tonic in diarrhoea and cough.

Active Constituents : Air dried roots contain upto 1.2% of alkaloids chiefly acotine and pseudoacotin

References : Agarwal (2003), Asolkar *et al.* (1992), Chauhan (1999), Sharma (2003).

Aconitum falconeri Stapf.

Family : Ranunculaceae.

Distribution : Sub-Alpine & Alpine Zones of Himalaya: 3,000-4,000 m.

Part Used : Root.

Utilization : Roots used in the treatment of fever, rheumatism, diarrhoea, skin diseases, neuralgia, nervous and digestive disorders.

Active Constituents : Falaconitine, mithaconitine, pseudaconitine, indaconitine, veratroylpseudaconitine, bishatisine and bishaconitine are the major components.

References : Ambasta (1986), Asolkar *et al.* (1992), Bhattacharjee (2001), Nautiyal *et al.* (2003), Rastogi & Mehrotra (1999a), Samant *et al.* (2001).

Aconitum ferox Wall.

Family : Ranunculaceae.

English, Hindi & Sanskrit Names:

Eng. : Indian Aconite, Monkshood, Wolfsbane;

Hindi : Bachhang, Bikh, Mith Visha, Mohra, Singya, Teliabis;

Sans. : Aheya, Amrita, Brahamaputra, Darada, Halahala, Jivanagata, Kalakuta, Kasakula, Kshveda, Nila, Pradipana, Raktashringika, Shringi.

Distribution : Alpine and E. Himalaya (3,000-4,000 m).

Part Used : Root.

Utilization : A pinch of dried root chewed along with saline water as an emetic in case of food poisoning. Powdered root paste good for scabies and neuralgia. The decoction given for the mad dog bite whereas tincture of the root (1 in 8 of alcohol in doses of 2-5 drops) used in combination with other drugs for the treatment of fever and rheumatism. Also used in diabetes.

Active Constituents : Chief constituents are aconitine, chasmanine, pseudoaconitine, chasmaconitine, indaconitine, bikhaconitine, veratroylpseudaconitine, diacetyl-pseudaconitine, picro-aconine, aconine, benzyl aconine, homo-napalline, veratroyl-ψ-aconitine, 2-(1H)-quinolinone, 3, 4-dihydro-6-hydrxy -2-(1H)-quinolinone, vakognavine, crassicauline, falconericine, bikhaconine, pseudaconine, neoline, senbusine a, isotalatizidine, columbianine, lipopseudaconitine, lipoyunaconitine, lipoind-aconitine and lipobikhaconitine, C-8 linoleoyl, palmitoyl and steroyl esters of veratroylpseudaconine, anisoylyunaconine, benzoylindaconine and veratroylbikhaconine.

Biological activity : Considered to be cardiac stimulant, hypoglycaemic, diaphoretic, antiphlogestic, antidiabetic, antiperiodic, antipyretic in small doses. In large doses it acts as a powerful sedative.

References : Agarwal (2003), Ambasta (1986), Anonymous (1948, 1968), Asolkar *et al.* (1992), Caius (2003), Chatterjee & Pakrashi (1997), Chhetri (2005), Drury (1978), Kirtikar & Basu (1935), Patil (2000), Rai & Sharma (1994), Rastogi & Mehrotra (1998), Singh *et al.* (2003), Sood *et al.* (2005a), Thakur *et al.* (1989).

Aconitum heterophylloides (Bruhl.) Stapf.

Family : Ranunculaceae.

Distribution : Arunachal Pradesh, Sub-Alpine Regions of N.E. States Upto 3,600 m.

Part Used : Dried Tubrous Roots.

Biological activity : Root considered diaphoretic, diuretic, aniperiodic, anodyne, antidiabetic, antiphlogestic and antiperiodic in very small doses. In large doses it is a virulent poison, narcotic and powerful sedative. It reduces the frequency and tension of the pulse and paralyses the respiratory centre.

References : Kumar (2002), Sood *et al.* (2005a).

Aconitum heterophyllum Wall. *ex* Royal

Syn : *A. atees* Royal; *A. cordatum* Royal; *A. ovatum* Lindle.

Family : Ranunculaceae.

English, Hindi & Sanskrit Names:

Eng. : Atis Root, Indian Atees;

Hindi : Atis, Ativika, Ativisha, Patis;

Sans. : Amrita, Aruna, Atisaraghini, Bhangura, Bringi, Mahoshadha, Mridvi, Prativisha, Pravisha, Shringi, Shringika, Shuklakanda.

Distribution : N.W. Himalaya (Kumaon): 3,500-4,200 m.

Part Used : Root.

Utilization : Regarded as an antiflatulent, antifertility agent, aphrodisiac, emetic, stomachic, anthelmintic and also used for treating cough, cold, fever, diarrhoea, dysentery, piles, digestive and gall bladder complaints, poisonous bites and stings, hysteria, throat infections, dyspepsia, diabetes and abdominal pain. Its paste applied externally in the treatment of gout and rheumatism. A pinch of powder given to treat fever and stomachache in animals.

Active Constituents : These are atisine (0.4%), dehydroatisine, heteratisine, hetisine, heterophyllisine, heterophylline, heterophyllidine, atidine, hetidine, benzolheteratisine, F-dihydroatisine, hetisinone, isoatisine, acotinic acid, tannic acid pectin and starch.

Biological Activity : Exhibits antiviral activity against spinach mosaic virus.

References : Agarwal (2003), Ambasta (1986), Anonymous (1948, 1968), Asolkar *et al.* (1992), Aswal (1996), Aswal & Mehrotra (1994), Ballabh & Chaurasia (2006), Beig *et al.* (2003), Bhattacharjee (2001), Caius (2003), Chandrasekar & Srivastava (2003), Chauhan (1999), Chatterjee & Pakrashi (1997), Chhetri (2005), Chopra *et al.* (1956), Dey (1973), Drury (1978), Jain (1991), Jain & Goel (2005), Hooker (1872), Kapur & Singh (1996), Kaul (1997), Khare (2004), Kirtikar & Basu (1935), Lal *et al.* (1996), Nautiyal *et al.* (2003), Negi *et al.* (1999), Rastogi & Mehrotra (1999b), Said (1997), Samant *et al.* (2001), Sarin (1996), Sharma (2003), Singh (1996, 1999), Singh & Kumar (2000b), Sinha & Sinha (2001), Sood *et al.* (2001), Sood *et al.* (2005), Sood & Prakash (2007), Thakur *et al.* (1989), Tiwari & Pande (2006).

Aconitum palmatum D. Don

Syn : *A. bisma* Rap.; *Caltha bisma* Ham.

Family : Ranunculaceae.

Hindi Names:

Hindi : Bikhma, Bishawa, Bishma, Bisma.

Distribution : C. & E. Himalaya.

Part Used : Root.

Utilization : Considered as tonic, aphrodisiac, astringent, antiperiodic and given for rheumatism and diarrhoea. Mixed with pepper it is recommended for cough, dyspepsia, enteretis, pains in the abdomen, bowel complaints, vomiting and intestinal worms.

Active Constituents : These are pamatisine, vakognavine, vakatisine, vakatisinine, vakatidine and atisine.

References : Agarwal (2003), Ambasta (1986), Caius (2003), Rastogi & Mehrotra (1999b).

Aconitum rotundifolium Kar. & Kir.

Family : Ranunculaceae.

Hindi Name:

Hindi : Mithazahar.

Distribution : Wet Shady Places & Stony Slopes of Sub-Alpine N. W. Himalaya: 3,000-5,000 m, J. & K. & H. P.

Part Used : Root.

Utilization : Water extr. of root prescribed against severe headache (5 drops twice a day). Also given for blood purification and cough.

References : Kala & Manjrekar (2000), Kaul (1997), Kayal *et al.* (2003).

Aconitum violaceum Jacq. *ex* Stapf.

Syn : *A. multidum* Royal; *A. multidum* var. *rigidum* Hook.f. & Th; *A. napellus* Clegh.; *A. napellus* var. *dissectum* Duthie.

Family : Ranunculaceae.

Hindi & Sanskrit Names:

Hindi : Dudhi-Bish, Tilia Kachnag;

Sans. : Visha.

Distribution : Alpine Zones of the Himalaya from Gilgit to Kumaon: 3,300-5,000 m.

Part Used : Root.

Utilization : Boiled dried roots used in small doses to cure cough, cold, fever, piles, stomachache, pain, cystitis, and diarrhoea. Roots dissolved in mother's milk said to be good for infants suffering from severe diarrhoea, stomatitis and allied gastric complaints. Extr. of the root useful for renal pain, rheumatism, high fever, abdominal pain, bronchitis, epilepsy, headache, inflammation, neck pain, snake bite and lice. Dried root pieces are fitted in tooth cavities to check toothache.

Active Constituents : Aconitine, atidine, mesaconitine, neoline, songoramine, deoxyaconitine, hypaconitine, nepelline, luciculine, 12-epidehydronapelline, 12-epiacetyldehydronapelline, 1, 14, diacetylneoline, indaconitine and vitamin C are the main ingredients of the root.

References : Agarwal (2003), Ambasta (1986), Anonymous (1948), Ballabh & Chaurasia (2006), Bhattacharjee (2001), Chatterjee & Pakrashi (1997), Chauhan (1999), Chaurasia *et al.* (2000b), Kala & Manjrekar (2000), Jain (1991), Kaul (1997), Kirtikar & Basu (1935), Samant *et al.* (2001), Sharma (2003).

Acorus calamus L. (Pl. 2B)

Syn : *A. calamus* var. *verus L.*

Family : Araceae.

English, Hindi & Sanskrit Names:

Eng. : Bastard Water Flag, Calamus, Myrtle Grass, Sweet Flag, Sweet Myrtle;

Hindi : Bach, Bhoja, Gorbach, Igir, Jatila, Safed Bach, Vacha;

Sans. : Bhadra, Bhutanashini, Galani, Ikshuparni, Shadgrantha, Ugragantha, Vijaya.

Distribution : Marshy Places Upto 2,000 m in Himalaya, Manipur, Naga Hills & S. India.

Parts Used : Root. Rootstock.

Utilization : Roots prescribed for easy delivery, pneumonia, flatulence, colic, chronic diarrhoea, dyspepsia, abdominal pains, epilepsy, malaria, bronchitis, cough, irritation of throat, snake bite and calculus affections, and considered anthelmintic, emetic, stomachic and carminative. Locally, its paste applied directly on head during fever. In Kwashiorkar disease of children, small pieces of roots are tied around the neck as an amulet. Also dried roots chewed to treat cough, cold and toothache. Its poultice usefully applied to paralysed limbs, cuts and rheumatic swellings whereas its paste mixed with honey given to cure stammering in children (½ teaspoon in the morning for 20 days). Paste of rootstock in mustard oil makes a good lotion for body massage in children.

Active Constituents : Chief ingredients are acorin, calamine, tannin, mucilage, starch, vit. c, fatty acid, sugar, calcium oxalate, calamenone, calamendiol and isocalamendiol.

Biological Activity : Root extr. exhibits antimicrobial activity against *Staphylococcus aureus, E. coli* and *Asparagus niger.*

References : Agarwal (2003), Caius (2003), Chauhan (1999), Dastur (1970), Dey (1973), Dhiman (2003), Drury (1978), Dwivedi (1993), Gopal & Chandra (2003), Jain (1968, 1991), Jain & Saklani (1992), Kharkongor & Joseph (1997), Kirtikar & Basu (1984), Kumar & Jain (1998), Mao (1993), Patil (2000), Rastogi & Mehrotra (1998, 2000), Said (1997), Saklani & Jain (1994), Sarin (1996), Sharma (2003), Sharma *et al.* (2003), Singh & Prakash (1994), Sood & Prakash (2007), Sood & Thakur (2004), Watt (1889).

Acrocephalus indicus (Burm.f.) O. Ktz.

Syn : *Acrocephalus capitatus* (Roth.) Benth.; *A. hispidus* (L.) Nicolson & Sivadason; *Gomphrena hispida* L.

Family : Lamiaceae.

Hindi Name:

Hindi : Aurea.

Distribution : Warmer Regions of India.

Part Used : Roots.

Utilization : Decoction of 5g root along with *Andrographis nallamalayana* given orally from third day for four days in leucorrhoea.

References : Agarwal (2003), Chopra *et al.* (1956), Venkataratnam & Raju (2004).

Actiniopteris radiata (Sw.) Link.

Syn : *A. dicotoma* Bedd

Family : Actinopteridaceae.

English & Sanskrit Names:

Eng. : Peacock's Tail;

Sans. : Mayurashikhaa, Neelkantha, Shikhao, Varhishikhaa.

Distribution : Common in the Kumaon Hills & Nilgiri.

Part Used : Roots.

Utilization : Root pounded with milk prescribed to the woman in season for conceiving in Gadnigraha.

Reference : Jain *et al.* (1991).

Actinocarya tibetica Benth.

Family : Boraginaceae.

Distribution : W. Tibet, Nubra (near Kassar village).

Part Used : Roots.

Utilization : Decoction good against fever.

References : Chaurasia *et al.* (2000b), Hooker (1885).

Adanosia digitata Linn.

Family : Bombacaceae.

English & Hindi Names:

Eng. : Boobab, Monkey Bread Tree;

Hindi : Gorak Amli.

Part Used : Root.

Utilization : Considered as febrifuge.

Reference : Dastur (1970).

Adenanthera pavonina Linn.

Family : Fabaceae.

English, Hindi & Sanskrit Names:

Eng. : Coral Wood Tree, Red Wood, Sandal Wood;
Hindi : Rakta Chandan, Raktakanchan;
Sans. : Kunchandana.

Distribution : E. Sub-Himalayan Tract, W. Ghats & Andamans.

Part Used : Roots.

Utilization : Considered as an emetic.

Active Constituents : These are oleanolic and echinocystic acid.

References : Agarwal (2003), Ambasta (1986), Dhiman (2003), Rastogi & Mehrotra (2001).

Adiantum lunulatum Burm.

Syn : *A. philippense* Burm.
Family : Polypodiaceae.

English, Hindi & Sanskrit Names:

Eng. : Walking Maiden Hair Fern;
Hindi : Hansapadi, Kalijhant, Paresiyarasan;
Sans. : Hamsapadi, Hangsarati.

Distribution : Throughout India Upto 1,200 m.

Part Used : Root.

Utilization : Decoction given once to cure discharge of blood in urine (single root of a plant boiled in 500 ml water till it becomes one fourth), throat infections, epilepsy and febrile conditions in children. Its poultice bandaged for 8-10 days on bone fracture.

References : Asolkar *et al.* (1992), Chhetri (2005), Girach (1992), Parrotta (2001), Varghese (1996).

Adiantum venusetum D. Don

Family : Polypodiaceae.

English Names:

Eng. : Black Maiden Hair Fern, Venus Hair.

Distribution : N. E. Himalaya, Kashmir, Shimla: 1,350-3,150 m.

Part Used : Root.

Utilization : Paste of root mixed with sugar considered effective to cure stomachache.

References : Chatterjee & Pakrashi (2001), Nautiyal *et al.* (2003).

Aegle marmelos (L.) Correa *ex* Roxb.

Syn : *Belou marmelos* (L.) Lyons; *Bilacus marmelos* (L.) Kuntze.; *Crataeva marmelos* L.; *C. religiosa* Ainslie.

Family : Rutaceae.

English, Hindi & Sanskrit Names:

Eng. : Bael Tree, Bengal Quince;
Hindi : Bael, Bel, Bilwa, Siriphal;
Sans. : Bilva, Shiva Druma, Malura, Vilva.

Distribution : Scattered in Dry Deciduous Forest of Peninsular India Upto 1,200 m. W. Himalaya.

Parts Used : Root. Root bark.

Utilization : Root as well as bark used in the form of a decoction as a remedy in melancholia, palpitation of the heart, high blood pressure, hypochondrias, inflammation of the heart, intermittent fevers, high fevers, diarrhoea, dysentery, stomach pain and urinary troubles. Root extr. taken orally about a week to cure piles and gastric troubles; and once a day for three days in jaundice. 25 g root taken with 2-5 pieces of black pepper and 25 ml water every 3 hrs as an antidote for snake venom. Root pills considered as a brain tonic.

Active Constituents : Main constituents are coumarins-scoparone, scopoletin, umbelliferone, marmesin, β-sitosterol, γ-sitosterol, xanthotoxin, 6, 7-dimethoxycoumarin, tembamide, marmin, skimmianine, skimming, aurapten, lupeol and 6′, 7′epoxyaurapten.

Biological Activity : Root is antiinflammatory and hypoglycaemic.

References : Asolkar *et al.* (1992), Chatterjee & Pakrashi (1997), Chopra *et al.* (1956), Dastur (1970), Dhiman (2003), Dwivedi (1993), Jain (1968, 1991), Khare (2004), Kirtikar & Basu (1935), Kurian (1995), Maiti & Mishra (2000), Mehrotra *et al.* (1995), Parmar & Kaushal (1982), Parrotta (2001), Patil & Patil (2002), Rastogi & Mehrotra (1998, 1999a, b), Reddi *et al.* (2005), Said (1997), Saklani & Jain (1996), Sarin (1996), Sharma (2003), Singh & Pandey (1998), Thakur *et al.* (1989), Thomas & Britto (2003), Venkataratnam & Raju (2004), Varghese (1996), Watt (1889).

Aerva aspera L.

Family : Amarantaceae.

Part Used : Root.

Utilization : 25 g root paste given orally every 2 hrs as an antidote for snake venom.

Reference : Maiti & Mishra (2000).

Aerva lanata (L.) Juss. *ex* Schult.

Syn : *Achyranthes lanata* L.; *A. villosa* Forssk.; *Aerva elegans* Moq.; *Illecebrum lanatum* L.Murr.

Family : Amarantaceae.

English, Hindi & Sanskrit Names :

Eng. : Polpala;
Hindi : Chaya, Gorkhabundi, Kapurajadi, Papurijadi;
Sans. : Aadaanpaaki, Astambayda, Gorakhagaanjaa, Shatkabhedi.

Distribution : Throughout the Plains of India.

Part Used : Root.

Utilization : Used to cure boils, burning micturition, headache, white leucorrhoea, jaundice and diarrhoea. Root juice allays liver congestion, biliousness and dyspepsia. Decoction of root is cooling, emulcent, diuretic, lithotriptic and used in haemetemesis, diabetes, lithiasis and arresting haemorrhage associated with pregnancy. Its paste fed to stop abnormal high bleeding in menstruation. Decoction of the root boiled along with the root of *Cyperus rotundus* L. and fennel (*Foeniculum vulgare* Geartn.) prescribed 2 or 3 times a day to cure diarrhoea in children.

References : Bhatt *et al.* (2003), Bhattacharjee (2001), Chatterjee & Pakrashi (1997), Chopra *et al.* (1956), Hooker (1885), Jadeja (1979), Khare (2004), Mohanty & Padhy (1996), Parrotta (2001), Reddi *et al.* (2005), Singh & Pandey (1998), Sur & Halder (2002), Varghese (1996).

Aerva sanguinolenta (L.) Blume

Syn : *Achyranthes sanguinolenta* L.;
Aerva scandens (Roxb.) Moq. in DC.
Family : Amarantaceae.

Distribution : Throughout India, ascending Upto 1,800 m in Himalaya.

Part Used : Root.

Utilization : Good against dysentery. Fresh roots collected after sunrise are crushed and mixed with those of *Smilax ovalifolia* Roxb. and *Asparagus racemosus* Willd. in the ratio of 2:1:1 for oral administration against leucorrhoea (twice a day, for a week).

References : Ambasta (1986), Asolkar *et al.* (1992), Sahoo & Mudgal (1995).

Aeschynanthus sikkimensis Stapf.

Syn : *A. maculatus* auct. non Lindl.;
A. maculatus var. *sikkimensis* Cl. in DC.
Family : Gesneriaceae.

Part Used : Root.

Utilization : Decoction used to cure fever.

References : Chhetri (2005), Lalramghinglova & Jha (1999).

Aeschynomene aspera L.

Family : Fabaceae.

English & Hindi Names:

Eng. : Solapith Plant;
Hindi : Sola.

Distribution : Plentiful in Bengal, Assam & S. India.

Part Used : Root.

Utilization : 3g dose of powdered root with 3 black peppers (*Piper nigrum*) given twice a day to cure gastric pain.

References : Agarwal (2003), Girach (1992).

Aganosma marginata G. Don

Family : Apocynaceae.

Distribution : Scattered in Bengal.

Part Used : Root.

Utilization : Decoction checks urinary troubles, anaemic condition and as an emmenagogue.

Reference : Agarwal (2003).

Agave americana L.

Syn : *A. cantula* Roxb.
Family : Amaryllidaceae.

English, Hindi & Sanskrit Names:

Eng. : American Aloe, Carata, Century Plant;
Hindi : Banskeora, Barakanwar, Hathisengar, Kantala, Ramkanta;
Sans. : Kalakantala, Kantala.

Distribution : Cultivated Throughout India as a Hedge Plant.

Part Used : Root.

Utilization : Considered stomachic, diuretic, diaphoretic and antisyphilitic.

Active Constituents : Hecogenin, tigogenin, gitogenin, β-sitosterol-3-O-β-D-arabinofuranosyl (1→4)-O-α-L-rhamnopyranosy (1→4)-β-D-glucopyranoside are the chief ingredients of roots.

Biological Activity : Roots possess diuretic properties.

References : Agarwal (2003), Ambasta (1986), Anonymous (1968), Chopra *et al.* (1956), Drury (1978), Parrotta (2001), Rastogi & Mehrotra (1998, 2002), Singh (2000), Sood & Thakur (2004), Watt (1889).

Ageratum conyzoids L.

Syn : *A. cordifoium* Roxb.

Family : Asteraceae.

English, Hindi & Sanskrit Names:

Eng. : Goat Weed, White Weed;
Hindi : Koobhi, Visadodi;
Sans. : Visamustih.

Distribution : Throughout India.

Part Used : Root.

Utilization : Pounded root taken for 7 days from the days of menstruation for two months for dysmenorrhoea and its juice said to possess antilithic, anthelmintic, antidysenteric properties. Cold decoction used as a lotion in purulent ophthalmia.

References : Caius (2003), Chatterjee & Pakrashi (1997), Chhetri (2005), Chopra *et al.* (1956), Dwivedi (2003), Jain & Sharma (2000), Jha & Kumar (2003), Parrotta (2001), Reddi *et al.* (2005), Sahoo & Mudgal (1995), Sood & Thakur (2004), Varghese (1996).

Agrimonia pilosa Ledeb.

Family : Rosaceae.

Distribution : Most Common Throughout Manipur, Nagaland, Assam & Meghalaya Upto 2,000 m.

Part Used : Root.

Utilization : Chewed to ameliorate toothache.

Reference : Kumar (2002).

Agropyron repens Beaur.

Syn : *Triticum repens* Linn.

Family : Poaceae.

English Names:

Eng. : Couch Grass, Dog Grass, Quack Grass, Quilch, Twich Grass, Witch Grass.

Distribution : Kashmir & W. Tibet, ascending Upto 4,600 m in the Karakoram Range.

Part Used : Root.

Utilization : Juice of the roots used for cirrous liver, tumours, cancerous growths, skin eruptions and rheumatic complaints. Infusion generally regarded as a soothing diuretic and useful to the bladder and kidneys.

References : Agarwal (2003), Ambasta (1986), Asolkar *et al.* (1992), Caius (2003).

Ailanthus altissima (Mill.) Swingle

Syn : *A. glandulosa* Desf.; *Toxicodendron altissima* Mill.

Family : Simarubaceae.

English & Hindi Names:

Eng. : Ailanto, Tree of Heaven;

Hindi : Vilayati Madala.

Distribution : Occurs Throughout Punjab.

Part Used : Root bark.

Utilization : Used to cure epilepsy, heart trouble and asthma.

Biological Activity : These are 1-methoxycanthin-6-one, canthin-6-one, canthin-6-one-3-oxide, β-sitosterol, scopoletin, 1-OH-canthin-6-one and shinjudilactone.

References : Ambasta (1986), Asolkar *et al.* (1992), Bhattacharjee (2001), Rastogi & Mehrotra (1999b).

Ainsliaea aptera DC.

Family : Asteraceae.

Distribution : Throughout H. P.: 2,000-3,200 m.

Part Used : Root.

Utilization : 2 g root powder taken with luke warm water as an effective cure for acute gastric pain. Fresh roots chewed in emergency to allay acute stomach pain, diarrhoea, and as wormicide.

References : Ambasta (1986), Asolkar *et al.* (1992), Bhattacharjee (2001), Chauhan (1999).

Ajuga parviflora Benth.

Syn : *Teucrium hispidum* Wall.

Family : Lamiaceae.

Distribution : W. Himalaya, Kashmir-Nepal.

Part Used : Root.

Utilization : Powdered roots taken to suppress hunger (1-2 teaspoons once every day for 3-5 days), and its extr. given to cure typhoid fever.

References : Nautiyal *et al.* (2003), Sood & Thakur (2004).

Alangium salvifolium (L.f.) Wang.

Syn : *A. decapetalum* Lam.; *A. hexapetalum* Roxb.; *A. lamarkii* Th.; *A. salvifolium* ssp. *decapetalum* (Lam.) Wangerin; *A. tomentosum* Lamk.; *Grewia salvifolia* L.f.

Family : Alangiaceae.

English, Hindi & Sanskrit Names:

Eng. : Sage-Leaved Alangium;
Hindi : Akarkanta, Akola, Anedhera, Ankola, Dhera;
Sans. : Ankola, Ankot.

Distribution : Plains & Foot Hills Throughout the Greater Part o India.

Parts Used : Root. Root bark.

Utilization : Roots considered anthelmintic, astringent, antipyretic, laxative and good for piles, body swellings, menstrual disorders, epilepsy, fever, rib muscle pain and cat bite. Its poultice applied on dog bite and its juice taken to relieve fever. Root paste along with the fruits of 'Bichati' (*Tragia involucrata*) in equal quantities given to cure paralysis (twice a day for two months). Also taken orally to check food poison. The root bark taken with butter milk as an efficacious remedy for fever (doses: 0.13-0.32 g), diarrhoea, dysentery, leprosy, skin diseases, syphilis, rheumatism, dropsy and also considered diaphoretic, diuretic and emetic. Decoction prescribed to treat rabies.

Active Constituents : Principle constitutes are alangin A, marckin, marckidin, tubulosine, cephaeline, isotubulosine, N-methycephaeline, deoxytubulosine, psychotrine, alangicine and demethylpsychotrine (root bark).

Biological Activity : Root extr. showed hypotensive and antidepressant activities.

References : Agarwal (2003), Ambasta (1986), Anonymous (1948), Asolkar *et al.* (1992), Brahmam & Saxena (1990), Chatterjee & Pakrashi (1997), Chopra *et al.*(1956), Dastur (1970), Drury (1978), Dwivedi (1993), Goud & Pullaiah (1996), Goud *et al.* (2000), Girach (1992), Jadeja (1999), Jain (1991, 1996), Khanna (2002b), Punjani (2003), Parrotta (2001), Pullaiah & Kumar (1996), Rao & Henry (1996), Rastogi & Mehrotra (1999a), Reddi *et al.* (2005), Sinha & Sinha (2001), Thomas & Britto (2003), Varghese (1996), Watt (1889).

Albizia lebbeck (L.) Benth.

Syn : *Acacia lebbeck* (L.)Willd.; *A. sirissa* Hamilton; *A. speciosa* (Jacq.) Willd.; *Albizia latifolia* Boirin.; *Inga leucoxylon* Hassk.; *Mimosa lebbeck* L.; *M. serissa* Roxb.

Family : Fabaceae.

English, Hindi & Sanskrit Names:

Eng. : East India Walnut, Lebbeck, Siris Tree;
Hindi : Siris, Sirisha;
Sans. : Bhandi, Sirisa.

Distribution : Common Throughout India, from the plains to 900 m in the Himalaya.

Parts Used : Root. Root Gum.

Utilization : Astringent roots used to treat hemicrania and ophthalmia. Powdered root and root gum used as a dental powder for strengthening gums and externally applied as plaster in leprous ulcers.

Active Constituents : These are echinocystic acid-3-O-L-rhamnopyranosyl (1→5)-β-D-xylopyranosyl (1→5)-β-D-arabinofuranosyl and (1→4)-β-D-glucopyranoside.

Biological Activity : Exhibits anticancerous, hypotensive and spermicidal properties.

References : Agarwal (2003), Chopra *et al.* (1956), Dastur (1970), Dhiman (2003), Kothari & Moorthy (1996), Parrotta (2001), Rastogi & Mehrotra (2002), Varghese (1996).

Alcea rosea L.

Syn : *Althaea rosea* (L.) Cav.
Family : Malvaceae.

English, Hindi & Sanskrit Names:

Eng. : Althaea Rose, Hock Herb, Hollyhock, Round Dock;
Hindi : Gulkhairu, Gulkhirri;
Sans. : Khatami.

Distribution : Common Ornamental Plant of Indian Gardens.

Part Used : Root.

Utilization : Given to cure ulcers and as a demulcent. Its decoction prepared by boiling in milk given for treating goitre, and also applied locally.

References : Beig *et al.* (2003), Caius (2003), Jain *et al.* (1991), Thakur *et al.* (1989).

Alocasia cucullata Schott.

Syn : *Arum cucullatum* Lour.

Family : Araceae.

Distribution : Bengal. Assam. N. E. Himalaya (Manipur).

Part Used : Rootstock.

Utilization : Consumed to cure rheumatism and gout.

References : Rastogi & Mehrotra (2001), Sasikala *et al.* (2003), Sood & Prakash (2007), Watt (1889).

Alocasia indica (Roxb.) Schott.

Family : Araceae.

English, Hindi & Sanskrit Names:

Eng. : Taro, Yam;
Hindi : Mankanda;
Sans. : Manaka.

Distribution : Grows in N. E. Himalaya.

Part Used : Rootstock.

Utilization : Considered diuretic, laxative and also employed in the treatment of anaemia, enervation, leprosy, dropsy, abdominal and spleen disorders.

References : Agarwal (2003), Ambasta (1986), Sasikala *et al.* (2003).

Alocasia macrorhiza Schott.

Syn : *A. indica* (Lour.) Spach.; *Arum macrorhizum* L.

Family : Araceae.

English, Hindi & Sanskrit Names:

Eng. : Acrid Taro, Giant Taro;
Hindi : Baramankachu, Mankadana;
Sans. : Manaka.

Distribution : Found Wild & Cultivated All Over India.

Part Used : Rootstock.

Utilization : Useful in inflammation, diseases of abdomen, constipation, piles, scorpion sting, skin irritation and as diuretic.

References : Ambasta (1986), Asolkar *et al.* (1992), Caius (2003), Chopra *et al.* (1956), Kumar (2002), Sood & Prakash (2007).

Aloe barbadensis Mill. (Pl. 2C)

Family : Liliaceae.

English Name:

Eng. : Indian Aloe.

Distribution : Semi-Wild in Almost All Drier Parts of the Country.

Part Used : Root.

Utilization : Used in colic pain.

Active Constituents : Contains aloin, isobarbaloin and emodin as active principles.

References : Dwivedi (1993), Varghese (1996).

Alpinia bracteata Roxb.

Family : Zingiberaceae.

Part Used : Root.

Utilization : Taken in case of food poisoning.

Reference : Jamir & Rao (1990).

Alstonia scholaris (L.) Br. *ex* DC.

Family : Apocynaceae.

English, Hindi & Sanskrit Names:

Eng. : Devil's Tree, Dita Bark Tree;
Hindi : Chatian, Chhatim, Satium;
Sans. : Saptaparna.

Distribution : Moist Regions Throughout the Country.

Parts Used : Root. Root bark.

Utilization : About 2 teaspoonful of root juice taken in severe stomach pain. One teaspoonful of the root bark powder prescribed once daily orally as a tonic for improving health. Also, bark useful in stomach ulcers (5 spoonfuls twice a day for 10 days).

Active Constituents : Root bark contains α-amyrin, lupeol acetate, stigmasterol, campesterol, α-sitosterol and isokanin-7-O-rhamnoside (8, 3', 4'-trihydroxyflavonone-7-O-α-L-rhamnopyranoside.

References : Ambasta (1986), Asolkar *et al.* (1992), Jain (1968), Molla & Roy (1996), Rao & Henry (1996), Rastogi & Mehrotra (1999b, 2002), Reddi *et al.* (2005).

Alternanthera sessilis (L.) R.Br. *ex* DC.

Syn : *A. denticulata* R.Br.; *A. nodiflora* R.Br.; *A. repens* Gmel. non Link.; *A. triandra* Lam.; *Gomphrena sessilis* L.

Family : Amarantaceae.

Hindi & Sanskrit Names:

Hindi : Kanchari;
Sans. : Gojihra.

Distribution : Throughout the Warmer Parts of the Country Upto 1,200 m in the Himalaya.

Part Used : Root.

Utilization : Considered useful for vomiting. 3-4 teaspoonfuls of its extr. crushed with stem bark of *Alstonia scholaris* (L.) Br. taken with cow milk for spermatorrhoea. Poultice of the roots applied externally to relieve inflamed wounds.

Active Constituent : Lupeol.

References : Ambasta (1986), Asolkar *et al.* (1992), Kumar & Goel (1998), Parrotta (2001), Rastogi & Mehrotra (1999b), Varghese (1996).

Althaea officinalis Linn. (Pl. 2D)

Family : Malvaceae.

English & Hindi Names:

Eng. : Bull's Eyes, Marsh Mallow, White Mallow, Wymote;

Hindi : Gulkhairo, Khaira, Khairakajhor, Khitmikajhar.

Distribution : Himalaya: Kashmir-Punjab.

Part Used : Root.

Utilization : Considered useful against irritation of alimentary canal, respiratory and urinary organs, bruises, cough, whooping cough, diabetes, gum troubles, hoarseness of voice, asthma bronchitis, pleurisy, dysentery, muscular and nerve injury, poisoning, gastric and intestinal inflammations, toothache, and for lubricating and relaxing the passage in nephritic and calculus complaints. Its poultice cures abrasions and eruptions externally. Powdered root boiled in milk regarded useful for haemorrhage from urinary organs and in dysentery.

Active Constituents : Contains about 37 % starch, 11 % mucilage, 11 % pectin, flavonoids, phenolic acids, sucrose, asparagines, althcin, sulphate-isoscutellarein-4'-methyl ether-8-(2"-30-3K) glucoside (I) and hypoletin.

References : Asolkar *et al.* (1992), Caius (2003), Kaul (1997), Khare (2004), Kirtikar & Basu (1935), Rastogi & Mehrotra (1998), Said (1997), Sood *et al.* (2005a), Watt (1889).

Alysicarpus bupleurifolius (L.) DC.

Family : Fabaceae.

Distribution : Throughout Plains of India, ascending to 1,200 m in Hills.

Part Used : Root.

Utilization : Root extr. (10-30 ml) given twice a day for 7- 9 days to cure asthma.

References : Rastogi & Mehrotra (2002), Sharma & Singh (1999).

Alysicarpus vaginalis (L.) DC.

Syn : *A. nummularifolius* DC.; *A. vaginalis* var. *nummularifolius* Baker; *Hedysarum vaginalis* L.

Family : Fabaceae.

English Names:

Eng. : Alyce Clover, One-Leaved Clover.

Distribution : Common Throughout India on Roadsides, Lawns & in Pastures Upto 1,200 m.

Part Used : Root.

Utilization : Used for treating leprosy, urinary troubles, cough, cold and as expectorant. 10g of its juice given once a day from third day of menstruation for 15 days to check conception. 5 g root powder given with water at hourly intervals till recovery from snake bite.

References : Bhalla & Bhalla (2002), Chopra *et al.* (1956), Jain *et al.* (1991), Jha & Verma (1996), Khan *et al.* (2003), Parrotta (2001), Varghese (1996).

Amaranthus caudatus L.

Syn : *A. edulis* Speg.; *A. paniculatus* Linn.

Family : Amarantaceae.

English & Hindi Names:

Eng. : Love-Lies-Bleeding, Tassel Flower;
Hindi : Ramdana.

Distribution : Grows Throughout Plains of India; often Cultivated.

Part Used : Root.

Utilization : One teaspoonful of powder taken with luke warm water to relieve gastric prolems.

References : Asolkar *et al.* (1992), Chopra *et al.* (1956), Khanna & Kumar (2000).

Amaranthus spinosus L. (Pl. 3A)

Family : Amarantaceae.

English, Hindi & Sanskrit Names:

Eng. : Prickly Amaranth, Spiny Pigweed, Thorny Amaranth;

Hindi : Choli, Janumarak, Kantabhaji, Kantanatia, Katailichaulai;

Sans. : Alpamarisha, Marish, Tandula, Tanduliyah.

Distribution : Very Common in Undisturbed, Open Habitats & Roadsides Throughout the Country.

Part Used : Roots.

Utilization : Considered anthelmintic, diuretic, laxative and also used for leucorrhoea, onychia, gonorrhoea, dysuria, weakness after child birth, eczema, colic pain, menorrhagia and as lactagogue. Fresh roots collected on Saturday given to the patient on empty stomach daily in the morning for three days to cure spermatorrhoea. Its decoction with a pinch of common salt given as vermicide against roundworms, especially for children (two spoonfuls twice a day for a week); and for urinary troubles and abnormal urine colour. Root paste useful against scabies, indigestion, haemorrhage, chronic diarrhoea, boils, wounds and inflammations and its extr. administered orally to reduce chest pain.

Active Constituents : It comprises of oleanolic acid, D-glucose & D-glucuronic acid along with α-spinasterol, α-spinasterol octacosanoate, and a new saponin – β-D-glucopyranosyl (1→4)-β-D-glucopyranosyl (1→3)-oleanolic acid.

References : Agarwal (2003), Bhattacharjee (2001), Chatterjee & Pakrashi (1997), Chowdhury *et al.* (2001), Das & Sharma (2003), Dey (1973), Gogoi & Borthakur (2001), Gogoi & Das (2003), Jain (1996), Kirtikar & Basu (1984), Maliya (2005), Maya *et al.* (2003), Parrotta (2001), Patil (2004), Patil & Patil (2002), Rahman (2000), Rastogi & Mehrotra (1999b), Rao & Henry (1996), Reddi *et al.* (2005), Sarma *et al.* (2002),

Shil & Sharma (2003), Shukla *et al.* (2003), Singh *et al.* (1996), Singh & Pandey (1996), Tiwari & Tiwari (1996), Upadhyay & Singh (2005), Varghese (1996).

Amaranthus tricolor L.

Syn : *A. gangeticus* L.; *A. mangostanus* L.; *A. melancholicous* Linn.; *A. polygamus* Linn., Hook.f.; *A. tristis* L.

Family : Amarantaceae.

English, Hindi & Sanskrit Names:

Eng. : Chinese Spinach;
Hindi : Lalchowlai, Lalsag;
Sans. : Tanduliya.

Distribution : Throughout India.

Part Used : Root.

Utilization : Root extr. given to check vomiting, on mixing it with corm extr. of banana (*Musa paradisiaca* L.) recommended orally for menstrual disorders (5 spoonfuls twice a day for 30 days). Its paste in combination with honey and rice water useful in leucorrhoea. Powdered root also effective in onychia.

References : Agarwal (2003), Asolkar *et al.* (1992), Chaterjee & Pakrashi (197), Rao & Henry (1996), Sharma & Singh (1999).

Amaranthus viridis L. (Pl. 3B)

Syn : *Alberua caudata* Boiss.; *Amaranthus fasciatus* Roxb.; *A. gracilis* Desf.; *A. polystachyus* Ham.; *Chenopodium caudatum* Jacq. *Euxolus caudatus* Moq.; *E. viridis* Moq.;

Family : Amarantaceae.

English, Hindi & Sanskrit Names:

Eng. : Chinese Spinach, Green Amaranth, Green Pigweed, Wild Amaranth;
Hindi : Tandulaya;
Sans. : Tanduliya, Vishaghna.

Distribution : Waste Places Throughout India.

Part Used : Root.

Utilization : Root chewing useful against kidney stones, conception, snake bite and as an appetiser. Also, its juice taken with 'amala' (*Emblica officinalis* Gaertn.) juice twice a day for required days to control excessive bleeding during menstruation and in urinary problem.

Active Constituent : Amasterol.

Biological Activity : Amasterol inhibited seed germination and growth of seedlings in lettuce. It also inhibited growth of *Helminthosporum oryzi*.

References : Ambasta (1986), Asolkar *et al.* (1992), Hooker (1885), Negi *et al.* (1999), Panthi & Chaudhary (2003), Rastogi & Mehrotra (2001), Reddi *et al.* (2005), Sood & Thakur (2004).

Amorphophallus campanulatus (Roxb.) Bl. *ex* Decne.

Syn : *A. paeoniifolius* (Dennst.) Nicolson; *Arum campanulatum* Roxb.

Family : Araceae.

English, Hindi & Sanskrit Names:

Eng. : Elephant Foot Yam;

Hindi : Arsaghana, Balukand, Kanda, Ol, Zaminkand;

Sans. : Arsaghana, Bahukanda, Olla, Sukandi, Vatari.

Distribution : Cultivated Throughout Plains.

Parts Used : Root. Rootstock.

Utilization : Fresh roots act as an acrid stimulant, emmenagogue, expectorant and beneficial in acute rheumatism, boils and ophthalmia. Rootstock useful as an appetiser, stomachic, constipating and in piles, spleen enlargement, tumours, asthma, vomiting, abdominal pain, blood diseases and elephantiasis.

References : Agarwal (2003), Caius (2003), Drury (1978), Gogoi & Das (2003), Sood & Prakash (2007), Watt (1889).

Ampelocissus barbata (Wall.) Planch.

Syn : *Vitis barbata* Wall.

Family : Vitaceae.

Distribution : Hills of Assam & Meghalaya Upto 1,200 m.

Part Used : Root.

Utilization : Root paste with mustard oil applied over the affected area to cure eczema.

References : Ambasta (1986), Asolkar *et al.* (1992), Jain (1991), Jain *et al.* (1991), Shukla & Verma (1996).

Ampelocissus latifolia (Vahl.) Planch.

Syn : *Vitis latifolia* Roxb.; *V. montana* Lawson.

Family : Vitaceae.

English, Hindi & Sanskrit Names:

Eng. : Jangli Angoor;
Hindi : Dokar Bel, Pani Bel;
Sans. : Amlavetsah, Amrtasrava, Ganavalli, Paniyavalli, Sitalata.

Distribution : Throughout India Upto 1,200 m in the Sub-Himalayan Hills & W. Ghats.

Part Used : Root.

Utilization : Considered astringent and its decoction taken to treat chronic dysentery, inflammation, menstrual complaints, stomach pain, pneumonia, skin diseases, snake bite, sores and as galactagogue. Fresh root paste applied topically twice a day for a week on affected part of the skin to cure fungal diseases, such as eczema and ringworm and also for promiting healing of wounds, bone fracture and wound in human beings and animals. Extr. also given orally to cure dyspepsia and indigestion. Root powder said to be good for high fever and to clear yellow colour of the urine (1 cup twice daily for three days on empty stomach).

References : Ambasta (1986), Asolkar *et 1al.* (1992), Chopra *et al.* (1956), Hooker (1872), Jain (1991), Katewa *et al.*

(2001), Parrotta (2001), Punjani (2004), Sikarwar (1996), Singh (2006), Singh & Pandey (1998), Varghese (1996).

Anacyclus pyrethrum DC.

Syn : *A. officinarum* Hayne.

Family : Asteraceae.

English, Hindi & Sanskrit Names:

Eng. : Pellitory, Spanish Pellitory;
Hindi : Akarkara;
Sans. : Aakallaka, Aakulakrit, Agragrahi, Arkakarbha.

Distribution : Katra (J.K.).

Part Used : Root.

Utilization : Known to be astringent, sialogogue, tonic, anti-inflammatory, irritant, stimulant, pungent, rubifacient, depurative and useful in cases of dental caries, odontalgia, pharyngitis, tonsillitis, hemiplegia, paralysis, epilepsy, cholera, rheumatism, typhoid fever, colic, chronic bowl complaints and sore throat. Alc. infusion of roots and an ointment prepared from them used for mange disease of horses.

Active Constituents : Anacyclin, pellitorine, enetriyne alcohol, hydrocarolin, inulin, sesamin, N-(2′-p-hydroxy-phenylethyl)-deca-dodeca-, tetradeca trans-2, trans-4-dienamides and polyacetylenic amides I-IV are the chief constituents of roots.

References : Ambasta (1986), Asolkar *et al.* (1992), Chatterjee & Pakrashi (1997), Khare (2004), Reddi *et al.* (2004).

Andrographis nallamalayana Ellis.

Family : Acanthaceae.

Part Used : Root.

Utilization : Decoction of 5g of its root along with those of *Acrocephalus indicus* given orally for four days in leucorrhoea.

Reference : Venkataratnam & Raju (2003).

Andrographis paniculata Nees.

Syn : *A. subspathulata* C.B.Clarke; *Justicia paniculata* Burm.f.

Family : Acanthaceae.

English, Hindi & Sanskrit Names:

Eng. : Creat, King of Bitters;

Hindi : Bhuineem, Chirayata, Kalmegh, Kirata, Kiryat, Mahatita;

Sans. : Kalmegh, Kiratatiktah.

Distribution : Mainly in the Plains from H.P. to Assam & Mizoram Southwards Through Peninsular India & Sri Lanka.

Part Used : Root.

Utilization : 3g powdered root given with water or its decoction given to children once a day for expelling worms, whooping cough (one teaspoonful of root with rhizome of *Zingiber zerumbent* (L.) Rosc. *ex* Smith in equal parts by weight, thrice a day for 15 days), dysentery, liver complaints, stomach complaints, dyspepsia and as febrifuge, aperient, stimulant, and tonic. Paste of root with turmeric (*Curcuma longa*) applied to relieve itching and skin rashes.

Active Constituents : Contains andrographin (5-hydroxy-7,7,2′-trimethoxy flavone), panicolin (5, 2′-dihydroxy-7, 8-dimethoxy flavone), apigenin-4′,7-dimethylether,mono-O-methyl wightin and 5 hydroxy – 7, 8, 2′, 3′-tetramethoxy flavone, β-sitosterol, diterpene glucoside-neoandrographolide, α-sitosterol, 7-O-methylwogonin, andrographidine A and andrographidines B, C, D, E, and F.

References : Agarwal (2003), Asolkar *et al.* (1992), Borthakur (1993), Chatterjee & Pakrashi (1997), Dwivedi (1993), Girach (1992), Jain (1991), Rastogi & Mehrotra (1999b, 2002), Parrotta (2001), Patil & Patil (2002), Thakur *et al.* (1989).

Aneilema scapiflorum Wight.

Family : Commelinaceae.

Hindi Name :

Hindi : Siyah Musli.

Distribution : Temperate & Tropical Himalaya, Gangetic Plains, W. Peninsula.

Parts Used : Root. Root bark.

Utilization : Root said to be astringent and considered useful in headache, giddiness, fever, jaundice, deafness, antidote to poisons and snake bite. Shade dried root bark employed with benefit in asthma, colic, piles, infantile convulsions, impotence, spermatorrhoea and for incontinent urine.

References : Agarwal (2003), Caius (2003), Chopra *et al.* (1956), Kirtikar & Basu (1984), Watt (1889).

Anemone obtusiloba D. Don

Family : Ranunculaceae.

Hindi Name:

Hindi : Ratanjot.

Distribution : Temperate Himalaya from Kashmir to Sikkim.

Parts Used : Root. Rootstock.

Utilization : Rootstocks used for concussions whereas roots given internally with milk for contusions and externally as blistering agent. Also root paste applied for healing wounds, sores and boils.

References : Agarwal (2003), Ambasta (1986), Chhetri (2005), Chopra *et al.* (1956), Kapur *et al.* (1996), Kaul (1997), Watt (1889).

Anemone rupicola Camb.

Family : Ranunculaceae.

Distribution : Inner Alpine Valleys of the Himalaya from Kashmir to Sikkim: 4,000-5,000 m.

Part Used : Roots.

Utilization : Roots mixed with milk given internally for contusions and externally as a blistering agent.

References : Hooker (1872), Kapur (1996a).

Anemone vitifolia Ham.

Family : Ranunculaceae.

English Name:

Eng. : Vine Leaved Anemone.

Distribution : Temperate Himalaya from Kashmir to Arunachal Pradesh: 1,300-2,700 m.

Part Used : Root.

Utilization : Decoction prescribed for pain in lumbar region.

References : Ambasta (1986), Asolkar *et al.* (1992), Kapur (1996a).

Angelica glauca Edgew.

Family : Apiaceae.

English, Hindi & Sanskrit Names:

Eng. : Angelica, Long Wort;
Hindi : Chorak, Churu;
Sans. : Choraka, Parnachoraka, Taskara, Tilaparnika.

Distribution : High Altitudes Upto 4,000 m on W. Himalaya, Kashmir to Shimla & Garhwal.

Part Used : Roots.

Utilization : Root powder prescribed for headache, insanity, epilepsy, coryza, cough, bronchial asthma, dysentery, menorrhoea, vomiting, biliousness and as cardiac stimulant, rubifacient, diaphoretic, stomachic, diuretic, purgative, carminative and tonic. Its powder with black salt given in flatulence and dyspepsia and extr. given to livestock for severe food poisoning and internal injury.

Active Constituents : Main constituents are valerianic acid, lactones, sesquiterpenes, pinene, phellandrane, α -& β-phellandrenes, α-pinene, macrocyclic lactones, penta- and heptadecanolide, furocaumarins including chlorogenic acid, angelicolide, new furocaumarin-2″-O-acetyloxypeucedanin hydrate, archangelin, oxypeucedanin, 2,2-dimethyloxirane, formic acid, 6-methyl-bi-cyclo[3, 2, 0]-hepten-2-one, 6-propyl-bicyclo [3,2.0]-hept-6-dien-2-one, 6-propylbicyclo [3, 2, 0]-hept-6-one, [1(S)-(1α,3a, 4α,8aβ)]-1, 4-methanoazulene, β-chamigrene, dehydroledol and elemol.

Biological Activity : Phellandrene has a stimulating action on the nervous system. The bitterness of furocaumarins accounts for digestive stimulant effects.

References : Agarwal (2003), Ambasta (1986), Aswal (1996), Bhattacharjee (2001), Chauhan (1999), Jain (1991), Kaul (1997), Khare (2004), Nadeem *et al.* (2001), Nautiyal *et al.* (2003), Rastogi & Mehrotra (1998, 2001, 2002), Samant *et al.* (2001), Sharma (2003), Sharma & Rana (2000), Singh (1999), Singh & Kumar (2000b), Tiwari & Pande (2006).

Angiopteris erecta (Forssk.) Hoffm.

Family : Marattiaceae.

Distribution : Throughout Hilly Regions of India; N.E. India Upto 2,000 m.

Part Used : Root.

Utilization : Its extr. mixed with milk prescribed orally for chest pain and heart ailments (3 spoonfuls twice a day for 3 days).

References : Jain (1991), Rajendran *et al.* (1997), Rastogi & Mehrotra (1999a, b), Sood & Prakash (2007).

Anona squamosa L.

Family : Annonaceae.

English & Hindi Names:

Eng. : Custard Apple Tree;
Hindi : Sharifa, Sitaphal.

Distribution : Wild as well as Cultivated Throughout Drier Areas of India.

Part Used : Root.

Utilization : Used for abdominal pain, melancholia, spinal disorder, scorpion sting and obstinate constipation. 25 g of root paste in 1 glass of water taken once on empty stomach for inducing abortion upto 4 months pregnancy.

Active Constituents : Hydrocyanic acid.

References : Dastur (1970), Dwivedi (1993), Varghese (1996).

Anthocephalus chinensis (Lamk.) A. Rich. *ex* Walp.

Syn : *A. cadamba* (Roxb.) Miq.; *A. indicus* Rich.; *Cephalanthus chinensis* Lamk.

Family : Rubiaceae.

Hindi & Sanskrit Names:

Hindi : Kadamb;
Sans. : Kadamba.

Distribution : Sub-Himalayan Tract from Sikkim Eastwards to Bengal. N. Cicars. W. Ghats.

Part Used : Roots.

Utilization : Used as an abortifaciant.

Active Constituents : Contains quinoric acid.

References : Agarwal (2003), Asolkar *et al.* (1992), Dhiman (2003).

Antidesma diandrum (Roxb.) Roth.

Family : Euphorbiaceae.

English & Hindi Names:

Eng. : Matha;
Hindi : Amta, Dhakki, Katma.

Distribution : Garhwal, Kumaon, Oudh, Bengal, S. India & Myanmar: Common in the Hill Forests of the Santal Pergunnahs.

Part Used : Root.

Utilization : Pieces of roots tied to both legs and hands to cure polio.

References : Ambasta (1986), Hembrom (1996), Watt (1889).

Apium graveolense L.

Family : Apiaceae.

English, Hindi & Sanskrit Names:

Eng. : Celery, Cultivated Celery, Wild Celery;
Hindi : Ajmoda, Bari Ajmod, Shalari;
Sans. : Ajmoda.

Distribution : Naturalised in the Foot Hills of N.W. Himalaya & the Outlying Hills of J. & K., H.P. & U.P.

Part Used : Root.

Utilization : Known to be used as an emmenagogue, expectorant, antirheumatic, carminative, diuretic, apertive, colic and for curing flatulence, dyspepsia, torminia, cystitis, anasarca, dysuria, bronchitis cough, dysmenorrhoea, functional amenorrhoea, jaundice and obstruction in urinary passages.

Active Constituents : Contains a few acetylenic compounds in the concentration of 0.0001-0.001%.

Biological Activity : Antimicrobial and spasmolytic activities confirmed.

References : Agarwal (2003), Ambasta (1986), Anonymous (1968), Bhattacharjee (2001), Dastur (1970), Karnick (1994), Said (1997), Thakur *et al.* (1989), Watt (1889).

Aporusa lindeyana Baill.

Family : Euphorbiaceae.

Hindi Name:

Hindi : Valaka.

Distribution : Grows in the Forest of Karnataka.

Part Used : Root.

Utilization : Decoction given in jaundice, fever, headache, insanity and seminal loss.

References : Agarwal (2003), Ambasta (1986), Caius (2003), Maya *et al.* (2003).

Aquilegia vulgaris L. ssp. *pyrenaica* DC.

Syn : *A. karelini* Baker; *A. moorcroftiana* Wall. *ex* Royle; *A. olympica* Boiss.

Family : Ranunculaceae.

English Name:

Eng. : Common Columbine.

Distribution : Himalaya: (Kashmir to Nepal): 3,800-4,200 m.

Part Used : Root.

Utilization : Root extr. good against inflammations and removal of kidney stones.

Active Constituents : Berberine and megnoflorine (20-.75 %) are the active principles of the roots.

References : Ambasta (1986), Kaul (1997), Rastogi & Mehrotra (1999a, b).

Arctium lappa Linn.

Syn : *Lappa major* Geartn.; *L. tomentosa* Lamk.

Family : Asteraceae.

English Names:

Eng. : Bengal's Buttons, Burdock, Cuckle-Buttons, Hardock, Lappa, Personata, Turky-Bur.

Distribution : Native to W. Himalaya from Kashmir to Shimla.

Part Used : Root.

Utilization : Regarded as diuretic, antiphlogistic, depurative, diaphoretic, antidiabetic and anticancerous and also used to cure cutaneous eruptions, rheumatism, cystitis and anorexia nervosa.

Active Constituents : Contains inulin, mucilage, lappin, mixtures of (4E, 6E, 12E) 4,6,12-tetradecatrien-8,10-diyn-1,3-diyldiacetate, (4E,6Z,12E) 12-tetradecatrien-8, 10-diyn-1, 3-diyldiacetate (2:1) & of (4E, 6E) 4, 6-tetradecadien-8,10,12-triyn-1, 3-diyldiacetate & (4E, 6Z) 4, 6-tetradecadien-8, 12-triyn-1, 2-diyldiacetate (5:2), 11(E)1, 11-tridecadien-3, 5, 7, 9-tettrayne, (3E,

11E01, 3, 11-tridecatriene-5, 7, 9-triyne, 3 (E) 3-tridecen-5, 7, 9,11-tetrayne-1, 2-epoxide & (8Z,15Z) heptadeca-1, 8, 15-trien-11, 13-diyne, S-containing acetylenic acid, acetic acid and γ-guanidino-n-butyric acids.

Biological Activity : Root extr. inhibits tumour growth.

References : Agarwal (2003), Ambasta (1986), Asolkar *et al.* (1992), Bhattacharjee (2001), Caius (2003), Chauhan (1999), Chopra *et al.* (1956), Hooker (1887), Kapur (1996a), Karnick (1994), Rastogi & Mehrotra (2002), Sood *et al.* (2001, 2005a, b).

Arctium minus (Hill.) Bernh.

Family : Asteraceae.

Part Used : Root.

Utilization : Tincture of fresh root beneficial for rheumatism, stomachache, lung diseases, cough, asthma and menstruation.

Reference : Bhattacharjee (2001).

Ardisia colorata Roxb.

Family : Myrsinaceae.

Distribution : Assam & Cachar to Myanmar & the Nilgiris.

Part Used : Root.

Utilization : Used for allaying fever, diarrhoea and rheumatism.

References : Ambasta (1986), Chopra *et al.* (1956).

Ardisia crispa (Thunb.) DC.

Syn : *A. crenata* Roxb.; *A. crenulata* Lodd.; *A. glandulosa* Blume.

Family : Myrsinaceae.

Distribution : E. Himalaya: 1,000-2,600 m.

Part Used : Root.

Utilization : Useful against fever, diarrhoea and rheumatism.

Reference : Ambasta (1986).

Ardisia odontophylla Wall.

Family : Myrsinaceae.

Distribution : Khasia & Jantia hills.

Part Used : Root.

Utilization : Prescribed for fever, diarrhoea and rheumatism.

References : Ambasta (1986), Asolkar *et al.* (1992).

Ardisia paniculata Roxb.

Family : Myrsinaceae.

Part Used : Root.

Utilization : Decoction said to cure rheumatism and pain in venereal diseases. About 1 cup decoction of the root in combination with those of *Smilax ovalifolia* Roxb. and *Bridelia tomentosa* Bl. taken twice daily for jaundice.

Reference : Lalramnghinglova & Jha (1999).

Ardisia polycephala Wall. *ex* A. DC.

Family : Myrsinaceae.

Part Used : Root.

Utilization : Root paste prepared in cold water good for excessive bleeding and that with hot water for normal bleeding after child birth.

Reference : Lalramnghinglova & Jha (1999).

Ardisia solanacea Roxb.

Syn : *A. humilis* auct. non Vahl.

Family : Myrsinaceae.

Hindi Name:

Hindi : Banjan.

Distribution : Throughout India.

Part Used : Root.

Utilization : About two tablespoon of root extr. given thrice a day for arresting internal haemorrhage. Root paste applied locally on muscular swelling, and also said to be useful in diarrhoea, rheumatism and fever. The roots along with flowers of *Bauhinia purpurea* mixed in cow's ghee, given to check asthma.

Active Constituents : Contains bauerenol.

References : Agarwal (2003), Ambasta (1986), Aminuddin & Girach (1996), Asolkar *et al.* (1992), Chopra *et al.* (1956), Girach & Aminuddin (1995), Hosagoudar & Henry (1996a), Jha & Verma (1996), Tiwari & Tiwari (1996).

Ardisia villosa Roxb.

Syn : *A. mollis* Blume; *A. glabrata* Blume

Family : Myrsinaceae.

Distribution : From Mergui to Malacca & Penang.

Part Used : Root.

Utilization : Known to cure fever, diarrhoea and rheumatism.

References : Ambasta (1986), Hooker (1888).

Areca catechu Linn.

Family : Arecaceae.

English, Hindi & Sanskrit Names:

Eng. : Areca Nut Palm, Betel Nut Palm, Catechu Palm, Fasel Nut, Indian Nut Tree, Pinang Palm, Supari Palm;

Hindi : Supari;

Sans. : Akota, Chikkana, Kapitana, Puga, Rajatala, Tambula, Valkataru.

Distribution : Widely Distributed in the Coastal Regions from Maharashtra to Kerala & Tamil Nadu, Deccan Plateau Assam, Meghalaya, W. Bengal, Andaman & Nicobar Islands & Sri Lanka.

Part Used : Root.

Utilization : Decoction given to cure sore lips and liver ailments.

References : Caius (2003), Jain (1991), Parrotta (2001).

Arenga saccharifera Labill.

Syn : *A. pinnata* (Wurmb.) Merrill; *Borassus gomutus* Lour.; *Gomutus saccharifera* Spr.; *Sagurus rumphii* Roxb.

Family : Arecaceae.

English Names:

Eng. : Areng Palm, Gomati Palm, Malay Sago Palm, Sago Palm.

Distribution : Rare in India; Assam.

Part Used : Root.

Utilization : Considered stomachic, pectoral and for bronchitis.

References : Ambasta (1986), Anonymous (1948), Caius (2003), Watt (1889).

Argemone maxicana L. (Pl. 3C)

Family : Papaveraceae.

English, Hindi & Sanskrit Names:

Eng. : Gamboge Thistle, Jamaica Yellow Thistle, Mexican Poppy, Prickly Poppy;

Hindi : Bharbhand, Biladhutura, Kutaila, Satyanashi, Shialkanta, Ujarkanta;

Sans. : Hemadugdha, Kanchani, Katuparni, Pitapushpa, Suvarna, Tiktadugdha.

Distribution : Throughout India.

Parts Used : Root. Root bark.

Utilization : Root useful in scabies, leucoderma, menstrual disorders, gonorrhoea, piles, cough, fever, malaria, regulating fertility, stomachache, inflammation and as mouthwash, anthelmintic and blood purifier. Its paste applied as an antidote for scorpion sting and ant bite. One teaspoonful of root powder given three times a day along with cow's ghee (2:1) once in morning for two weeks for spermatorrhoea; and its juice used as an eye drops to cure inflammation. Fried paste of 6g root bark taken for 11 days to treat syphilis.

Active Constituents : Contains protopine, berberine, resin and a toxic principle.

References : Agarwal (2003), Anonymous (1968), Caius (2003), Chopra *et al.* (1956), Dash & Mishra (2000), Dastur (1970), Drury (1978), Girach *et al.* (1997), Gogoi & Das (2003), Jain (1991, 1996), Khan *et al.* (2003), Khanna & Kumar (2000), Kumar & Jain (1998), Kurian (1995), Panda & Das (2000), Parrota (2001), Patil (2004), Patil & Patil (2002), Rao & Henry (1996), Reddi *et al.* (2005), Sahoo & Mudgal (1995), Sarangi & Sahu (2003), Sikarwar (2002), Singh (2000), Srivastava *et al.* (2003), Upadhyay & Singh (2005), Varghese (1996), Verma & Chauhan (2006).

Argyreia cuneata (Willd.) Ker-Gawl.

Syn : *Convolvulus cuneatus* Willd.; *Lettsomia cuneata* Roxb.; *Rivea cuneata* Wight.
Family : Convolvulaceae.

English Name :

Eng. : Purple Convolvulus.

Distribution : Upper Hills of Peninsular India: 950-1,400 m.

Part Used : Root.

Utilization : 200 ml decoction of root and cumin given orally to initiate labour pain, easy delivery and expelling the placenta. Root paste applied on scabies.

References : Hosagoudar & Henry (1996a), Parrotta (2001).

Argyreia cymosa (Roxb.) Sweet.

Syn : *Convolvulus glomeratus* Wall.; *Lettsonia cymosa* Roxb.
Family : Convolvulaceae.

Distribution : Deccan Peninsula.

Parts Used : Root.

Utilization : Fresh root juice (1-2 drops) used as an eye drops to check redness of eyes and decoction (1 teaspoon, every morning with desired quantity of milk) given for 3-4 days to treat burning micturition.

References : Girach (2001), Hooker (1885).

Argyreia nervosa (Burm.f.) Boj. (Pl. 3D)

Syn : *A. speciosa* (L.f.) Sweet; *Convolvulus speciosa* L.; *C. speciosus* L.f.; *C. nervosus* Burm.f.; *Ipomoea speciosa* (L.f.) Pers.; *Lettsomia nervosa* (Burm.f.) Roxb.; *Rivea nervosa* (Burm.f.) Hall.f.

Family : Convolvulaceae.

English, Hindi & Sanskrit Names:

Eng. : The Elephant Creeper, Woolly Morning Glory;
Hindi : Samandarsokh;
Sans. : Ajantri, Bastaantri, Dridhadaru, Murva, Samudrapalaka, Vriddhadara.

Distribution : Assam. Bengal. Karnataka. Tamil Nadu to an elevation of 900 m on the Eastern Slopes of the W. Ghats.

Part Used : Root.

Utilization : Known to be bitter, diuretic, nervine tonic, alterative, aphrodisiac and useful in gleet, gonorrhoea, anaemia, syphilis, strangury, diseases of nervous system, migraine, healing wounds, muscular pain, skin diseases, chronic ulcers and venereal diseases. Its paste applied locally in snake bite, and also on mixing it with those of *Asparagus racemosus, Grewia hirsuta* and *Hemidesmus indicus* given in cold, cough and fever.

References : Asolkar *et al.* (1992), Chatterjee & Pakrashi (1997), Dastur (1970), Dey (1973), Dutta & Nath (2000), Dwivedi (1993), Jadeja (1999), Khare (2004), Kumar (2002), Maheshwari *et al.* (1997), Parrotta (2001), Reddi *et al.* (2005), Sarangi & Sahu (2004), Tirkey (2004), Watt (1889).

Argyreia sericea Dalz.

Family : Convolvulaceae.

Part Used : Root.

Utilization : Given in vertigo.

Reference : Kothari & Moorthy (1996).

Argyreia strigosa (Roth.) Sant & Patel

Syn : *A. setosa* (Roxb.) Choisy; *Convolvulus setosus* Wall.; *C. strigosus* Spreng.; *Lettsonia setosa* Roxb.

Family : Convolvulaceae.

Distribution : Throughout Tropical & Sub-Tropical India.

Part Used : Root.

Utilization : Powder taken orally to cure burning sensation in abdominal region and urinary duct.

References : Ambasta (1986), Hooker (1885), Sharma & Singh (1999), Singh & Pandey (1996).

Arisaema intermedium Bl.

Family : Araceae.

English Name:

Eng. : Indian Turnip.

Distribution : W. Himalaya (Shimla, 800 m).

Part Used : Root.

Utilization : Paste applied on ulcer to remove the puss.

Reference : Bhattacharjee (2001).

Arisaema tortuosum (Wall.) Schott.

Syn : *Ariesaema curvatum* Kunth. Sensu Hook.f. ; *A. helleboriefolium* Schott.; *Arum tortuosum* Wall.

Family : Araceae.

English & Hindi Names:

Eng. : Cobra Flower;

Hindi : Birbanka, Gurien, Kiri-Ki-Kuri, Samp-Ki-Khumb.

Distribution : Temperate Himalaya from Shimla to Assam & on W. Ghats Upto 2,800 m.

Part Used : Roots.

Utilization : Given to kill worms infesting cattle in the rains. Its paste applied externally to reduce body swelling in domestic animals.

References : Agarwal (2003), Ambasta (1986, Chopra *et al.* (1956), Sood & Prakash (2007), Sood & Thakur (2004), Watt (1889).

Arisaema wallichianum Hook.f.

Syn : *A. propinquum* Schott.

Family : Araceae.

Part Used : Root.

Utilization : Its extr. applied to hasten healing of cuts and wounds caused by leopard.

References : Kaul (1997), Negi *et al.* (1993).

Aristolochia bracteata Retz.

Syn : *A. bracteolata* Lamk.

Family : Aristolochiaceae.

English, Hindi & Sanskrit Names:

Eng. : Bracteated Birthwort;
Hindi : Bhringi, Gandan, Kiramar;
Sans. : Dhumrapatra, Kitamari.

Distribution : Upper Gangetic Plains, W. Bengal, Gujarat & Peninsular India.

Part Used : Root.

Utilization : Dried roots used internally as an abortifacient. 1/2 cup water extr. of roots given orally in dysmenorrhoea, and its decoction useful for expelling round worms and curing ophthous ulcers. Root paste applied externally for curing scorpion stings and snake bites.

Active Constituents : Aristolochic acid, KCl and KNO_3.

References : Agarwal (2003), Ambasta (1986), Chopra *et al.* (1956), Dastur (1970), Jain (1991), Kumar & Pullaiah (2000), Parrotta (2001), Rao *et al.* (1996), Rastogi & Mehrotra (1999a, b), Sarangi & Sahu (2004), Singh & Pandey (1998), Thomas & Britto (2000).

Aristolochia indica L.

Syn : *A. lanceolata* Wight.
Family : Aristolochiaceae.

English, Hindi & Sanskrit Names:

Eng. : The Indian Birthwort;
Hindi : Isharmul, Ishvarimul, Ishwari;
Sans. : Arkmuula, Jata, Iswari, Naakuli.

Distribution : Throughout the Low Hills & Plains of India from Nepal to W. Bengal & S. India.

Part Used : Root.

Utilization : Known to be useful as an abortifacient, gastric stimulant, bitter tonic, blood purifier, emmenagogue, antiarthritic and for checking headache, vertigo, skin diseases, dysentery, diarrhoea, white leprosy, body pain, snake bite, rashes, dropsy, rheumatism, fever, impotency, bowel complaints of children and intermittent fever. Smelling fresh root believed to expedite delivery.

Active Constituents : Active principles are nitrophenanthrene, aristolochic acid, aristolic, p-coumaric acids, aristolactam, aristolactam glucoside, aristolic acid, aristolamide, aristolochic acid II methyl ether, aristolochene, seline -4 (14), ishwarol, ledol, ishwarone, stigmast-4en-3-one, ceryl alcohol, d-camphor, ishwarane, aristolochene , two sesquiterpene alcohols-(128)-7,12-secoishwaran-12-ol and (+) ledol, 2-hydroxy-1-methoxy-4H-dibenzo[de,g]quinoline 4,5-(6H)-dione(I), aristolide, 5α-stigmastan-3, 6-dione, (-) cubebin, (-) hinokinin and two aristolindiquinone.

Biological Activity : Aristolochic acid has been found to possess anticancer, antiestrogenic and antiimplantation, antispermatogenic properties.

References : Agarwal (2003), Anonymous (1948, 1968), Asolkar *et al.* (1992), Chatterjee & Pakrashi (1997), Chopra *et al.* (1956), Dastur (1970), Dey (1973), Drury (1978), Goud *et al.* (2000), Girach (1992), Hosagoudar & Henry (1996b), Goud & Pullaiah (1996), Jain (1991), Jha & Verma (1996), Khare (2004), Kothari & Moorthy (1996), Kumar & Jain (1998), Kumar & Pullaiah (2000), Painuli & Maheshwari (1996), Parrotta (2001), Patil & Patil (2002), Ragupathy & Mahadevan (1991), Rajendran *et al.* (2001), Rao & Henry (1996), Rastogi & Mehrotra (1999a, b, 2001), Sarin (1996), Saxena *et al.* (1997), Singh & Maheshwari (1992), Singh & Pandey (1998), Sur & Halder (2002), Thakur *et al.* (1989),Watt (1889).

Aristolochia saccata Wall.

Family : Aristolochiaceae.

Distribution : Entire N.E. Region, E. & C. Himalaya.

Part Used : Root.

Utilization : Used as tonic, emmenagogue, stimulant, alexiteric, antiarthritic and considered valuable antidote for bites of poisonous snakes and insects, and in diseases of the spleen.

References : Jain (1991), Jain *et al.* (1991).

Aristolochia serpentina L.

Family : Aristolochiaceae.

English & Hindi Names:

Eng. : Birthwort;
Hindi : Sarpmoola.

Distribution : Indian Gardens.

Part Used : Root.

Utilization : Snake-like aromatic roots used in cobra bite and for curing malaria, typhous fever, small pox and pneumonia. Poultice of roots employed for healing open wounds and skin ulcers.

References : Agarwal (2003), Bhattacharjee (2001).

Armoracia rusticana P. Gaertner, B. Meyer & Scherb.

Syn : *A. lapathifolia* Gillb.; *Cochlearia armoracia* Linn.

Family : Brassicaceae.

English Name:

Eng. : Horse Radish.

Part Used : Roots.

Utilization : Reported to be useful in albuminuria, aphonia, asthma, gonorrhoea, athlete foot and urinary disorders and as stimulant, diaphoretic, diuretic, digestive, antiscorbutic and appetizer.

Active Constituents : Contains a volatile oil and singirin.

Biological Activity : Root extr. exhibits B.P. depressant and spasmolytic properties.

References : Ambasta (1986), Asolkar *et al.* (1992), Baburaj *et al.* (2000).

Arnebia benthamii (Wall. *ex* G. Don) J.M. Johanston

Syn : *Mactrotomia benthamii* (Wallich) A.DC.

Family : Boraginaceae.

Hindi Names:

Hindi : Balchari, Ratanjot.

Part Used : Root.

Utilization : Useful for the treatment of asthma, rheumatic pains, swellings of organs, heart and hysterial disease, headache, fever, boils, ulcers, cuts, wounds, eye complaints, and as antiseptic. Powdered purple coloured root mixed with hair oils considered as a red dye and hair tonic by the local people.

References : Asolkar *et al.* (1992), Aswal & Mehrotra (1994), Arya & Prakash (2000), Beigh *et al.* (2003), Bhattacharjee (2001), Jain (1991), Kaul (1997), Negi & Pant (1990).

Arnebia euchroma (Royle) Johnston.

Syn : *Macrotomia perennis* Boiss.

Family : Boraginaceae.

Hindi Name:

Hindi : Ratanjot.

Distribution : Alpine W. Himalaya, W. Tibet, Nepal: 3,300-4,500 m.

Part Used : Root.

Utilization : Known to be used for toothache, eye diseases, cuts, wounds, fire burns, tumours, cough, blood purification and as an antipyretic, antimicrobial, antiinflammatory and tonic for brain. The purple roots used as hair dye and tonic by local people.

Active Constituents : Active properties are due to alkanin β, β'-dimethylacrylate, shikanofurans B & C, de-O-methyl-lasiodiplodin, arnebinone,shikonin, deoxyshikonin, β-hydroxyisovalerylalkanin, acetylshikonin, β, β-simcthylacrylshikonin, β, β-dimethylacrylalkanin, â-hydroxyisovalerylshikonin, â-acetoxyvalerylalkanin, tetraacrylshikonin, arnebinol, shikanofuran and de-O- methyllasiodiplodin, arnebinone, arnebifuranone, polysaccharides-LR-2IId-Ib, LR-2IIb-3a & LR-2IId-5a- consisting mainly of mannose, galactose, glucose, glucuronic acid, rhamnose, fucose and arabinose.

Biological Activity : Anticancer and antimicrobial activities positive.

References : Ambasta (1986), Aswal & Mehrotra (1997), Chauhan (1999), Jain (1991), Kaul (1997), Rastogi & Mehrotra (1998, 2001, 2002), Sharma (2003), Sood *et al.* (2001, 2005c).

Arnebia guttata Bunge

Syn : *A. tibetana* Kurz.
Family : Boraginaceae.

Distribution : Kashmir: 2,100-3,600m.

Part Used : Root.

Utilization : Used as blood purifier and a source of red dye.

Active Constituents : Deoxyshikonin, acetylshikonin, shikonin, β-hydroxyisovalerylshikonin, β, β-dimethylacryl-shikonin, and tetraacryshikonin are the active constituents.

References : Ambasta (1986), Asolkar *et al.* (1992), Kaul (1997), Rastogi & Mehrotra (2001, 2002).

Artanema sesamoides (Vahl.) Benth.

Family : Scrophulariaceae.

Hindi & Sanskrit Names:

Hindi : Neermulli;
Sans. : Kokilaksha.

Distribution : W. Peninsula.

Part Used : Root.

Utilization : Decoction of root effectively employed for rheumatism, siarrhoea, kidney stones, syphilis and ophthalmia.

References : Agarwal (2003), Kirtikar & Basu (1984), Maya *et al.* (2003).

Artemisia japonica Thunb.

Syn : *A. parviflora* Roxb. *ex* D. Don
Family : Asteraceae.

Hindi Name:

Hindi : Pati.

Distribution : Himalaya (Kashmir to Sikkim): 2,100-3,500 m, Meghalaya 1,500-2,100 m), Parasnath Hill in Bihar,

1,200 m, W. Ghats from Konkan Southwards to Pulney Hills.

Part Used : Root.

Utilization : Roots good for throat disorders.

References : Jain (1991), Jain & Goel (2005), Rastogi & Mehrotra (1999a, b).

Artemisia nilgirica (Clarke) Pamp.

Syn : *A. indica* Willd.; *A. vulgaris* auct (non L.);
A. vulgaris var. *nilagirica* C.B.Clarke.
Family : Asteraceae.

English, Hindi & Sanskrit Names:

Eng. : Indian Worm Weed, Mugwort;
Hindi : Nagdana;
Sans. : Damanaka, Naagdamani, Valamota.

Distribution : Throughout the Hilly Regions of India, ascending to 3,600 m in the W. Himalaya & 1,500-2,400 m in Sikkim & Khasia hills. Mount Abu in Rajasthan. W. Ghats. Konkan Southwards to Kerala.

Part Used : Root.

Utilization : Decoction used as tonic, antiseptic and analgesic.

Active Constituents : These are trideca-1, 3, 5-trien-9, 9, 11-triyne, artemisia ketone and 1-acetoxytetradeca-4, 6-dien-8, 10, 12-triyn.

References : Agarwal (2003), Ambasta (1986), Chatterjee & Pakrashi (1997), Dhiman (2003), Jain (1991), Khare (2004), Negi *et al.* (1999), Rastogi & Mehrotra (1999a, b).

Artemisia sacrorum Ledeb.

Syn : *A. gmelini* Web. *ex* Stechm.
Family : Asteraceae.

Distribution : W. Tibet, Kanawar, & Kumaon: 3,000-5,600 m.

Part Used : Root.

Utilization : Poultice used for relieving headache and healing sores.

References : Negi *et al.* (1999), Watt (1889).

Artocarpus heterophyllus Lam.

Syn : *A. integra* (Thunb.) Merill.; *A. integrifolia* L.f.

Family : Moraceae.

English, Hindi & Sanskrit Names:

Eng. : The Jack-Fruit Tree;

Hindi : Kathal;

Sans. : Panasa.

Distribution : Evergreen Forest of the W. Ghats: 450-1,200 m. Hotter Parts of the Country (Assam being the largest producer).

Part Used : Root.

Utilization : Decoction of the root useful in diarrhoea, skin diseases, asthma, fever and stomach complaints.

Active Constituents : Active ingredients are β-sitosterol, ursolic acid, cycloartenone, artoflavonone, artonins A & B, heterophyllin, cycloheterophyllin, artonins J, K & L, cyclocarpin A, heteroflavone C & 9-hydroxytridecyldocosanoate, betulin and betulinic acid

References : Asolkar *et al.* (1992), Chatterjee & Pakrashi (1997), Chopra *et al.* (1956), Jain (1991), Kurian (1995), Rastogi & Mehrotra (1998).

Asarum canadens L.

Family: Aristolochiaceae.

Part Used : Root.

Utilization : Considered stimulant and used to treat heart palpitation, nervous problems and fevers. Its powdere inhaled as snuff to relieve aching head and eyes. A tea prepared from root used to relieve stomach gas.

Reference : Bhattacharjee (2001).

Asclepias curassavica Linn.

Family: Asclepiadaceae.

English & Hindi Names:

Eng. : Blood Flower, Curassavian Swallow-Wort, West Indian Ipecacuanha;

Hindi : Kakatundi.

Distribution : S. India & Bengal.

Part Used : Root.

Utilization : Useful in India as purgative, cathartic, emetic, irritant, astringent, depurative, vermifuge, haemostatic and in the treatment of cancerous growth, warts, piles, dysentery, pneumonia and gonorrhoea.

References : Ambasta (1986), Anonymous (1948, 1968), Chopra *et al.* (1956), Kurian (1995), Subramanium (2000), Watt (1889).

Asparagus adscendens Buch-Ham. *ex* Roxb.

Syn : *Asparagopsis adscendens* Kunth.; *Asparagus volubilis* Wall.

Family : Liliaceae.

Hindi Names:

Hindi : Satavar, Sufedmusli.

Distribution : Tropical & Sub-Tropical Parts of N.W. Himalaya Upto 1,600 m.

Parts Used : Root. Root bark.

Utilization : Root bark with cow's milk beneficial as tonic Root used as appetizing, demulcent in veterinary medicines, refrigerant, galactagogue, diuretic, aphrodisiac, astringent and also useful in dysentery, diarrhoea, general debility, throat complaints and leprosy.

Active Constituents : Contains asparagines, saponin (steroidal), asparanins C & D, asparosides C & D, 3β-O-β-D-(2'-O-tetracosyl) xylopyranosylstimasterol and 3β -O-β-D-glucopyranosyl (1→2)-α-L-arabinopyranosylstigmasterol.

References : Agarwal (2003), Anonymous (1948), Asolkar *et al.* (1992), Aswal (1996), Chauhan (1999), Chopra *et al.* (1956), Dwivedi (1993), Kirtikar & Basu (1984), Negi *et al.* (1993), Rastogi & Mehrotra (1998, 2001), Samant *et al.* (2001), Singh (1999), Sood & Prakash (2007), Sood & Thakur (2004), Watt (1889).

Asparagus gonocladus Baker

Family : Liliaceae.

English & Hindi Names:

Eng. : Piri Kari Jeri;
Hindi : Shakakul.

Distribution : Peninsular India from Konkan Southwards in the Ghats: 1,200-1,800. Plains of India.

Part Used : Root.

Utilization : Useful in cutaneous diseases, gonorrhoea and as aphrodisiac; and also for treating fever in children by hanging around the neck.

References : Chopra *et al.* (1956), Hembrom (1996), Sood & Prakash (2007).

Asparagus gracillus Royle

Family : Liliaceae.

Part Used : Root.

Utilization : Crushed root applied as poultice on the head to check headache. A snuff of dried root given to cattle to stop recurrent urinary bleeding.

Reference : Negi *et al.* (1999).

Asparagus racemosus Willd.

Family : Liliaceae.

English, Hindi & Sanskrit Names:

Eng. : Indian Asparagus, Wild Asparagus;
Hindi : Chatwal, Satawari, Satmuli;
Sans. : Atirasa, Bahusutaa, Indivari, Satavarai.

Distribution : Common in India, ascending Upto 1,300 m in the Himalaya & Sri Lanka.

Part Used : Root.

Utilization : Known to be useful in diabetes, acidity, diarrhoea, dropsy, dizziness, excessive menstrual flow, post-delivery complaints, bodyache, abdominal pain, leucorrhoea, stomach ulcers, nephrocalcinosis, lactation, gastric trouble, flatulence, gonorrhoea, urinary trouble, skin itching, joint pains, jaundice, hypertension, harmorrhage, malaria, as galactagogue and febrifuge. 5g dose of dried/fresh root powdered with black peppers (*P. nigrum*) given twice a day to check frequent nocturnal emissions. The boiled milk extr. of fresh roots (tuber) taken orally to cure blood dysentery. To treat hydrocele, root pounded together with black pepper applied as paste on affected part. About 10g of root powder along with cow's butter given as an antacid twice daily as ulcer protective. Extr. of roots grounded with those of *Bombax ceiba* and sugar candy taken daily to check premature ejaculation. Juice of the roots (10 ml) with equal quantity of milk given twice daily for a fortnight in chronic liver disorders associated with alcoholism and loss of appetite. Roots are also kept below the pillow for inducing sleep in children. Poultice of the root paste applied on forehead to get relief from headache and for healing cattle wounds. Pounded roots along with aerial red root of *Smilax prolifera* prescribed as a drink to cure urinary disorders as well as discharge of blood in urine.

Active Constituents : These are satavarin, a glycoside of sarsasapogenin, shatavarin I, shatavarin II and shatavarin IV.

Biological Activity : Shatavarin-I showed specific antioxytocic activities *in vitro* and *in vivo*. Alc. extr. of roots found to possess antiamoebic effect against *E. histolica* and antiviral effect against Ranikhet disease of virus.

References : Agarwal (2003), Ambasta (1986), Aminuddin & Girach (1991, 1996), Anonymous (1948, 1968),

Asolkar *et al.* (1992), Aswal (1996), Bhatt (2002), Bhatt *et al.* (2003), Bhattacharjee (2001), Borthakur (1993), Borthakur *et al.* (2004), Chaudhari & Hutke (2002), Chopra *et al.* (1956), Dwivedi (1993), Girach (1992), Girach & Aminuddin (1995), Gogoi & Das (2003), Gogoi *et al.* (2003 a, b), Hembrom (1996), Islam & Jha (2003), Jain (1991), Kadel and Jain (2006), Kumar & Lal (2003), Khare (2004), Kirtikar & Basu (1984), Kumar & Pullaiah (1998), Maheshwari *et al.* (1997), Maliya (2005), Mohanty & Padhy (1996), Nayak *et al.* (2003), Panda & Das (2000), Pandey & Rout (2006), Panthi & Chaudhary (2003), Parrotta (2001), Punjani (2006), Rastogi & Mehrotra (2002), Reddi *et al.* (2005), Sarangi & Sahu (2004), Saren *et al.* (2000), Sarma *et al.* (2001), Sharma (2003), Singh (1999), Singh *et al.* (2003), Thakur *et al.* (1989), Varghese (1996), Watt (1889).

Asphodelus fistulosus Linn.

Syn : *A. tenuifolius* Cav.

Family : Liliaceae.

English & Hindi Names:

Eng. : Asphodel;

Hindi : Piazi.

Distribution : Throughout Gangetic Plains from Punjab to Bengal.

Part Used : Root.

Utilization : Ash of the root mixed with oil, applied externally in baldness, skin blotches, pigmentation and ringworm. Internally, the ash is diuretic, antiflammatory and emmenagogue.

References : Agarwal (2003), Khare (2004).

Aster amellus Linn.

Syn : *A. trinervius* Roxb.

Family : Asteraceae.

Distribution : C. & W. Himalaya & Khasia Hills.

Part Used : Root.

Utilization : Used for cough, pulmonary affections, malarial fever and haemorrhage.

References : Ambasta (1986), Anonymous (1948), Chopra *et al.* (1956).

Aster molliussculus Wall.

Family : Asteraceae.

Distribution : W. Himalaya (Kashmir-Kumaon): 1,500-2,400 m. Sikkim.

Part Used : Root.

Utilization : Decoction rubbed on the chest to relieve muscular pain.

References : Asolkar *et al.* (1992), Negi *et al.* (1999).

Asteracantha longifolia (L.) Nees.

Syn : *Hygrophilla auriculata* (Schumach) Heime *H. spinosa* T. Anders.

Family : Acanthaceae.

Hindi & Sanskrit Names:

Hindi : Gokhula-Kanta, Gokura, Talimakhana;

Sans. : Kokilaksha.

Distribution : Wasteplaces, Hedges, Rice Fields, Roadsides in Tropical Parts of India.

Part Used : Root.

Utilization : Used as demulcent, refrigerant, diuretic and tonic.

Active Constituents : Contains betulin and lupeol.

References : Ambasta (1986), Anonymous (1968), Chatterjee & Pakrashi (1997), Chopra *et al.* (1992), Dastur (1970), Drury (1978).

Astilbe rivularis D. Don

Family : Saxifragaceae.

Distribution : Himalaya (Kashmir to Bhutan): 1,500-2,700 m. Meghalaya: 1,200-1,800 m.

Part Used : Root.

Utilization : Dried root chewed directly in case of dysentery diarrhoea. Its paste along with root of *Berginia ciliata* used in post-natal recovery of mother after child birth.

References : Chhetri (2005), Rastogi & Mehrotra (1999b), Sharma (2003).

Astragalus chlorostachys Lindl.

Family : Fabaceae.

Distribution : Chamoli, U.P.

Part Used : Root.

Utilization : Powdered root taken with milk as a vital tonic for longevity and strength.

References : Asolkar *et al.* (1992), Negi *et al.* (1999).

Astragalus polycanthus Royle *ex* Benth.

Syn : *A. grahamianus* Royle; *A. scariosus* Benth.

Family : Fabaceae.

Distribution : Ascending from plains of the Punjab to 4,000 m, reaching to Kumaon & Gharwal.

Part Used : Root.

Utilization : Powdered roots used for ophthalmic ailments and as toothbrush to remove toothache.

References : Kapur *et al.* (1996), Hooker (1879).

Astragalus rhizanthus Royle *ex* Benth.

Family : Fabaceae.

Distribution : W. Himalaya, Temperate & Alpine Regions.

Part Used : Root.

Utilization : Used as heart stimulant. Paste applied externally against skin infection.

References : Ballabh & Chaurasia (2006), Kaul (1997), Sood *et al.* (2001).

Atropa belladonna Linn.

Syn : *A. acuminata* Royle *ex* Lindl.
Family : Solanaceae.

English, Hindi & Sanskrit Names:

Eng. : Belladona, Deadly Nightshade;
Hindi : Sag-Angur;
Sans. : Suchi.

Distribution : W. Himalaya from Kashmir-Shimla: 2,000-4,000 m.

Part Used : Root. .

Utilization : Recorded to be known as anodyne, antispasmodic, emetic, mydriatic, narcoticsedative, antiseptic, stimulant to respiration and blood circulation, and also for wounds, ulcers, giddiness, dilation of pupils, asthma, whooping cough, external treatments of gout neuralgic pains and rheumatism and healing fractured and dislocated parts.

Active Constituents : Main ingredients are chief alkaloids, (-)-hyocyamine, atropine, apotropine, scopolamine, tropine, hydroxycaumarins (including scopoline, scopoletine), tannins, hyosine (scopolamine), bellaradine, cuscohygrine, N-methylputrescine oxidase, putrescine, N-methyl transferase, scopoletin, starch, hyoscine-N-oxide, nicotine, asparagines, choline, succinic acid, 7-monoglucoside of kaempferol and belladonnine.

References : Agarwal (2003), Bhattacharjee (2001), Chatterjee & Pakrashi (1997), Chopra *et al.* (1956), Dey (1973), Khare (2004), Rastogi & Mehrotra (1999b), Sharma (2003), Singh (1996), Sinha & Sinha (2001).

Atylosia scrabaeoides (L.) Benth.

Syn : *Cantharospermum pauciflorum* Wight & Arn.; *Dolichos scrabaeoides* L.
Family : Fabaceae.

Hindi Name:

Hindi : Walkollu.

Distribution : W. Himalaya & Sub-Tropical Regions of India.

Part Used : Root.

Utilization : 3g powdered root per dose given thrice a day to cure gastric trouble and stomachache. Its poultice checks hair fall or cure baldness and rheumatism. Aq extr. of roots used as ear drops (5-10 drops once a day) to cure deafness.

References : Agarwal (2003), Girach (1992), Parrotta (2001), Reddi *et al.* (2005), Sahoo & Bahali (2003).

Atylosia volubilis (Blanco) Gamble.

Syn : *A. mollis* Baker (FBI) in Part non Benth.; *A. crassa* Prain.

Family : Fabaceae.

Distribution : Almost Throughout India Upto 900 m.

Part Used : Root.

Utilization : Decoction (half teaspoonful) taken orally with a cup of cow's milk to treat piles for 3-4 days consecutively. Roots chewed in Bihar for mad dog bite and toothache.

References : Asolkar *et al.* (1992), Jain (1996).

Averrhoa carambola Linn.

Family : Oxalidaceae.

Hindi Name:

Hindi : Kamrakh.

Distribution : Cultivated in the Hotter Parts of India.

Part Used : Root.

Utilization : Paste of the root mixed with those of *Eclipta alba* effectively applied in gum bleeding.

References : Agarwal (2003), Saikia & Nath (2003).

Azima tetracantha Lam.

Syn : *Fagonia montana* Miq.; *Monetia barleriodes* L'Herit; Roxb.
Family : Salvadoraceae.

English, Hindi & Sanskrit Names:

Eng.	: Mistletoe Berry Thorn;
Hindi	: Kantagurkamai;
Sans.	: Kundali.

Distribution : Orissa, W. Bengal & Peninsular India.

Parts Used : Root. Root bark.

Utilization : Known to be useful in dropsy, rheumatism and as diuretic. Root bark lessens pain in joints and muscular rheumatism.

Active Constituents : Roots yield friedelin, lupeol, glutinol and β-sitosterol.

References : Agarwal (2003), Asolkar *et al.* (1992), Chatterjee & Pakrashi (1997), Chopra *et al.* (1956), Rao & Henry (1996), Watt (1889).

Balanites aegyptiaca (L.) Del.

Syn : *B. roxburghii* Planch.; *Ximenia aegyptiaca* L.
Family : Zygophyllaceae.

English, Hindi & Sanskrit Names:

Eng.	: Desert Date;
Hindi	: Hingan, Hingoli, Hingu;
Sans.	: Angavriksha, Hingot, Ingudi.

Distribution : Wild in Semi-Arid Areas of M. P., Rajasthan & A.P.

Part Used : Root.

Utilization : Decoction reportedly emetic, anti-inflammatory and also cures dysentery. Root paste applied in eye infection of cattle, and also rubbed locally in internal haemorrhage following accidental fall.

Active Constituents : These are yamogenin-O-α-L-(1→2)-β-D-glucopyranosyl (1→4)[α-L-rhamnopyranosyl (1→2)]-β-D-glucopyranoside, yamogenin-O-β-D-xylopyranosyl

(1>3)-[α-L-rhamnopyranosyl (1>2)]glucopyranoside and yamogenin-O-α-L-rhamnopyranosyl (1>2)-β-D-glucopyranoside.

Biological Activity : Recommended as a potent molluscicide.

References : Meena *et al.* (2003), Parrotta (2001), Rao & Henry (1996), Singh & Pandey (1998), Thakur *et al.* (1989).

Baliospermum montanum (Willd.) Muell. Arg.

Syn : *B. axillare* Bl.; *B. polyandrum* Wight; *Jatropha montana* Willd.

Family : Euphorbiaceae.

English, Hindi & Sanskrit Names:

Eng. : Wild Castor, Wild Croton;

Hindi : Danti, Dantiamool, Hakum, Jangli Jamalgota;

Sans. : Amukheti, Anukula, Bhadra, Dantika, Gunapriya, Jayapala, Madhupushpa, Makunaka.

Distribution : Damp Places in Humid Areas of India, Particularly in E. & S. India.

Part Used : Root.

Utilization : Recorded to be known as purgative, cathartic, anthelmintic, diuretic, alexiteric and also useful in pains, diseases of the skin and abdomen, piles, wounds, enlarged spleen, itching, inflammations, anaemia, leucoderma, jaundice, oedema, constipation, internal abscesses, dropsy, anasarca and flatulence. Filtrate of the root crushed with jaggery (*Saccharum officinarum*) given orally for rib muscle pain and stomach pain.

Active Constituents : Contains baliospermin, montanin, phorbol-12-deoxy-13-O- palmitate, phorbol-12-deoxy-16 hydrowy-13-O-palmitate and phorbol-12-deoxy-5β-hydroxy-13-myristate.

Biological Activity : Antileukaemic and cytotoxic activities have been demonstrated in the esters of both 12-deoxyphorbol and 12- deoxy-16-hydroxyphorbol.

References : Agarwal (2003), Ambasta (1986), Aminuddin & Girach (1996), Anonymous (1948), Asolkar *et al.*

(1992), Caius (2003), Chatterjee & Pakrashi (1997), Chauhan (1999), Chopra *et al.* (1956), Dey (1973), Khare (2004), Kirtikar & Basu (1984), Mitaliya & Bhatt (2003), Nayak *et al.* (2003), Parrotta (2001), Rao & Henry (1996), Rastogi & Mehrotra (1999a, b), Reddi *et al.* (2005), Sarin (1996), Sharma (2003), Thakur *et al.* (1989), Watt (1889).

Bambusa arundinacea Willd.

Syn : *B. bambos* Druce.
Family : Poaceae.

English, Hindi & Sanskrit Names:

Eng. : Spiny Bamboo, The Thorny Bamboo;
Hindi : Bans;
Sans. : Shatparvaa, Trinadhwaja, Vamsha, Vansalochana, Venu.

Distribution : Hill Forests of W. & S. India, Orissa, W. Bengal, Assam, ascending Upto 1,000 m.

Part Used : Root.

Utilization : Credited with astringent, tonic, and refrigerant properties and also used to treat jont pain, general debility, ringworm infections of the skin, bleeding gums and painful conditions of joints.

References : Agarwal (2003), Khare (2004), Said (1997).

Barleria cristata L.

Syn : *B. ciliata* Roxb.
Family : Acanthaceae.

Hindi & Sanskrit Names:

Hindi : Raktajinti, Tadrelu;
Sans. : Jhintika, Jhiti, Sahacarah.

Distribution : Throughout India to 200 m in the Himalaya. Dry Deciduous Forests Throughout Peninsular India.

Part Used : Root.

Utilization : Decoction given to check anaemia, cough, pneumonia, rheumatism, toothache and headache.

Root paste with jaggery good for chronic stomach pain.

Active Constituents : Main constituents are anthraquinins, barlacristone and cristabarlone.

References : Agarwal (2003), Asolkar *et al.* (1992), Chatterjee & Pakrashi (1997), Chopra *et al.* (1956), Jain (1991), Parrotta (2001), Varghese (1996).

Barleria cuspidata Heyne *ex* Nees.

Family : Acanthaceae.

Part Used : Root.

Utilization : Used for dizziness and vertigo.

Reference : Kothari & Moorthy (1996).

Barleria prionitis L.

Family : Acanthaceae.

English, Hindi & Sanskrit Names:

Eng. : Thornnail Dye;
Hindi : Catserina, Jhinti, Kat-Sereya;
Sans. : Kantakuranta, Karunta, Kurantaka, Sahacara, Vajradanti.

Distribution : Naturally Throughout Hotter Parts of India in Ravine & Shady Habitats in Deciduous Forests.

Part Used : Root.

Utilization : Useful in fever, bronchitis and infected teeth, and its paste applied externally to reduce boils and glandular swellings. Its decoction used as mouthwash in toothache and extr. obtained by crushing the root taken orally twice a day for a week in leucorrhoea. Also, its decoction prescribed with *Capsicum* 'Mirch' powder to cure fever after delivery.

Active Constituents : Contains a new flavonoid luteolin-7-O-(2″-O-P-coumaryl)-β-D glycopyranoside.

References : Agarwal (2003), Ambasta (1986), Anonymous (1968), Asolkar *et al.* (1992), Chatterjee & Pakrashi

(1997), Chopra *et al.* (1956), Dwivedi (1993), Parrotta (2001), Rao & Henry (1996), Rastogi & Mehrotra (2002), Sahu & Mudgal (1995), Saklani & Jain (1994), Sarangi & Sahu (2004).

Barringtonia acutangula (L.) Gaertn.

Family : Lecythidaceae.

English, Hindi & Sanskrit Names:

Eng. : Indian Oak;
Hindi : Hijjal, Ijal, Ingar, Panniari, Samundar-Phal;
Sans. : Hijjala, Nichula.

Distribution : Sub-Himalayan Tracts & in the Plains of Bengal.

Part Used : Root.

Utilization : Root bitter and similar in properties to cinchona, i.e. cooling, aperient and expectorant.

Biological Activity : Root extr. exhibits hypoglycaemic activity in albino rats.

References : Agarwal (2003), Anonymous (1968), Chopra *et al.* (1956), Dastur (1970), Drury (1978), Kumar (2002), Watt (1889).

Basella alba Linn.

Syn : *B. rubra* Linn.
Family : Basellaceae.

English, Hindi & Sanskrit Names:

Eng. : Indian Spinach, Malabar Nightshade;
Hindi : Bon-Poi (Wild), Myal-Ki-Bhaji, Poi (Cultivated), Sufed-Bachla;
Sans. : Potaki.

Distribution : Cultivated Throughout N.E. States; Sometimes Growing Wild near Hedges.

Part Used : Root.

Utilization : Known as rubifacient, emetic and blood purifier. Its poultice reduces local swelling and decoction relieves vomiting.

Biological Activity : Root poisonous.

References : Asolkar *et al.* (1992), Kumar (2002), Kurian (1995), Watt (1889).

Bauhinia purpurea L.

Syn : *B. triandra* Roxb.

Family : Fabaceae.

English, Hindi & Sanskrit Names:

Eng. : Butterfly Tree, Geranium Tree, Orchid Tree, Pink Bauhinia;

Hindi : Kandan, Koliar, Khairwal, Kwillar, Lal Kachnar, Sona;

Sans. : Kanchan.

Distribution : Throughout Peninsular India in Deciduous Forests. Planted as a Garden Ornamental throughout India Upto 1,300 m.

Part Used : Root.

Utilization : Known to be useful in snake bite, piles, sore throat, tympanic gland trouble, arthritis, expelling gases, griping pains from the stomach and bowels, and as carminative, tonic and haemostatic agent. Its paste applied locally for oedema and pain. To treat cold and cough about 10 ml decoction with one glass of lukewarm water given twice a day for 5-7 days.

References : Agarwal (2003), Ambasta (1986), Asolkar *et al.* (1992), Bondya *et al.* (2006), Kumar & Pullaiah (2000), Kurian (1995), Parrotta (2001), Reddi *et al.* (2005), Varghese (1996), Watt (1889).

Bauhinia vahlii (Wt. & Arn.) Benth.

Syn : *B. racemosa* Vahl.; *Phanera vahlii* Benth.

Family : Fabaceae.

English, Hindi & Sanskrit Names:

Eng. : Camel's Foot Climber;

Hindi : Jallur, Mahul, Maldan, Malu;

Sans. : Asmantaka, Malanjhana, Murva.

Distribution : Deciduous Forests of India from Gujarat Southwards to Maharashtra & N. A. P.

Part Used : Root.

Utilization : Reported to be used as an antidote for scorpion sting, in bone fracture, dysentery, stomachache, fever and carminative. Its aq. extr. (1 teaspoon thrice a day) given orally for three days for the treatment of syphilis. Medicated oil obtained by boiling 500g fresh crushed root with 1 litre 'til' (*Sesamum indicum*) oil good for massaging (twice a day for one month) to treat facial paralysis.

References : Chopra *et al*. (1956), Bhattacharjee (2001), Khanna (2002b), Maheshwari *et al*. (1997), Maliya (2005), Parrotta (2001), Reddi *et al*. (2005), Sharma (2003), Singh (1996), Singh & Prakash (1994), Varghese (1996).

Bauhinia variegata L.

Syn : *B. candida* Roxb.; *B. purpurea* Wall.; *Phanera variegata* Benth.

Family : Fabaceae.

English, Hindi & Sanskrit Names:

Eng. : Buddhist Bauhinia, Mountain Ebony, Orchid Tree, Paderian, Variegated Bauhinia;

Hindi : Barial, Gurial, Kachnar, Kandan, Kovidara, Raktakanchan;

Sans. : Ashmantaka, Chamrika, Gandari, Girija, Kanchanara, Kudara, Mahapushpa, Tamrapushpa, Yugapatraka.

Distribution : Throughout India.

Parts Used : Root. Root bark.

Utilization : Root decoction useful in stomachache, sore throat, ulcers, tympanic gland trouble, dyspepsia, flatulence and as antidote to snake poisoning. Root paste heated and applied externally on boils to drive out pus. Decoction of root bark useful for lessening fatness and against tumours.

References : Agarwal (2003), Anonymous (1968), Asolkar *et al*. (1992), Chauhan (1999), Chopra *et al*. (1956),

Deshmukh *et al.* (2000), Dhiman (2003), Kirtikar & Basu (1984), Kumar & Lal (2003), Khare (2004), Kurian (1995), Panthi & Chaudhary (2003), Parrotta (2001), Reddi *et al.* (2005), Sood & Thakur (2004), Thakur *et al.* (1989), Watt (1889).

Begonia inflata Cl.

Family : Begoniaceae.

Distribution : Darjeeling, 1,000 m.

Part Used : Root.

Utilization : Decoction taken as effective remedy for genito-urinary problems.

References : Hooker (1879), Laramnghinglova & Jha (1999).

Belamcanda chinensis (L.) Redoute

Family : Iridaceae.

English & Hindi Names:

Eng. : Black Berry Lily, Leopard Lily;
Hindi : Surajkanti

Distribution : Cultivated Throughout India.

Part Used : Root.

Utilization : 20 ml of its juice taken orally during bed time for 3-4 days to cure constipation. Also, known to be used as aperient and resolvent.

References : Agarwal (2003), Chopra *et al.* (1956), Dash *et al.* (2003).

Berberis aristata DC.

Syn : *B. coriaria* Royle *ex* Lindle.

Family : Berberidaceae.

English, Hindi & Sanskrit Names:

Eng. : Indian Berberry, Ophthalmic Berberry;
Hindi : Chotra, Dar-Hald, Kashmal, Rasaunt, Rasvat;
Sans. : Daaruharidra, Darvi, Kata, Kateri, Pitadru.

Distribution : Temperate Himalaya (Indus Eastwards to Bhutan): 1,800-3,500 m.

Part Used : Root. Root bark.

Utilization : Decoction of root along with 'haridra' (*Terminalia chebula*) used for diabetes. Root bark useful to cure rat and snake bites, boils, rheumatic pains, eye complaints, jaundice, chronic uterine troubles, piles, cholera, diarrhoea, menorrhagia, leucorrhoea, debility, periodic neuralgia, enlarged spleen, intermittent fevers, skin diseases, and also as alterative, deobstruent, and powerful antiseptic. Tincture made from the root bark also used as a bitter tonic, stomachic, anti-inflammatory, cholagogue, antiperiodic, alterative in cases of remittent as well as intermittent fevers, and application in affections of the eyelids, debility, chronic ophthalmia, unhealthy ulcers and sores.

Active Constituents : Active ingredients are berberine hydrochloride, palmatine hydrochloride, karachine, aromatine, oxyberberine, oxycanthine, berberine and taxilamine.

Biological Activity : Antibacterial and antiinflammatory activities positive.

References : Agarwal (2003), Ambasta (1986), Anonymous (1948, 1968), Caius (2003), Chatterjee & Pakrashi (1997), Chopra *et al.* (1956), Jain (1968, 1991), Jha & Kumar (2003), Kaul (1997), Khare (2004), Kirtikar & Basu (1935), Nautiyal *et al.* (2003), Samant *et al.* (2001), Said (1997), Sinha & Sinha (2001), Sood *et al.* (2005a), Thakur *et al.* (1989), Watt (1889).

Berberis asiatica Roxb. *ex* DC.

Family : Berberidaceae.

Hindi & Sanskrit Names:

Hindi : Daruhaldi, Kilmora, Kingora, Sumlu;

Sans. : Daruharidra, Darvi, Haimavati, Hemkranta, Nirdishta, Parjani, Pita, Pitika, Sthirraga.

Distribution : Dry Outer Himalaya (600-2,550 m) from Kumaon Eastwards & Assam also.

Part Used : Root.

Utilization : Known to be useful in conjunctivitis, leucorrhoea, diseases of mouth, fevers, healing ulcers, treatment of malaria, liver disorders, urethral discharges, ophthalmia, jaundice, and as an antidote to snake venom, purgative for children, blood purifier and antiseptic. Extr. of fresh root (after filtration with a clean cloth) applied externally for the treatment of eye diseases and its poultice applied externally in wounds and inflammations.

Active Constituents : Chief constituents are berberine, palmatine, jatrorrhizine, columbamine, tetrahydropalmitine, berbamine, oxyberberine and oxycanthine.

Biological Activity : Alc. extr. showed hypoglycaemic activity.

References : Agarwal (2003), Anonymous (1948), Arya & Prakash (2000), Asolkar *et al.* (1992), Bhattacharjee (2001), Caius (2003), Chhetri (2005), Chopra *et al.* (1956), Jain *et al.* (2003), Jain & Patole (2001), Kirtikar & Basu (1935), Thakur *et al.* (1989).

Berberis chitria Ham.

Family : Berberidaceae.

Hindi Name:

Hindi : Daruhaldi.

Distribution : Himalaya (Kashmir-Bhutan): 1,600-3,700 m.

Part Used : Root.

Utilization : Decoction given orally as a blood purifier and for conjuctivitis, and its extr. from the root also used for the treatment of skin diseases.

Active Constituents : Contains hentriacontane, triacontane, cetyl alcohol, β-sitosterol, γ-sitosterol, dihydrokaempferol, quercetin, oleic, stearic, palmitic, linoleic acids, three alkaloids-chitriana A, B & C, and dihydropalmatine N-oxide.

Biological Activity : Exhibits CNS depressant activity in albino mice and antimicrobial activity against *E. coli, Staphylococcus* *aureus, Bacillus proteus, Salmonella typhi* and *Pseudomonia pyocyanae.*

References : Negi *et al.* (1999), Lal *et al.* (1996), Rastogi & Mehrotra (2002), Singh & Kumar (2000b).

Berberis lycium Royle

Syn : *B. parkeriana* Schn.

Family : Berberidaceae.

English, Hindi & Sanskrit Names:

Eng. : Raisin Berberry;
Hindi : Chitra, Kushmul;
Sans. : Daruharidra.

Distribution : W. Himalaya (1,000-3,000 m), M.P. & Tamil Nadu (Nilgiris).

Parts Used : Root. Root bark.

Utilization : Root extr. used in ophthalmia, chronic diarrhoea, piles, skin diseases, sores, leprosy, food poisoning, ear troubles, remedy for oral ulcers, disorders of the urinary tract and enlarged spleen. 1-2g powdered root given with water once every morning for 20 days to cure nervous disorders. A simple decoction of it with honey given in jaundice. Root bark yields 'Rasount' of commerce, and useful in curing fever, ulcers, constipation and as tonic.

Active Constituents : Berberine, berbericine, berbericinine, beerberine, berbamine, umbellatine, berbenine, berbericine and oxyberberine are the main root alkaloids.

Biological Activity : Beberine acts as a cardiac depressant. Berberine produced sedation in mice and conscious cats and potentiated phenobarbitone sleeping time. Antibacterial and anti-inflammatory activities found to be positive.

References : Agarwal (2003), Ambasta (1986), Anonymous (1968), Caius (2003), Chauhan (1999), Chopra *et al.* (1956), Jain (1991), Kaul (1997), Kayal *et al.* (2003), Kirtikar & Basu (1935), Rastogi & Mehrotra (1999a, b), Sarin (1996), Singh (1999, 2000), Sood & Thakur (2004), Verma & Chauhan (2006), Watt (1889).

Berginia ciliata (Haw.) Stenb. (Pl. 4A)

Syn : *B. ligulata* Engl.; *Saxifraga ciliata* Wall.; *S. ligulata* Wall.

Family : Saxifragaceae.

English, Hindi & Sanskrit Names:

Eng. : Rock Foil;
Hindi : Pashanbhed;
Sans. : Paashaanbheda.

Distribution : Throughout Temperate Himalaya (Kashmir-Bhutan) 2,300-3,300 m & Khasia Hills at 1,300 m.

Part Used : Root.

Utilization : Useful in India for fever, burns, urinary troubles, toothache, piles, backache, bronchitis, dysuria, cough, cold, dysentery, ophthalmia, cuts, boils, swelling, sores, dropsy and as abortifacient. Tea prepared from its roots taken as cardiac tonic. A mixture comprising its root and that of *Astilbe rivularis* is made into paste for relieving body pain after child birth. Root mixed with soyabean (*Glycine max*) seeds, given to cattle in diarrhoea. About 5g root paste taken with one glass of hot water (twice or thrice a day for a week) for curing stomach pain caused by stone.

Active Constituents : Contains β-sitosterol, bergenin, galloylated leucoanthocyanidin-4-(2-O-galloyl)glucoside, gallic acid, tannic acid (14.2 %), glucose (5.6 %), mucilage, wax, (-) afzelechin, c-glycoside of bergenin, and (+) catechin-3-gallate.

Biological Activity : Alc. extr. of roots, antilithic, diuretic whereas acetone extr. of root bark exhibited cardiotoxic, CNS depressant and antiinflammatory properties.

References : Agarwal (2003), Ambasta (1986), Anonymous (1948, 1968), Arya & Prakash (2000), Asolkar *et al.* (1992), Chauhan (1999), Chhetri (2005), Chopra *et al.* (1956),

Jain *et al.* (1991), Kaul (1997), Khare (2004), Kumar (2002), Kumar & Rao (2001), Lal *et al.* (1996), Nadeem *et al.* (2001), Negi *et al.* (1993), Pandey & Tiwari (2003), Rastogi & Mehrotra (1999a, b), Samant *et al.* (2001), Sharma & Sood (1997), Sharma *et al.* (2003), Singh (1996), Singh *et al.* (2003), Singh & Kumar (2000b).

Berginia purpurascens (Hook.f. & Thoms) Engl.

Syn : *Saxifraga purpurascens* Hook.f. & Thoms.

Family : Saxifragaceae.

Distribution : Himalaya (Nepal-Arunachal Pradesh): 3, 000-5,000 m.

Part Used : Root.

Utilization : Poultice of its paste applied over the fracture or taken orally 2-3 times in a day to get relief from pain.

Active Constituents : These are polyphenols-4, 6-di-O-galloylarbutin, 2, 4, 6-tri-O-galloyglucose, 6-O-galloylarbutin, bergenin and its gallates.

References : Ambasta (1986), Maity *et al.* (2003), Rastogi & Mehrotra (2002).

Berginia stracheyi (Hook.f. *et* Thomas) Engl.

Syn : *Saxifraga ciliata* Lindl. non Royle; *S. stracheyi* Hook.f. Thomas

Family : Saxifragaceae.

Distribution : Kashmir-U.P.: 2,400-4,800 m.

Part Used : Root.

Utilization : Powder used for abdomen and joint pain, toothache, swellings, kidney stones, sores, and considered diuretic, febrifuge, ophthalmic and tonic. Its paste heals cuts, ulcers, swelling of limbs, and stiff joints.

Active Constituents : Found to contain a new catechin derivative, (+) catechin-3-gallate.

References : Anonymous (1968), Aswal & Mehrotra (1994), Asolkar *et al.* (1992), Ballabh & Chaurasia (2006),

Bhattacharjee (2001), Jain (1991), Kaul (1997), Negi *et al.* (1999), Rastogi & Mehrotra (1999a, b), Singh (2000), Sood *et al.* (2000).

Beta vulgaris Linn.

Family : Chenopodiaceae.

English, Hindi & Sanskrit Names:

Eng. : Sugar Beet, The Beet Root;
Hindi : Chukkandar;
Sans. : Plakya.

Distribution : Widely Cultivated Throughout India.

Part Used : Root (Tuber).

Utilization : Useful in chronic catarrhal conditions, tuberculosis, psycopathy, cough, inflammation, general debility and as expectorant and tonic.

Active Constituents : Possesses phytoalexins-3, 5-dihydroxy-6, 7-methylene dioxy flavone, 5, 2'-dihydroxy-6, 7-methylenedioxyflavone and 5-hydroxy-6, 7-methylene dioxyflavone.

References : Baburaj *et al.* (2000), Rastogi & Mehrotra (2002), Raddi *et al.* (2005).

Bidens biternata (Lour.) Merrill & Sheriff.

Syn : *B. chinensis* Willd.; *B. pilosa* auct. non Linn.
Family : Asteraceae.

Distribution : Throughout India as weed Upto 2,000m.

Part Used : Root.

Utilization : Juice given to check toothache.

References : Asolkar *et al.* (1992), Trivedi (2002).

Biophytum sensitivum (L.) Don

Syn : *Oxalis sensitivum* L.
Family : Oxalidaceae.

Hindi & Sanskrit Names:

Hindi : Lajjalu, Lakshana, Zarer;

Sans. : Amlika, Lajalu, Lajjaluka.

Distribution : Open Habitats During the Rainy Season Throughout the Hotter Parts of India, ascending to 1,800m in the Himalaya.

Part Used : Root.

Utilization : Known to be useful in gastric troubles, gonorrhoea, impotency, fevers, lithiasis and urinary calculi.

References : Agarwal (2003), Asolkar *et al.* (1992), Chatterjee & Pakrashi (1997), Chopra *et al.* (1956), Parrotta (2001), Prasad *et al.* (2002), Reddi *et al.* (2005).

Bistorta vivipara (L.) S.F.Gray

Family : Polygonaceae.

Part Used : Root.

Utilization : Used in diarrhoea, dysentery, haemoptysis, gout, and leucorrhoea. It makes an excellent gargle for sore throat, spongy gums and a good lotion for ulcers.

References : Kala & Manjrekar (2000), Maity *et al.* (2003).

Bixa orellana L.

Family : Bixaceae.

English, Hindi & Sanskrit Names:

Eng. : Arnatto Plant, Chinese Dye Tree, Monkey Turmeric;

Hindi : Sindhuriyan, Latkan, Vatkana;

Sans. : Karachchada, Raktabija, Raktapushpa, Shonapushpi, Sinduri, Trivapushpi, Virpushpa.

Distribution : Largely Cultivated Throughout India.

Part Used : Root bark.

Utilization : Two teaspoons decoction of root bark given in fever twice daily for three days. Also considered antiperiodic, antipyretic, and of great use in uncomplicated intermittent, remittent and continued fevers, and to be a good remedy for gonorrhoea.

Active Constituents : Contains triterpene tomentosic acid.

References : Agarwal (2003), Anonymous (1968), Asolkar *et al.* (1992), Chaudhary & Hutke (2002), Chopra *et al.* (1956), Dhiman (2003), Dwivedi (1993), Kirtikar & Basu (1935), Rao & Henry (1996), Reddi *et al.* (2005), Yoganarasimhan *et al.* (1981), Watt (1889).

Blepharis linariaefolia Pers.

Syn : *B. sidica* T.Anders

Family : Acanthaceae.

Distribution : Punjab, Rajputana Desert, Gujarat, Sind, and Baluchistan.

Part Used : Root.

Utilization : One teaspoonful powder along with honey given twice a day to cure liver disorders. Paste of roots applied over body to treat leucoderma.

References : Chopra *et al.* (1956), Khanna & Kumar (2000).

Blepharis persica (Burm.f.)Kuntze.

Syn : *B. edulis* Pers.

Family : Acanthaceae.

Hindi & Sanskrit Names:

Hindi : Chaupatia, Uttanjan;

Sans. : Shiki, Utamganasya.

Distribution : Punjab & W. Rajasthan.

Part Used : Root.

Utilization : Considered diuretic, dysmenorrhagic and useful in irregular menstruation and urinary discharges.

References : Agarwal (2003), Ambasta (1986), Asolkar *et al.* (1992), Chatterjee & Pakrashi (1997).

Blumea eriantha DC.

Family : Asteraceae.

Part Used : Root.

Utilization : Decoction taken for rabies (mad dog bite).

Reference : Goud & Pullaiah (1996).

Blumea lacera (Burm.f.) DC.

Syn : *B. subcapitata* DC.; *Conyza lacera* Burm.f.

Family : Acanthaceae.

Hindi & Sanskrit Names:

Hindi : Jangli-Muli, Kakronda, Kukur-Benda, Nimrudi;

Sans. : Kukundara.

Distribution : Wasteplaces, Gardens, Fields, Railway Tracks, Roadsides in Plains of India.

Part Used : Root.

Utilization : Useful in cholera and as astringent and febrifuge.

Active Constituents : Contains α-amyrin, campesterol, lupeol, stigmasta-5, 9 (11)-3β-ol and β-sitosterol.

References : Agarwal (2003), Asolkar *et al.* (1992), Chatterjee & Pakrashi (1997), Watt (1889).

Blumea laciniata (Roxb.) DC.

Syn : *B. lacera* var. *blumei* DC.

Family : Asteraceae.

Distribution : Punjab to Arunachal Pradesh, Mizoram & Peninsular India Upto 1,350 m.

Part Used : Root.

Utilization : Root paste collected in a cup of water drunk for snake bite.

References : Asolkar *et al.* (1992), Lalramnghinglova & Jha (1999).

Boehmeria clidemioides Miq.

Syn : *B. macrophylla* Don, *B. platyphylla* D. Don; *Urtica pendulifera* Wall.

Family : Urticaceae.

Hindi Names:

Hindi : Gargela, Rhea Ghas.

Distribution : Moist & Shady Places Throughout N.E. Region.

Part Used : Root.

Utilization : Paste applied to heal wounds and cures eczema.

References : Agarwal (2003), Kumar (2002), Rastogi & Mehrotra (1999a, b), Sood & Thakur (2004), Watt (1889).

Boerhaavia diffusa L.

Syn : *B. repens* L.; *B. repens* var. *diffusa* (L.) Hook.f.; *B. repens* var. *procumbens* (Roxb.) Hook.f.

Family : Nyctaginaceae.

English, Hindi & Sanskrit Names:

Eng. : Pigweed, Spreading Hogweed;

Hindi : Beshakapore, Biskhafra, Punernava, Santhi, Thikri;

Sans. : Punarnava, Raktakanda, Raktapunarnava, Sophagni.

Distribution : Drier Part of the Country in Open Spaces Upto 2,000 m in the Himalaya.

Part Used : Root.

Utilization : Reported in India for asthma, cough, dropsy, heart diseases, rheumatism, jaundice, paralysis, epilepsy, eye diseases, sore throat, piles, gonorrhoea, diarrhoea, flatulence, acidity, blood dysentery, anaemia, cuts, wounds, haemoptysis, urinary diseases, respiratory diseases, fever, stomach pain, enlargement of spleen, general debility, gastric and liver diseases, oedema, swelling, calculi, obesity, and also used as purgative, anthelmintic, febrifuge, diuretic, diaphoretic, expectorant and stomachic. Ointment made from it prescribed for the treatment of leprosy, other skin diseases and as an inflammatory agent in inhibiting rheumatic pains, ascites, anasarca, internal inflammation and chronic peritonitis. Root extr. along with black pepper (200 ml twice a day for 4-7 days) beneficial for spermatorrhoea.

Active Constituents : Yields triacontane, β-ecdysone(0.05%), hentriacontane, β-sitosterol, ursolic acid, 5, 7-dihydroxy-3′, 4′-dimethoxy-6, 8-dimethylflavone, punarnavaine, rotenoid analogues- boeravinone A-F, punarnavoside, amino acids, fatty acids, isofuroxanthone, boeravine, lignans, liriodendrin, hypoxanthine-9-L-arabinofuranoside, 12a-hydroxyrotenoid-boeravinone-c, repenone and repenol borahavine.

Biological Activity : Aq. & acetone extrs. in a dose of 4 mg/100 g of body weight showed significant anti-inflammatory activity in carragenin induced oedema and formaldehyde-induced arthritis in albino rats. Liriodendrin exhibited significant Ca^{2+} channel antagonistic effect in frog heart single cell.

References : Agarwal (2003), Asolkar *et al.* (1992), Beg *et al.* (2005), Bhatt *et al.* (1999), Bhatt & Mitaliya (1999), Bhattacharjee (2001), Binu *et al.* (2003), Chatterjee & Pakrashi (1997), Chopra *et al.* (1956), Dastur (1970), Drury (1978), Dwivedi (2003), Hosagoudar & Henry (1996b), Jain (1968), Jain *et al.* (1991), Khare (2004), Kumar & Lal (2003), Maya *et al.* (2003), Mehrotra *et al.* (1995), Parrotta (2001), Patil (2000), Rao & Henry (1996), Rastogi & Mehrotra (1998, 1999a, b, 2001, 2002), Reddi *et al.* (2005), Said (1997), Saini (1996b), Samvatsar & Diwanji (1996), Sen & Behra (2003), Sharma (2003), Singh (1999, 2000), Singh & Maheshwari (1992), Singh & Prakash (1996), Sinha & Sinha (2001), Thakur *et al.* (1989), Watt (1889).

Bolahaavia variegata L.

Family : Fabaceae.

Part Used : Root.

Utilization : Half cup juice given in snake bite.

Reference : Jain & Patole (2001).

Bombax ceiba L.

Syn : *B. hexaphyllum* Roxb.; *B. malabaricum* DC.; *Gossampinus malabarica* (DC.) Merill.; *Salmalia malabarica* (DC.) Schott & Endl.

Family : Bombacaceae.

English, Hindi & Sanskrit Names:

Eng. : Cotton Tree, Red Cotton Tree, Red Silk-Cotton Tree, Silk-Cotton Tree;

Hindi : Kantisembal, Pagun, Ragatsemal, Simal, Somr;

Sans. : Apurani, Bahuvirya, Chirjivi, Kantakudruma, Mochani, Nissara, Pichhala, Ramyapushpa, Yamadruma.

Distribution : Common Upto 1,500 m Throughout India.

Parts Used : Root. Root bark.

Utilization : Roots known to be useful for biliousness, heat of the body, inflammations, urinary troubles, cystitis, pimples, dysentery, menstrual disorders, and as diuretic, tonic, stimulant, astringent to the bowels, stomachic and coolant. To stop urination of children during night, it is given with puffed rice for one week on empty stomach in the morning. In case of liver enlargement and loss of vigour, roasted tap root of tender plant taken (about 50g twice daily) till cure. Sap extracted from the root consumed on empty stomach in leucorrhoea. 6g powdered root mixed with the pulse, *Phaseolus* sp. and is given with cow's milk once at bed time in spermatorrhoea. Pounded root bark boiled with mustard oil and garlic, applied for bodyache and rheumatism.

Active Constituents : Root bark contains hemigossypolone, hemigossypolone-7-methylether, isohemigossypol-1-methylether (hemigossypol-6-methylether), iso-hemigossypol-1, 2-dimethylether, 8-formyl-7-hydroxy-5-isopropyl-2-methoxy-3-methyl-1, 4-nepthaquinone, 7-hydroxycadalene, lupeol, β-sitosterol, 3′, 4′, 5, 7-tetrahydroxy-6-methoxyflavon-3-O-β-D-glucopyranosyl-α-D-xylopyranoside (I), triacontanol, and 1, 6-dihydroxy-3-methyl-5-isopropyl-7-methoxy-8-naphthalene carboxylic acid (8→1) lactone (VI).

References : Agarwal (2003), Ambasta (1986), Aminuddin & Girach (1991), Anonymous (1968), Aswal (1996), Caius (2003), Chatterjee & Pakrashi (1997), Das & Sharma (2003), Deshmukh & Rothe (2003), Gogoi & Borthakur (2001), Gogoi *et al.* (2003a), Gopal & Chandra (2003), Gosh & Sensharma (1997), Jamir (1997), Katewa *et al.* (2001), Khanna (2002b), Khare (2004), Kirtikar & Basu (1935), Kumar & Lal (2003), Maheshwari *et al.* (1997), Mukherjee & Namhata (1990), Parrotta (2001), Patil (2000), Rao & Henry (1996), Rastogi & Mehrotra (1999a, b, 2001), Sen & Behera (2003), Sharma (2000), Thakur *et al.* (1989), Varghese (1996).

Borassus flabellifer Linn.

Syn : *B. flabelliformis* Roxb.

Family : Arecaceae.

English, Hindi & Sanskrit Names:

Eng. : Brab Tree, Palmyrah Palm;

Hindi : Tal, Tala, Tar, Tari;

Sans. : Talah.

Distribution : Coastal Areas of Peninsular India, Bihar & Bengal.

Part Used : Root.

Utilization : Juice of the young root used in cases of gastric catarrah, hiccups, burning sensation, inflammation, gonorrhoea, and considered as stimulant, diuretic, anthelmintic, cooling and restorative.

References : Agarwal (2003), Asolkar *et al.* (1992), Caius (2003), Chopra *et al.* (1956), Dastur (1970), Kurian (1995), Rastogi & Mehrotra (2002), Reddi *et al.* (2005), Watt (1889).

Bragantia wallichii R.Br.

Syn : *Apama siliquosa* Lam.

Family : Aristolochiaceae.

Sanskrit Name:

Sans. : Chakrani.

Distribution : Deccan Peninsula & Sri Lanka.

Part Used : Root.

Utilization : Said to be efficacious in cholera, diarrhoea, dysentery, ulcers and carbuncles.

References : Agarwal (2003), Ambasta (1986), Anonymous (1948), Chopra *et al.* (1956).

Brassica napus L.

Syn : *B. campestris* Hook.f. & Thoms. ssp. *campestris* var. *dichotoma* Watt. in part; var. *toria* Duthie & Fuller.; *B. campestris* Hook.f. & Thoms. ssp. *napus* Hook.f. & T.Anders. in part; *Sinapsis dichotoma* Roxb.

Family : Brassicaceae.

English & Hindi Names:

Eng. : Rape, Toria;

Hindi : Kali Sarson, Lahi, Lutni, Magni, Toria, Toriva.

Distribution : Cultivated in Punjab, Bengal, U.P. & Bihar.

Part Used : Root.

Utilization : Useful as an emollient, diuretic and for checking cough and bronchial catarrh.

References : Agarwal (2003), Ambasta (1986), Chopra *et al.* (1956).

Bridelia retusa L. Spreng.

Syn : *B. cambodiana* Gagnep.; *B. crenulata* Roxb.; *B. spierrei* Gagnep.; *B. spinosa* (Roxb.) Willd.

Family : Euphorbiaceae.

Hindi & Sanskrit Names:

Hindi : Ekdania, Gauli, Gondai, Kaj, Kasai, Khaja;

Sans. : Asana, Ekadivi, Ekavira, Mahavira.

Distribution : Secondary Deciduous & Semi-Deciduous Forests Throughout India (Except in Dry Regions); Often Near Streams or Rivers to 1,800 m.

Part Used : Root.

Utilization : Decoction taken orally in dysentery and abdominal pain.

References : Painuli & Maheshwari (1996), Parrotta (2001), Watt (1889).

Bromus japonica Thunb.

Syn : *B. patulus* Mert & Koch.

Family : Poaceae.

Distribution : W. Himalaya: 3,000-4,600 m.

Part Used : Root.

Utilization : Root extr. known to cure eczema.

References : Hooker (1897), Nautiyal *et al.* (2003).

Bupleurum falcatum Linn.

Family : Apiaceae.

English & Hindi Names:

Eng. : Hare's Ear;
Hindi : Kalizewar, Tejraj.

Distribution : Wild in Himalaya (Kashmir-Bhutan): 1,000-4,000 m.

Part Used : Root.

Utilization : About 4 cm of root grounded with water and taken as aphrodisiac and tonic. This in combination with other drugs prescribed in liver and stomach disorders and as diaphoretic.

Active Constituents : Active properties are due to angelicin, saikosides ^{2}a, Ib & II, saikogenin A, triol-saikogeninB, saikogenins A, C, D, E, saikosaponins (I-V), saikosaponins e & f, saikosaponins-a,-b2,-b4,-c, & -d, α-spinasteryl-β-D-gluciside, saikosaponin b4, galacturonic acid, arabinose, rhamnose, galactose, α- & β-spinasterol, stigmasterol, stigmastenol & Δ^7-Δ^{22}-stigmastenol, bupleurumol, ribitol(adonitol), 10- nonacosanone, β-sitosterol, malonylsaikosaponins a and d.

Biological Activity : Root saponin mixture effective at 400 mg/kg in treatment of rat paw oedema induced by dextran.

References : Agarwal (2003), Anonymous (1948, 1968), Asolkar *et al.* (1992), Rastogi & Mehrotra (1999a, b, 2001, 2002), Shah & Singh (1990), Sood & Prakash (2007).

Bupleurum marginatum Wall. *ex* DC.

Syn : *B. falcatum* var. *marginatum* (Wall. *ex* DC.) Clarke, var. *scorzoneraefolium* (Willd.) Ledeb.; *B. linearifolium* DC.; *B. scorzoneraefolium* Willd.

Family : Apiaceae.

Hindi Names:

Hindi : Kalizewar.

Distribution : N. E. & N. W. Himalaya from Kashmir to Khasia Hills.

Part Used : Root.

Utilization : Root antipyretic and used in liver diseases and amenorrhoea.

Active Constituents : Contains saikosaponin (0.4 %), arabinose, glucose, sucrose, oleic acid and linoleic acid.

References : Asolkar *et al.* (1992), Kaul (1997), Rastogi & Mehrotra (2002), Sood & Prakash (2007).

Butea monosperma (Lam.) Taub. (Pl. 4B)

Syn : *B. frondosa* Koen. *ex* Roxb.; *Erythrina monosperma* Lam.

Family : Fabaceae.

English, Hindi & Sanskrit Names:

Eng. : Bastard Teak, Bengal Kino, Butea Gum, Flame of the Forest;
Hindi : Chinchra, Dhak, Kakria, Palas, Paras, Tesu, Tesu-Ka-Per;
Sans. : Kinsuka, Palasa, Palasha.

Distribution : All Over India Upto 1,000 m: Natural in the Deciduous Forests of Central India.

Parts Used : Root. Root bark.

Utilization : Root decoction used in urinary troubles, congested lungs, goitre, muscular swelling (root boiled in water and vapour directed to the face), night blindness and other defects of sight, tuberculosis (root paste taken with eggs) and elephantiasis. Root paste applied on wounds in foot and mouth diseases. Dried root bark powder mixed with cow ghee taken for 3 days from menstruation date to check barrenness. Also, bark considered as aphrodisiac, analgesic, anthelmintic and applied in sprue, piles, ulcers, tumours, dropsy and taken to check B.P.

Active Constituents : Found to contain glucose, glycine, protein, fat, carbohydrates and minerals.

References : Anonymous (1969), Dhiman (2003), Dwivedi (1993), Hosagoudar & Henry (1996a), Kadel & Jain (2006), Khare (2004), Kirtikar & Basu (1935), Kothari & Moorthy (1996), Kurian (1995), Maheshwari *et al.* (1997), Pandey & Rout (2006), Parrotta (2001), Reddi *et al.* (2005), Sarangi & Sahu (2004), Saren *et al.* (2000), Sinha & Sinha (2001), Sood & Thakur (2004), Thakur *et al.* (1989), Varghese (1996).

Butea superba Roxb.

Family : Fabaceae.

Hindi & Sanskrit Names:

Hindi : Bela Palas, Bodha, Chihumt, Palaslata;

Sans. : Latapalas, Latapalasha.

Distribution : Deciduous Hill Forests from U.P.-W. Bengal & Assam, Southwards Through Peninsular India to Tamil Nadu.

Part Used : Root.

Utilization : Poultice applied on weeping eczema.

References : Agarwal (2003), Parrotta (2001), Shah & Singh (1990).

Byttneria herbacea Roxb.

Syn : *Buetneria herbacea* Roxb.

Family : Sterculiaceae.

Distribution : Peninsular India from Gujarat Southwards to Tamil Nadu & Bihar.

Parts Used : Root. Root bark.

Utilization : Bark antidiabetic. Root extr. along with that of *Asparagus racemosus* given twice a day for 15 days in spermatorrhoea and leucorrhoea. Mixed with the roots of 'Kurchi' [*Holarrhena antidysentrica* (Roth.) A.DC.], it is given to prevent wind formation in the stomach and acidity. Root paste applied on maggot infested sores, cuts by iron tools, wounds and joints to relieve swelling.

References : Ambasta (1986), Bhatt *et al.* (2003), Chopra *et al.* (1956), Panda & Das (2000), Parrotta (2001), Reddi *et al.* (2005), Singh (1996), Singh & Prakash (1994), Varghese (1996).

Cadaba fruticosa (L.) Druce.

Syn : *C. farinosa* Forssk.; *C. indica* Lam.; *Cleome fruticosa* L.; *Stroemia tetandra* Vahl.
Family : Capparidaceae.

English & Hindi Names:

Eng. : Indian Cadaba;
Hindi : Dabi, Kodhab.

Distribution : Drier Low Land Regions of India & Sri Lanka.

Part Used : Root.

Utilization : Decoction used against chronic fever and for expelling intestinal worms.

References : Goud *et al.* (2000), Parrotta (2001), Reddi *et al.* (2005).

Caesalpinia bonduc (L.) Roxb.

Syn : *Bonduc minus* Medik.; *Caesalpinia bonducella* (L.) Fleming; *C. crista* L.; *C. jayabo* Naza.; *C. nuga* Ait.f.; *Guileandina bonduc* L.; *G. bonducella* L.
Family : Fabaceae.

English, Hindi & Sanskrit Names:

Eng. : Bonduc Nut, Fever Nut, Nickar, Physic Nut;
Hindi : Gatan, Gataram, Kanja, Karanju, Katkaranj, Sagagoti;
Sans. : Kuberaksi, Latakaranja, Putikaranja.

Distribution : Wild Throughout the Plains of India & Upto 850 m in hedges & Wastelands Near Villages. Coastal Regions of W. E. & S. India.

Parts Used : Root. Root bark.

Utilization : Root recommended as antipyretic and its decoction taken to treat bladder stones, leucorrhoea (mixed with honey), leprosy, calculus and dysentery. Root bark useful in epilepsy, tumours, eczema and as emmenagogue, anthelmintic and febrifuge.

References : Agarwal (2003), Anonymous (1948), Asolkar *et al.* (1992), Chopra *et al.* (1956), Dhiman (2003), Dastur (1970), Khanna (2002b), Nayak *et al.* (2003), Parrotta (2001), Rao & Henry (1996), Reddi *et al.* (2005), Said (1997), Varghese (1996), Watt (1889).

Caesalpinia digyna Rottl.

Syn : *C. oleosperma* Roxb.

Family : Fabaceae.

English, Hindi & Sanskrit Names:

Eng. : Teri pods;
Hindi : Vakarimul;
Sans. : Vakeri.

Distribution : E. Himalaya, E. & W. Peninsula & Sri Lanka.

Part Used : Root.

Utilization : Reported to be astringent and given in scrofulous affections, pthisis and diabetes.

Active Constituents : Bergenin and vikerin isolated from the roots.

Biological Activity : Ethanolic extr. (50%), active *in vitro* against *Mycobacterium tuberculosis* in guinea pigs.

References : Agarwal (2003), Ambasta (1986), Anonymous (1948, 1968), Asolkar *et al.* (1992), Chatterjee & Pakrashi (1997), Chopra *et al.* (1956), Watt (1889).

Cajanus cajan (Linn.) Millsp.

Syn : *C. bicolor* DC.; *C. flavus* DC.; *C. indicus* Spreng.; *C. obcordifolia* Singh; *Cyotisus cajan* L.

Family : Fabaceae.

English, Hindi & Sanskrit Names:

Eng. : Congo Pea, Pigeon Pea, Red Gram;
Hindi : Arhur, Tur, Tuver;
Sans. : Adhaki, Tuvarika, Tuwari.

Distribution : Cultivated Throughout Country & Upto 1,800 m in the Himalaya.

Part Used : Root.

Utilization : Juice given to new born babies to cure diarrhoea.

Active Constituents : Cajaquinone, cajaisoflavone, cajaflavanonegenistein, lupeol, α & β-amyrin, sitosterol, isoflavanone, 2′-O-Me-cajanone and prenylated isoflavon are the active constituents.

References : Agarwal (2003), Asolkar *et al*. (1992), Parrotta (2001), Rastogi & Mehrotra (1999a, b, 2001), Sur & Halder (2002), Watt (1889).

Caladium bicolour (Ait.) Vent.

Family : Araceae.

Part Used : Rootstock.

Utilization : Paste applied to cure facial paralysis.

Reference : Kumar (2002).

Callicarpa longifolia Lam.

Family : Verbenaceae.

Hindi Name:

Hindi : Daya.

Distribution : A Shrub of Nicobar Islands.

Part Used : Root.

Utilization : Useful against diarrhoea and syphilis.

References : Ambasta (1986), Anonymous (1948), Chopra *et al.* (1956), Tiwari & Tiwari (1996), Watt (1889).

Callicarpa macrophylla Vahl.

Syn : *C. incana* Roxb.; *C. roxburghii* Wall.

Family: Verbenaceae.

English, Hindi & Sanskrit Names:

Eng. : Perfumed Cherry;

Hindi : Daya, Priyamgu, Sumali;

Sans. : Priangu.

Distribution : Kashmir to Assam & Arunachal Pradesh Upto 1,800 m.

Part Used : Root.

Utilization : Powdered root taken for relief from stomach ailments, (1-2 teaspoons once every morning for 5-7 days).

References : Agarwal (2003), Ambasta (1986), Anonymous (1948, 1968), Chopra *et al.* (1956), Sood & Thakur (2004), Watt (1889).

Calotropis gigantea (L.) R.Br. *ex* Ait.

Syn : *Asclepias gigantea* L.

Family : Asclepiadaceae.

English, Hindi & Sanskrit Names:

Eng. : Bowstring Hemp, Gigantic Milk Weed, Gigantic Swallow-Wort, Madar;

Hindi : Ak, Akada, Arka, Madar, Mudar, Safed Ak;

Sans. : Aditya, Arkah, Mandara, Pratapasa, Ravi, Rupika.

Distribution : Common Throughout the Plains & Low Hills of India.

Parts Used : Root. Root bark.

Utilization : Known to be used in India as an expectorant, abortifacient, stomachic, purgative, and considered useful in ulcers, skin diseases (poultice), piles, syphilis, elephantiasis, hydrocele, cold, cough, asthma, arthritis, fever, constipation, inflammations, ear trouble, boils, swellings, pneumonia, and leprosy. Its paste applied as poultice for rheumatic pain and headache, and juice given in case of snake bite. Mixture of crushed roots along with fruits of *Terminalia chebula* prescribed for epilepsy (1 spoonful once a day for 30 days). Powdered root bark (a substitute of ipecac) given orally thrice a day for three days for jaundice, dysentery, bronchial troubles, and as diaphoretic and emetic.

Active Constituents : These are β-sitoterol, taraxasterol, α-& β-amyrin, valeric acid, acetic acid, ψ taraxasterol (root bark). α- & β-calotropeol, calotropin, calotroposides C, D, E, F & G, calotroposides A & B, designated as 12-O-benzyllineolon-3-O-β-D-cymaropyranosyl (1→4)-β-D-oleandropyranosyl (1→4)- β-D-cymarapyranoside, uscharidin and uzarigenin (root).

Biological Activity : Antidepressant activity positive.

References : Agarwal (2003), Ambasta (1986), Anonymous (1968), Asolkar *et al.* (1992), Chatterjee & Pakrashi (1997), Chauhan (1999), Chopra *et al.* (1956), Dwivedi (1993), Girach (1991), Kurian (1995), Negi *et al.* (1993), Panda & Das (2000), Parrotta (2001), Patil & Patil (2002), Rao & Henry (1996), Reddi *et al.* (2005), Sahoo & Bahali (2003), Said (1997), Sharma (2003), Singh & Maheshwari (1992), Singh & Rao (2003), Thakur *et al.* (1989), Varghese (1996), Watt (1889).

Calotropis procera (L.) Ait. (Pl. 4C)

Syn : *Asclepias procera* Ait.; *Calotropis hamiltonii* Wall.

Family : Asclepiadaceae.

English, Hindi & Sanskrit Names:

Eng. : Dead Sea Apple, Milkweed, Sodam Apple, Swallow Wort;
Hindi : Ak, Akada, Madar, Safed Ak;
Sans. : Alarka, Arka,.

Distribution : Common Throughout the Plains & Low Hills of India.

Part Used : Root. Root bark.

Utilization : Root bark known to be useful for treating chronic cases of dyspepsia, flatulence, constipation, loss of appetite, leprosy, indigestion, mucus in stool, malaria (decoction with black pepper), piles (smoke of root bark), syphilis, hydrocele (paste applied), skin diseases, elephantiasis, dysentery and worms, and also as antidysenteric, antispasmodic, emetic, expectorant and purgative. Extr. from root bark along with juice of *Allium cepa* L. and vinegar, sugarcane, 50 ml each, given orally twice a day for treating fever and cholera. Administration of the extr. at 3 minute interval causes abortion. Fresh root twig used as tooth brush in toothache. Root paste applied locally in scorpion sting, piles and boils. Necklace of root worn to cure jaundice. Honey mixed with the powder of roasted roots given orally before meal to cure asthma. Mixture of root powder and milk in the ratio of 1:100 taken internally to cure rheumatism, muscular swelling, cough and asthma.

Active Constituents : These are β-sitosterol, taraxasterol, α-& β-amyrin, valeric acid, acetic acid, benzoyllinedone, benzoylisollineolone, long-chain fatty acids and C (18) isoursane (root bark), beta-amyrin, quercetin-3-rutionside, α- and β-calotropeol (roots).

Biological Activity : Diuretic, vermicidal and antidepressant activities found to be positive.

References : Ambasta (1986), Anonymous (1968), Asolkar *et al.* (1992), Bhatt *et al.* (2002), Chatterjee & Pakrashi (1997), Dastur (1970), Dhiman (2003), Drury (1978), Jadeja (1999), Katewa *et al.* (2001), Maliya & Singh (2003), Parrotta (2001), Rastogi & Mehrotra (1999a, 2002), Reddi *et al.* (2005), Samvatsar & Diwanji (2000), Sarin (1996), Singh (2002, 2003), Singh & Maheshwari (1992) , Sinha & Sinha (2001), Thakur *et al.* (1989), Trivedi (2002), Watt (1889).

Caltha palustris Linn.

Family : Ranunculaceae.

English Name:

Eng. : The Marsh Marigold.

Distribution : Marshes of the W. Temperate Himalaya (Kashmir to Nepal): 2,600-3,300 m.

Part Used : Root.

Utilization : Known to cure gonorrhoea.

Active Constituents : Contains magnoflorine, helleborin and veratrin.

References : Asolkar *et al.* (1992), Aswal & Mehrotra (1994), Sharma (2003), Watt (1889).

Campylandra aurantiaca Wall. *ex* Baker

Family : Liliaceae.

Distribution : Himalaya (Nepal Eastwards to Meghalaya) & Nagaland: 1,200-2,400 m.

Part Used : Root.

Utilization : Paste massaged gently on head to prevent falling of hair and in the treatment of fever.

Active Constituents : 3-epiruscogenin, 3-epineoruscogenin, tupisgenin, aurantigenin, and ranmogenins A, B & C.

References : Maity *et al.* (2003), Rastogi & Mehrotra (2002).

Canavalia gladiata DC.

Syn : *C. ensiformis* Baker, non DC.

Family : Fabaceae.

English & Hindi Names:

Eng. : Jack Bean, Sword Bean;

Hindi : Bara Sem, Chaptisem, Makhan Sem.

Distribution : Warmer Parts of India.

Part Used : Root.

Utilization : Extr. given as an antidote for snake venom.

References : Agarwal (2003), Ambasta (1986), Maiti & Mishra (2000), Watt (1889).

Canavalia virosa (Roxb.) Wt. & Arn.

Family : Fabaceae.

Part Used : Root.

Utilization : Prescribed for stomach disorders.

Reference : Kothari & Moorthy (1996).

Canna indica Linn.

Syn : *C. indica* var. *orientalis* Rosc.; *C. orientalis* Rosc.

Family : Cannaceae.

English, Hindi & Sanskrit Names:

Eng. : Indian Bead, Indian Reed, Indian Shot;

Hindi : Kewara, Sabbajaya, Sarvajjya;

Sans. : Devakili, Shilarambha, Vanakadali.

Distribution : All Over Indian Gardens.

Part Used : Root.

Utilization : Decoction considered as diaphoretic, diuretic, demulcent, stimulant, febrifuge and for dropsy.

References : Agarwal (2003), Ambasta (1986), Anonymous (1948), Kirtikar & Basu (1984), Subramanium (2000), Watt (1889).

Cannabis sativa L.

Syn : *C. indica* ssp. *indica* var. *indica* (Lamk.) Wehmer.

Family : Cannabinaceae.

English, Hindi & Sanskrit Names:

Eng. : Indian Hemp, Marihuana, Marijuana, Soft Hemp, True Hemp;

Hindi : Bhang, Charas, Ganja, Gur, Kinnab, Phulganja, Siddhi;

Sans. : Bhanga, Chapola, Ganja, Ganjika, Harsini, Indrasana, Jaya, Ujaya, Vijaya.

Distribution : Throughout India.

Part Used : Root.

Utilization : Decoction checks stomach pain and headache (½ teaspoon twice daily for 3 days).

Active Constituents : Active constituents are capparine,F cappariline, capparinine, cannabidiol, cannabigerol, feruloyltyramine and p-coumaroyltyramine.

References : Ansari (1991), Rastogi & Mehrotra (1998, 2002), Sood & Thakur (2004).

Canthium parviflorum Lamk.

Family : Rubiaceae.

Part Used : Root.

Utilization : Juice good against intestinal disorders.

Reference : Goud & Pullaiah (1996).

Capparis decidua (Forssk.) Edgew.

Syn : *C. aphylla* Roth.

Family : Capparidaceae.

English, Hindi & Sanskrit Names:

Eng. : Caper Berry;

Hindi : Karel, Karer, Karu, Keril, Lete;

Sans. : Granthila, Gudhapatra, Kantaki, Kataphal, Nishpatrika, Shakapushpa, Tikshnasara, Ushanasundra, Vidahika, Vishvakpatra.

Distribution : Rajasthan, Punjab, Sindh; Southwards to Karnatka.

Parts Used : Root. Root bark.

Utilization : Two teaspoonful extr. of roots along with equal quantity of honey given thrice a day to get relief in dropsy. Paste of the root bark applied to malignant tumours, boils, eruptions, rheumatic swellings, gout, and also considered astringent, alterative, antitoxic agent and febrifuge.

Active Constituents : Mainly contains l-stachydrine, capparidisine, capparisine, capparisinine, n-pentacosane, n-triacotanol, beta-sitosterol, spermidine isocodonocarpine, 14-N-acetylisocodonocarpine and 15-N-acetylcapparisine (root bark), capparin, capparilin and capparinin (root).

References : Anonymous (1948, 1968), Caius (2003), Chopra *et al*. (1956), Khanna & Kumar (2000), Khare (2004), Rastogi & Mehrotra (1998, 2002).

Capparis sepiaria L.

Syn : *C. glauca* Wall. *ex* Hook.f. & Thoms.; *C. retusella* Thw.; *C. sepiaria* var. *incanescens* (DC.) Hook.f. & Thms.; *C. sepiaria* var. *retusella* (Thw.)Thw.; *C. sepiaria* var. *vulgaris* Hook.f. & Thoms.; *C. spinosa* Linn.

Family : Capparidaceae.

English, Hindi & Sanskrit Names:

Eng. : Caper Bush, Indian Caper;

Hindi : Ber, Hium, Kabra, Kanthari;

Sans. : Ahimsa, Amlaphala, Dupravesha, Gridhranakhi, Jali, Kakadani, Kanya, Vakrakantaki, Vayastinduka.

Distribution : Rocky & Stony Soil of N. W. India, Through Punjab, Rajasthan to the South.

Part Used : Root.

Utilization : Reported to cure earache, fever and gout. Its paste used in the form of vaginal pessary in disorders of the female genital tract, and its pieces stuck to the lock of hair of pregnant woman for inducing abortion. Simultaneously during the treatment, mixture of the powdered root of *Pueraria tuberosa* (Roxb.) DC. and 'daru' (*Punica granatum*) also given orally. Juice of the root effectively dropped in the ear to relieve earache and also taken orally to cure pneumonia and *Ascaris* worms. Its paste with opium applied to check swelling of eyes. 2-3 drops of root juice put in nasals to treat headache. To heal sores and mumps in Kangra, macerated roots are applied crushed with ginger and asafoetida. Root bark considered aperient, tonic, expectorant, anthelmintic, emmenagogue, analgesic, sedative, diuretic, stomachic and also good in enlarged spleen, tubercular glands, paralysis, cholera, toothache, dropsy and affections of liver and spleen.

Active Constituents : These are stachydrine, rutic acid and a volatile substance with garlic odour (root bark), indole glucosinolates, glucobrassicin, neoglucobrassicin and 4-methoxy-glucobrassicin (roots).

Biological Activity : Stachydrine when given to dogs, rabbits and rats quickened the coagulation of blood.

References : Agarwal (2003), Ambasta (1986), Anonymous (1948), Caius (2003), Jadeja (1999), Jain (1991), Jain *et al.* (1991), Khare (2004), Kirtikar & Basu (1935), Parrotta (2001), Punjani (2002), Reddi *et al.* (2005), Sharma & Singh (1999), Singh *et al.* (2003), Singh & Pandey (1998).

Capparis zeylanica L.

Syn : *C. hastigera* Linn.; *C. horrida* Linn.; *C. xanthophylla* Coll. & Hemsl.

Family : Capparidaceae.

English, Hindi & Sanskrit Names:

Eng. : Caper Plant, Ceylon Caper;

Hindi : Ardanda, Hins, Jakhambel, Jhiri, Karawa;

Sans. : Animrsa, Govindi, Granthila, Kantakalata, Karamba, Kinkani, Krishangi, Tapasapriya, Vartala, Vyaghraghanti, Vyaghrapada, Vyagranakhi.

Distribution : Lowland Localities in India from U.P. Eastwards & Southwards Through Peninsular India.

Part Used : Roots. Root bark.

Utilization : Roots effectively employed for curing rheumatism and dyspepsia. 3-5g dose of the mixture of its roots with 21 black peppers taken with water daily for getting relief in syphilis and its paste administered orally with sugar candy and 'kalimircha' (*Piper* nigrum) to cure snake bite. Root bark considered bitter, cholagogue, diuretic, sedative, aphrodisiac, stomachic, antihidrotic, and useful against earache, cholera, pus formation, tuberculosis, boils, swelling of testicles, paralysis, toothache and pyorrhoea and its extr. mixed with goat's milk considered useful aphrodisiac (2 spoonfuls once a day for 30 days). Pills made from root bark and black pepper (*Piper nigrum*) prescribed for tuberculosis (2 pills twice a day for 15 days); and with that of *Ficus benghalensis* and tail of garden lizard for paralysis (2 pills once a day for 40 days).

References : Agarwal (2003), Anonymous (1948, 1968), Balasubramanian & Prasad (1996), Caius (2003), Chatterjee & Pakrashi (1997), Chopra *et al.*(1956), Jain (1991, 1996), Kirtikar & Basu (1935), Parrotta (2001), Rao & Henry (1996), Reddi *et al.* (2005), Saini (1996b), Sarangi & Sahu (2004), Singh (2000), Sivakumar *et al.* (2003), Tewari *et al.* (2001).

Capsicum frutescens L.

Syn : *C. minimum* Roxb.

Family : Solanaceae.

English Name:

Eng. : Bird Chilli.

Distribution : Universally Cultivated Throughout India.

Part Used : Root.

Utilization : Prescribed for gastritis and burning sensation in stomach.

Reference : Varghese (1996).

Cardiospermum halicabum L.

Family : Sapindaceae.

English, Hindi & Sanskrit Names:

Eng. : Ballon Vine Heartseed, Blister Creeper, Winter Cherry;

Hindi : Kanphata, Kanphuti;

Sans. : Indravalli, Karnasphota, Parvatagni, Tejovati.

Distribution : Common Throughout India in Wastelands, Hedges & Thickets.

Part Used : Root.

Utilization : So far reported from India as diaphoretic, diuretic, aperient, emetic, laxative, abortifacient, rubifacient, febrifuge (decoction twice a day), stomachic, and also efficacious in nervous diseases, lumbago, piles (decoction), rheumatism (paste), arthritis, diarrhoea and dysentery (2-3 spoonfuls thrice a day for 1 to 3 days), menstrual disorder, white discharge-2 spoonfuls twice a day for 3 days after menstruation (root extr.).

Active Constituents : Phlobaptene, phlobatannin and sitosterol.

Biological Activity : Root extr. diuretic in male albino rats.

References : Agarwal (2003), Anonymous (1948, 1968), Asolkar *et al.* (1992), Chatterjee & Pakrashi (1997), Chopra *et al.* (1956), Dastur (1970), Drury (1978), Gogoi & Das (2003), Jain & Sharma (2000), Kurian (1995), Painuli & Maheshwari (1996), Patil (2000), Rao & Henry (1996), Ravishankar & Henry (1992), Reddi *et al.* (2005), Shukla & Verma (1996), Sikarwar (1996), Watt (1889).

Carica papaya L.

Family : Caricaceae.

English, Hindi & Sanskrit Names:

Eng. : Papaya Tree;
Hindi : Papaya, Pappita, Popai;
Sans. : Brahmairandah.

Distribution : Naturalized Throughout the Warmer Parts of India.

Parts Used : Root. Root bark.

Utilization : Roots beneficial as a tonic, medicine for excessive bleeding of the kidney and anthelmintic; and also said to cure yaws and piles. Its powder (6g) mixed with 100 ml of water (3 spoons twice daily) given to dissolve kidney stones, and infusion (10g) prescribed orally once a day to woman after child birth to promote secretion of breast milk. Root bark useful in constipation, piles and as galactagogue (two spoonfuls once a day for five consecutive days).

References : Agarwal (2003), Anonymous (1948), Borthakur (1993), Dutta & Nath (2000), Kurian (1995), Parrotta (2001), Reddi *et al.* (2005), Sinha & Sinha (2001).

Carissa carandas L.

Syn : *C. congesta* Wight.
Family : Apocynaceae.

English, Hindi & Sanskrit Names:

Eng. : Karaunda;
Hindi : Karonda;
Sans. : Avigna, Karamarda, Karamardika.

Distribution : Different Parts of India; Wild in Drier Forests, Cultivated in Gardens.

Part Used : Roots.

Utilization : Considered anthelmintic, bitter purgative, stomachic and locally used in scabies and antidote for snake bite, animal sores (pounded roots), skin diseases (paste in lime juice), cough and diarrhoea (decoction).

Active Constituents : Yields 4-cardioactive compounds, carissone, β-sitosterol, carindone and carinol.

Biological Activity : Extr. exhibits histamine releasing activity.

References : Agarwal (2003), Anonymous (1948, 1968), Asolkar *et al.* (1992), Chatterjee & Pakrashi (1997), Chopra *et al.* (1956), Kurian (1995), Rastogi & Mehrotra (1999a, b), Singh & Pandey (1998), Upadhye *et al.*

(1994), Verma & Chauhan (2006), Watt (1889).

Carissa spinarum L.

Syn : *C. diffusa* Roxb.; *C. opaca* Stapf. *ex* Haines.

Family : Apocynaceae.

English, Hindi & Sanskrit Names:

Eng. : Karaunda;
Hindi : Karaunda;
Sans. : Karamardika.

Distribution : Sub-Himalayan Tract Upto 1,000 m, Trans-Indus Territory & Coast of S. Andamans.

Parts Used : Root. Root bark.

Utilization : Root efficacious for treatment of rheumatism, worm-infested sores of animals (paste), stomachache (root extr.), fever (decoction), inflammation (paste) and as purgative. Root bark useful in malignant ulcers as antibiotic. Mixture of its root paste along with that of *Bombax ceiba* and rhizome paste of *Acorus calamus* applied on chest and back of the body for curing pneumonia.

Active Constituents : Roots contain 5 cardiac glycosides, including evomonoside, odoroside G and H.

References : Ambasta (1986), Anonymous (1948, 1968), Asolkar *et al.* (1992), Chopra *et al.* (1956), Maheshwari *et al.* (2001), Nayak *et al.* (2003), Parmar & Kaushal (1982), Rastogi & Mehrotra (1998), Singh & Kumar (2000a), Singh & Prakash (1994), Watt (1889).

Casearia esculenta Roxb.

Syn : *C. levigata* Dalz.; *C. ovata* (Lamk.) Willd.

Family : Flacourtiaceae.

Hindi Name:

Hindi : Kulkulta.

Distribution : Distributed in Maharashtra, Kerala.

Part Used : Root.

Utilization : Decoction given in liver troubles, diabetes, piles, indigestion, and as purgative, astringent and cathartic.

Active Constituents : Contains leucopelargonidin, tannin, resin and colouring matter.

Biological Activity : Administration of alc. extr. to rats led to significant reduction in blood sugar level after two hours. Glucose tolerance was also favourably influenced by chronic administration of its extr..

References : Agarwal (2003), Ambasta (1986), Chopra *et al.* (1956), Mitaliya & Bhatt (2003), Rastogi & Mehrotra (1999a, b), Sood *et al.* (2005a), Yoganarasimhan (1981), Watt (1889).

Casearia graveolens Dalz.

Family : Flacourtiaceae.

Part Used : Root.

Utilization : Paste used in piles.

Reference : Kothari & Moorthy (1996).

Casearia tomentosa Roxb.

Syn : *Anavinga lanceolata* Lam.; *Casearia elliptica* Willd.; *C. ovata* Roxb.

Family : Flacourtiaceae.

Hindi & Sanskrit Names:

Hindi : Bairi, Beri, Chilla;

Sans. : Kampillaka.

Distribution : Moist, Hilly Locations Upto 900 m from the Foot Hills of the Himalaya Southwards, Andaman Islands. Open Grazing Lands in S. India. Rare in W. Ghats.

Parts Used : Roots. Root bark.

Utilization : Root bark considered a reputed tonic for anaemic conditions and also stimulates liver function. Its paste applied on enlarged scrotum in children. A decoction of its root reported to be a remedy for diabetes and juice given for dysentery (2-3 teaspoons 2 times a day, for 1-2 days).

References : Kumar & Jain (1998), Kothari *et al.* (2003), Parrotta (2001), Reddi *et al.* (2005), Singh (1996).

Cassia alata L.

Family : Fabaceae.

English & Hindi Names:

Eng. : Ringworm Senna;

Hindi : Dat-Ka-Pat, Dadmurdan.

Distribution : Wild & Cultivated All Over Bengal & Many Other Parts of India.

Part Used : Root.

Utilization : Infusion beneficial in cases of irregular menses and liver incumbement. Roots also considered diuretic.

References : Ambasta (1986), Dey (1973), Jain & Sikarwar (1998), Sharma (2003), Watt (1889).

Cassia auriculata L.

Syn : *Senna auricalata* (L.) Roxb.

Family : Fabaceae.

English, Hindi & Sanskrit Names:

Eng. : Avaram, Tanner's Senna, The Tanner's Cassia;

Hindi : Aswal, Tarval, Taval;

Sans. : Avartaki.

Distribution : Dry Zones of Peninsular India. Particularly Common in A.P., Tamil Nadu & Karnataka.

Part Used : Root.

Utilization : Useful in diarrhoea, cough, cold, rheumatism and as nephroprotector. Its powder mixed with sugar and given orally to cure leucorrhoea and menorrhoea and its paste when mixed with mustard oil used against skin diseases and also applied on the abdomen against problems in urinary discharge. Root bark controls frequent urination.

References : Agarwal (2003), Ambasta (1986), Anonymous (1968), Chopra *et al.* (1956), Dastur (1970), Jain (1991), Jain *et al.* (1991), Parrotta (2001), Reddi *et al.*(2005), Singh & Pandey (1998), Venkataratnam & Raju (2004).

Cassia fistula L.

Syn : *C. rhombifolia* Roxb.; *Cathartocarpus fistula* (L.) Pers.

Family : Fabaceae.

English, Hindi & Sanskrit Names:

Eng. : Golden Shower, Indian Laburnum, Purging Cassia, The Cassia Fistula;

Hindi : Amaltas, Bandarlauri, Dhanbuhar, Girmalah, Kirwara, Warga.

Sans. : Aragvadha, Sampaka, Suvarnaka, Vyadhighata;

Distribution : Wild & Cultivated Throughout India in Mixed Deciduous Forests & Hilly Tracts in the Outer Himalaya. Common Throughout Gangetic Valley, C. & S. India.

Parts Used : Root. Root bark.

Utilization : Roots reported to be useful for common cold, fever, cardiac disorders, biliousness, rheumatic conditions, elephantiasis (extr.), haemorrhages, wounds, ulcers, boils, tubercular glands, various skin diseases, urinary troubles, syphilis, burning sensation, leprosy, gastric disorders (1/2 glass drink, 3-5 days), inflammation, constipation, asthma (root paste), and as tonic, febrifuge and strong purgative. Root bark useful as a laxative, tonic, and also employed in treatment of black fever. Powdered roots mixed with goat's milk (one teaspoonful taken daily twice) prescribed against breast diseases. Mixture of roots pounded together with *Cryptolepis buchananii* given as antinarcotics to relieve nervous system paralysed under the influence of liquor.

Active Constituents : Root bark contains tannin, phlobaphenes and oxyanthraquinone.

References : Beg *et al.* (2005), Chatterjee & Pakrashi (1997), Chauhan (1999), Chopra *et al.* (1956), Dastur (1970), Dey (1973), Dwivedi (1993), Girach & Aminuddin (1995), Jain (1991), Kirtikar & Basu (1935), Kurian (1995), Parrotta (2001), Reddi *et al.* (2005), Said (1997), Saren *et al.* (2000), Sarkar *et al.* (2000), Seetharam *et al.* (1999), Sharma (2003), Singh (1999).

Cassia hirsuta Linn.

Family : Fabaceae.

Distribution : Bihar, N. E. & Peninsular India in Plains.

Part Used : Root.

Utilization : Extr. efficacious against snake bite (*Vipera russelli*).

References : Lalramnghinglova & Jha (1997), Rastogi & Mehrotra (2002).

Cassia obtusifolia L.

Syn : *Cassia tora* auct.(non L.)

Family : Fabaceae.

English, Hindi & Sanskrit Names:

Eng. : Fetid Cassia, Ringworm Plant;

Hindi : Chakunda, Panevar;

Sans. : Avudham, Chakragaja, Chakramarda, Chakri, Dadrughna, Dridhabija, Edahasti, Gajakhya, Kusume, Meshakshi, Meshavhaya, Padmata, Taga, Tarkilla, Uranaksha, Vimardaka, Vyavartaka.

Part Used : Root.

Utilization : Powder taken orally as an antidote in case of scorpion stings, snake bite, and cures skin diseases, tuberculous glands, ringworm (paste). Also considered diuretic and antiperiodic.

Active Constituents : Chief constituents are crysophenol, 1, 7-dihydroxy-5-methoxy carboxy-1-3-methylxanthone, β-sitosterol, 1, 3, 5-trihydroxy-6, 7-dimethoxy -2-methylanthraquinone, leucopelargonidin-3-O-α-L-rhamnopyranoside, helminthosporin, methylchrysophanol, betulinic acid, 2, 5 dimethoxy benzoquinone and aloe-emodin.

Biological Activity : Aloe-emodin exhibits antimicrobial activity.

References : Bhattacharjee (2001), Chatterjee & Pakrashi (1997), Girach *et al.* (1997), Kirtikar & Basu (1935), Rastogi & Mehrotra (1999a, b, 2002), Upadhyay & Singh (2005).

Cassia occidentalis Linn.

Syn : *Senna iccidentalis* (L.) Link.

Family : Fabaceae.

English, Hindi & Sanskrit Names:

Eng. : Coffee-Senna, Fetid Senna, Negro Coffee;

Hindi : Barikasondi, Chakunda, Kasonda, Kasondi;

Sans. : Arimarda, Dipna, Kala, Kanaka, Kasamandah, Vimarda.

Distribution : Throughout India (Weed in Wasteplaces & Near Dwellings) from Himalaya to Cape Comarin.

Parts Used : Root. Root bark.

Utilization : Roots so far known to be used in skin diseases, stomach ache, whooping cough, filariasis, paralysis, chest pain, body swellings, jaundice, boils, burns, eye infections, snake bite, scorpion stings, pneumonia, malaria, aphonia, dysentery, sore throat (chewed), and as anthelmintic, diuretic, febrifuge (1g root powder twice daily with luke warm water) and antiemetic. Its powder mixed with the aril of *Pithecellobium* considered aphrodisiac. Mixture of the pounded root with black pepper checks headaches, and given to nursing women for purification of milk. Root bark used as a quinine substitute to cure malarial fever, and used as a stimulant and specific medicine for gonorrhoea and hepatic troubles (decoction).

Active Constituents : These are emodin, quercetin, chrysophanol, emodol, physcion, β-sitosterol, a phytosterol, aloe-emodin, occidentalol I and occidentalol II, islandicin, questin, chrysophanol-10, 10′-bianthrone germichrysone, methylgermitorosone and pinselin.

References : Barna *et al.* (2000), Chauhan (1999), Chopra *et al.* (1956), Gogoi & Das (2003), Hosagoudar & Henry (1996b), Jain (1991), Kirtikar & Basu (1935), Parrotta (2001), Rajendran & Henry (1994), Rastogi & Mehrotra (1999a, b, 2001), Reddi *et al.* (2005), Saxena *et al.* (1997), Sharma (2003), Upadhyay & Chauhan (2000), Varghese (1996).

Cassia sophera L.

Syn: *Cassia chinensis* Jacq.; *Senna esculenta* Roxb.; *S. purpurea* Roxb.; *S. sophera* Roxb.

Family: Fabaceae.

English, Hindi & Sanskrit Names:

Eng. : Dog Seena, Senna Sophera;

Hindi : Bas-Ki-Kasunda, Kasaunda, Kasundi;

Sans. : Kasamardan.

Distribution : Tropical Regions of India.

Part Used : Root.

Utilization : Given to cure cough and acute bronchitis. Its juice prescribed orally as tonic (10 ml, twice daily) to cure perpeurium.

Active Constituents : Anthraquinones-1, 8-dihydroxy-3, 6-dimethoxy-2-methyl-7-vinylanthraquinone and 1, 3- dihydroxy-5, 7, 8-trimethoxy-2-methylanthraquinone have been isolated.

References : Agarwal (2003), Ambasta (1986), Chatterjee & Pakrashi (2001), Rastogi & Mehrotra (2002), Sarma *et al.* (2001), Watt (1889).

Cassia tora L.

Syn : *Senna tora* (L.) Roxb.

Family : Fabaceae.

English, Hindi & Sanskrit Names:

Eng. : Foetid Cassia, Sickle Senna, Wild Senna;

Hindi : Chakavat, Chakunda, Chekour, Panevar, Puar, Tarata;

Sans. : Chakramardah, Dadmari.

Distribution : Throughout India to an elevation of 1,500 m in the Himalaya, Often Gregarious. Typically on Roadsides, Grazing Lands & Wasteplaces.

Part Used : Root.

Utilization : Recorded to be used in eczema, snake and scorpion bites (root paste), ringworms (root smashed in lime juice and applied) and fever. Its pieces tied as a necklace for child suffering from chronic fever associated with stiffness of muscles.

Active Constituents : Contains anthraquinones and beta-sitosterol.

References : Agarwal (2003), Chauhan (1999), Chopra *et al.* (1956), Dastur (1970), Jain (1991), Kothari & Londhe (2000), Parrotta (2001), Reddi *et al.* (2005), Sahoo & Bahali (2003), Singh & Pandey (1998), Subramanium (2000), Varghese (1996), Watt (1889).

Cassine glauca (Rottb.) Kuntze.

Syn : *Celastrus glaucus* (Rottb.) Vahl.; *Elaeodendron glaucum* (Rottb.) Pers.; *Mangifera glauca* Rottb.

Family : Celastraceae.

English, Hindi & Sanskrit Names:

Eng. : Ceylon Tea;

Hindi : Bakra, Chouri, Dhebri, Jamras, Jamrassi;

Sans. : Asitamuskaka, Bhutphal, Khavalli, Krishnamuskaka, Muskaka.

Distribution : Commonly in Dry & Mixed Deciduous Forests or Scattered in Sal (*Shorea robusta*) Dominated Forests in the Sub-Himalaya Upto 1,800 m particularly in N. W. U.P. & Throughout Peninsular India.

Parts Used : Root. Root bark.

Utilization : Root bark extr. given orally once a day for 3-4 days for curing dysentery, rheumatism, gout and as emetic. Root considered antidote to snake bite.

References : Jain (1991), Parrotta (2001), Singh (1996), Singh & Rao (2003), Sinha & Sinha (2001).

Catharanthus roseus (L.) Don

Syn : *Lochnera rosea* (L.) Reichb.; *Vinca rosea* L.

Family : Apocynaceae.

English, Hindi & Sanskrit Names:

Eng. : Madagascar Periwinkle, Periwinkle, Red Periwinkle;

Hindi : Sadabahar, Sada Suhagan, Sada Suhagi;

Sans. : Nityakalyani, Rasna, Sadampuspa.

Distribution : Planted in Gardens & Occasionally Naturalised to an altitude of 1,300 m.

Part Used : Root.

Utilization : Efficacious against diabetes and toothache; also found to have purgative, vermifugal, depurative, sedative and haemostatic properties.

Active Constituents : Active principles are ajmalicine, serpentine, tetrahydroalstonine, cavicine, dihydrositsirikine, isositsirikine; 2-acylindoles: perivine, perosine, akuammicine, lochnerine, lochnerivine, maandrosine, ammocalline, pericalline (apparicine, gomizine, tabunoschizine), virosine, vincapusine. L(+) bornesitol, venalstonine,vindolinine, lochnericine, catharanthine, vindoline, leurosine, ursolic acid, oleanolic acid and alstonine

Biological Activity : Hypotensive, sedative and tranquillizing properties found to be significant. Experiments on animals have shown that certain extr. of *Catharanthus* are useful in leukaemia.

References : Agarwal (2003), Ambasta (1986), Asolkar *et al.* (1992), Bhattacharjee (2001), Chatterjee & Pakrashi (1997), Jain (1991), Kurian (1995), Parrotta (2001), Rastogi & Mehrotra (1999a, 2001), Reddi *et al.* (2005), Sharma (2003), Sood *et al.* (2005a).

Catunaregam spinosa (Thnb.) Tirv.

Syn : *Gardenia dumetorum* Retz.; *G. spinosa* Thunb.; *Randia dumetorum* (Retz.) Poir.; *R. dumetorum* (Retz.) Poir. var. *floribunda* (DC.) Gamble; *R. brandisii* Gamble; *R. floribunda* DC.

Family : Rubiaceae.

English, Hindi & Sanskrit Names:

Eng. : Common Emetic Nut;
Hindi : Mainphal;
Sans. : Madana.

Distribution : A.P.

Part Used : Root.

Utilization : Recorded to be used for ulcer in intestine and stomach, and post delivery pains. Its paste containing turmeric (*Curcuma longa*) and goat's milk also prescribed for stomach ulcers (2 spoonfuls twice a day for 10 days).

References : Rao & Henry (1996), Reddi *et al.* (2005).

Cayratia auriculata (Roxb.) Gamble

Syn : *Cissus auriculata* Roxb.
Family : Vitaceae.

Distribution : Common in A.P.

Part Used : Root.

Utilization : Tuberous root applied externally to cure arthralgia.

References : Deshmukh & Rothe (2003), Rao & Henry (1996).

Cayratia carnosa (Lamk.) Gagnep.

Syn : *Cayratia trifolia* (L.) Domin.;
Vitis carnosa (Roxb.) Wall.; *Vitis trifolia* L.
Family : Vitaceae.

English, Hindi & Sanskrit Names:

Eng. : Fox Grape;
Hindi : Amalbel, Gidardrak, Kassar, Ramchana;
Sans. : Amlaparni, Amlavalli, Atyamlaparni.

Distribution : Open Habitats Throughout the Hotter Parts of India, including all of Peninsular India Upto 600 m.

Part Used : Root.

Utilization : Useful in India as an astringent, antiseptic, emetic, diaphoretic, febrifuge, and also given in anaemic and diabetic conditions. Root paste mixed with turmeric (*Curcuma domestica* Vahl.) given to cattle for foot and mouth diseases. Poultice of 2g root powder mixed with 10 ml cow urine considered good against boils, pimples, cuts, wounds, lymphatic swellings and animal bite to cure poisonous effect. Decoction effectively controls excessive bleeding from uterus, and 'Bhils' in Udaipur district use its fomentation to cure mumps.

References : Agarwal (2003), Ambasta (1986), Anonymous (1948), Asolkar *et al.* (1992), Bhogaonkar & Devarkar (2002), Chatterjee & Pakrashi (1997), Chopra *et al.* (1956), Painuli & Maheshwari (19960, Parrotta (2001), Punjani (2004), Ravisankar & Henry (1992), Sikarwar (1996), Singh & Pandey (1998).

Ceiba pentandra (L.) Gaertn.

Syn : *Bombax pentandrum* L.; *Eriodendron pentandrum* (L.) Kurz.

Family : Bombacaceae.

English, Hindi & Sanskrit Names:

Eng. : True Kapok Tree, White Silk-Cotton Tree;

Hindi : Hattian, Katan, Safed-Semul;

Sans. : Dhaba, Sveta Salmali.

Distribution : Hotter Parts of W. & S. India.

Part Used : Root.

Utilization : Useful for removing stones from bladder (decoction), diabetes (root juice), dysentery and gonorrhoea (tap root), snake bite, and as diuretic.

References : Agarwal (2003), Ambasta (1986), Anonymous (1948), Chatterjee & Pakrashi (1997), Chopra *et al.* (1956), Dastur (1970), Nayak *et al.* (2003).

Celastrus paniculatus Willd.

Syn : *C. multiflorus* Roxb.; *C. nutans* Roxb.

Family : Celastraceae.

English, Hindi & Sanskrit Names:

Eng.	:	Black-Oil Plant, Climbing Staff Plant;
Hindi	:	Apang, Jamrasi, Kondgaidh, Malkagni, Malkamini, Pheng, Sankhui;
Sans.	:	Avatapadi, Jyotishmati, Kanguni, Malkangui, Pitataila.

Distribution : Sub-Himalayan Tract Upto 2,000 m & Hilly Areas of Maharashtra, Gujarat & Tamil Nadu.

Part Used : Root.

Utilization : Considered efficacious in headache, snake bite, jaundice (paste applied over the body), piles (pills made from crushed roots; 2-3 times a day for 3 days), dysmenorrhoea, pneumonia, paralysis, ulcers (chewed), helmintic infection, eye diseases, diabetes (root juice), diarrhoea and kidney troubles. Powdered root (1 teaspoon twice a day, for 10 days) prescribed for the treatment of tumour cancer. Mixture of its paste and seeds applied externally in bodyache and joint pain, and its mixture comprising its root (5 cm long), 7 black peppers (*Piper nigrum*) and 'chawaldhua' water (water in which rice is washed) prescribed twice a day to cure spermatorrhoea, leucorrhoea and piles.

Active Constituents : These are mainly pristimerin, celastrol, zeylasterone and zeylasterol.

References : Binu *et al.* (2003), Brahmam & Saxena (1990), Jain (1991,1996), Kothari & Londhe (2000), Maheshwari *et al.* (1997), Nadanakunjidam (2003), Parrotta (2001), Patil (2000), Rastogi & Mehrotra (1998), Reddi *et al.* (2005), Singh (1996), Singh & Prakash (1994), Thakur *et al.* (1989), Varghese (1996).

Celosia argentea L. var. *cristata* (L.) Kuntze.

Syn : *C. cristata* L.

Family : Amarantaceae.

English, Hindi & Sanskrit Names:

Eng.	:	Cox Comb Plant, Quail Grass;
Hindi	:	Debkoti, Safed Murgha, Sarwari;
Sans.	:	Pothura, Vitunna.

Distribution : A Common Weed Throughout India.

Part Used : Root.

Utilization : Poultice applied all over the body to cure fever acccompanied by shivering, urinary stone (applied locally), anorexia, tastelessness, giddiness and as an antiemetic, and its extr. homogenised with five seeds of *Diospyros melanoxylon* given orally (1/2 cup) to women for birth control. Fresh root decoction (50 ml) used for one week as an antidote in snake bite.

Active Constituents : Hexadeca dienoic, tricosanoic, lignoceric, cerotic and α-aminobutyric acids together with 1, 1′-[3-(2-cyclopentyl-ethylidene)-1, 5-pentanediyl] bis-cyclopentane have been localized.

References : Agarwal (2003), Ambasta (1986), Anonymous (1948), Kshirsagar & Singh (2000a), Maheshwari *et al.* (1997), Pawar & Patil (2004), Ram *et al.* (2002), Rastogi & Mehrotra (2002), Reddi *et al.* (2005).

Centaurea behen Linn.

Family : Asteraceae.

English & Hindi Names:

Eng. : White Raphontic;

Hindi : Safed Bahman.

Distribution : Cultivated in India in Sub-Tropical Regions.

Part Used : Root.

Utilization : Efficaceous in jaundice and calculus affections.

Active Constituents : Contains behenin, taraxasterol, myristate, inulin and centaurea sterol A.

References : Agarwal (2003), Ambasta (1986), Anonymous (1948), Asolkar *et al.* (1992), Chopra *et al.* (1956), Khare (2004).

Centella asiatica (L.) Urb.

Syn : *Hydrocotyle asiatica* L.

Family : Apiaceae.

English, Hindi & Sanskrit Names:

Eng. : Indian Penny-Wort;
Hindi : Brahmi, Brahm-Manduki, Budhbrahmani;
Sans. : Mandukparni.

Distribution : Moist Places Throughout India; (Marshy Banks of Rivers, Streams & Ponds).

Part Used : Root.

Utilization : Decoction (half tea spoonful) taken orally for jaundice.

Active Constituents : These are polyacelenes I-V and nine other acetylenes.

References : Chatterjee & Pakrashi (1997), Jain (1968, 1996).

Centratherum anthelminticum (L.) Kuntze.

Syn : *Conyza anthelmintica* L.; *Veronia anthelmintica* L.
Family : Asteraceae.

Hindi & Sanskrit Names:

Hindi : Buckshi, Somraj;
Sans. : Bakuchee, Somraji.

Distribution : Wasteplaces, Hedges & Cultivated Lands Throughout India.

Part Used : Root.

Utilization : Useful in menorrhagia and post-partum haemorrhage.

Reference : Chatterjee & Pakrashi (1997).

Cephaelis ipecacuanha (Brot.) A.Rich.

Syn : *Callicocca ipecacuanha* Brot.; *Cephaelis emetica* Pers.; *Ipecacuanha officinalis* Arrunda; *Psychotria ipecacuanha* Stokes.
Family : Rubiaceae.

English & Hindi Names:

Eng. : Ipecac, Ipecacuanha Root;
Hindi : Ipecac.

Distribution : Commercially Cultivated in Darjeeling & Parts of Assam, Sikkim, S. India.

Part Used : Dried Root.

Utilization : So far known to cure amoebic dysentery, uterine haemorrhage, dyspepsia, morning sickness, whooping cough, asthma, menorrhagia, intermmitent fever, piles, cold, gonorrhoea, and as cholagogue, stimulant, diaphoretic, sedative, appetiser, emetic and antidysenteric.

Active Constituents : The major constituents are emetine, cephaline, psychotrine, methylether, emetamine, ipecacuanhic acid, ipecoside, erythrocephleine (colouring matter), saponin, resin, fat, calcium oxalate, starch, hydroipecamine, ipecamine, protoemetine, o-methylpsychotrine, choline, citric and malic acids and ipecacuanhin.

Biological Activity : Emetine (1mg/kg) showed tubocurarine-like effect on neuromuscular transmission in rat.

References : Agarwal (2003), Asolkar *et al.* (1992), Baburaj *et al.* (2000), Bhattacharjee (2001), Chatterjee & Pakrashi (1997), Chopra *et al.* (1956), Dey (1973), Rastogi & Mehrotra (1999a, b), Reddi *et al.* (2005), Sharma (2003), Watt (1889).

Chasalia chartacea Craeb.

Syn : *C. curviflora* Thw.; *Psychotria curviflora* Wall.

Family : Rubiaceae.

Hindi Name:

Hindi : Vellakurinji.

Distribution : E. Himalaya, S. India, W. India & Andamans.

Part Used : Root.

Utilization : Decoction used in rheumatism, pneumonia, head disorders, ear and eye diseases, malaria, throat complaints and cough. Its paste diluted in 10ml cow's milk administered orallly 2 to 3 times for stomach pain.

References : Agarwal (2003), Ambasta (1986), Anonymous (1948), Chopra *et al.* (1956), Rajendran & Henry (1994).

Chasalia ophioxyloides (Wall.) Craib.

Family : Rubiaceae.

Distribution : India, Sri Lanka, Myanmar & Malaysia.

Part Used : Root.

Utilization : Decoction efficacious against rheumatism, pneumonia, sore throat, head disorders, ear and eye diseases.

Reference : Yoganarasimhan (1981).

Cheilanthes farinosa (Forssk.) Kaul f.

Family : Polypodiaceae.

English & Hindi Names:

Eng. : Silver Fern;
Hindi : Shankhpushpi.

Distribution : All Over N.E. Region (Rock Crevices & Hill Cuttings in Shaded Situation).

Part Used : Root.

Utilization : Useful for eczema, menstrual disorders and stomachache.

References : Asolkar *et al.* (1992), Kumar (2002).

Cheilanthes tenuifolia (Burm.) Sweet

Family : Polypodiaceae.

Part Used : Root.

Utilization : Used for stomachache (1 pill thrice daily for 7 days with hot water).

Reference : Varghese (1996).

Chlorophytum arundinaceum Baker

Family : Liliaceae.

Hindi & Sanskrit Names:

Hindi : Biskandri, Safed Musli;
Sans. : Musali, Sweta-Musli.

Distribution : N.E. India (Bhutan to Manipur). Peninsular India from Orissa & Maharashtra Southwards to Tamil Nadu to an altitude of 1,200 m.

Part Used : Root.

Utilization : Known in India for dysmenorrhoea (root crushed in rice water and extr. taken twice a day for two days), arthritis (decoction in milk given to patient once a day for one month), stomachache (one teaspoonful of root paste taken with a cup of warm milk twice a day for 3 days), asthma (one glassful decoction of roots mixed with one spoonful of 'Mishri' (sugar pellets) and taken twice a day for one month) and also considered as tonic as it is said to improve vigour and vitality.

Active Constituents : Contains carbohydrate, proteins, root fibres, saponins and minerals.

References : Ambasta (1986), Asolkar *et al.* (1992), Bhattacharjee (2001), Dwivedi (1993), Parrotta (2001), Sahoo & Mudgal (1995), Sen *et al.* (2002), Sharma (2003), Tirkey (2004), Upadhyay (2003).

Chlorophytum borivilianum Sant. & Fern.

Family : Liliaceae.

Distribution : Sal & Moist-Teak Forests of M.P., Chatisgarh, Jharkhand.

Part Used : Root.

Utilization : Decoction prepared by boiling it with a glass of milk and taken daily at night for nearly a month as a health tonic. Its aq. extr. taken orally (twice daily for 7 days) to cure cough; and also for rheumatic and nervous complaints and in diabetes.

References : Chaudhari & Hutke (2002), Punjani (2006), Sharma (2003).

Chlorophytum glaucum Dalz.

Family : Liliaceae.

Part Used : Root.

Utilization : Root tubers nutritious and given to cure sexual impotency.

Reference : Bhogaonkar & Devarkar (2002).

Chonemorpha fragrans (Moon.) Alston

Syn : *C. macrophylla* (Roxb.) G.Don; *Echites macrophylla* Roxb.

Family : Apocynaceae.

Hindi & Sanskrit Names:

Hindi : Garbhedaro;

Sans. : Murva.

Distribution : Moist Forests Throughout India Including Andaman Islands; ascending to 1,400 m in Hills.

Part Used : Root.

Utilization : Found to possess astringent, laxative, thermogenic, depurative, carminative, antehelmintic, digestive, sudorifice, anodyne, expectorant, febrifuge properties, and also considered useful in leprosy, colic, constipation, dyspepsia, flatulence, hyperacidity, cough, ronchitis, intermittent fevers, cardiac debility, diabetes, jaundice and hyperdypsia. Its extr. efficacious against retained placenta (100 ml twice daily).

Active Constituents : Chief constituents are chonemorphine, funtumafrine C, deaminochonemorphinone, japindine, N-formylchonemorphine, glycerides of lauric (11.9), stearic (13.3), arachidic (17.0), myristic (20.2), palmitic (10.4), oleic (9.6), linolenic (11.8), ricinoleic acids (5.2%), octacosanol, ceryl alcohol, â-sitosterol and taraxasterol.

References : Agarwal (2003), Lalramnghinglova & Jha (1997), Rastogi & Mehrotra (1999a, b, 2001), Reddi *et al.* (2005), Watt (1889).

Chrozophora prostrata Dalz. & Gibs.

Syn : *C. plicata* (Vahl.) A.Juss. *ex* Spreng.

Family : Euphorbiaceae.

Hindi & Sanskrit Names:

Hindi	:	Shadevi, Subali;
Sans.	:	Suryavarta.

Distribution : Particularly Throughout India.

Part Used : Root.

Utilization : So far reported to cure gastric disorders (paste of a piece of root about 2.5 cm long), cough (root ash), cough and cold (10g root powder given with water), and also for asthma and bronchitis.

Active Constituents : Active principles are due to 3, 5, 6, 7, 8-pentamethoxyxanthone-1-O-rhamnopyranosyl (1→6)-glucopyranoside, 3, 5, 8-trimethoxyxanthone-1-O-glucopyranoside and 2-acetonyl-5-methyl-7-hydroxy-6C-glucopyranosyl-chromone-2″-O-glucopyranoside.

References : Agarwal (2003), Ambasta (1986), Jain *et al.* (1991), Rastogi & Mehrotra (2002), Saren *et al.* (2000), Shukla & Verma (1996).

Cichorium endiva Linn.

Family : Asteraceae.

English & Hindi Names:

Eng.	:	Endive, Garden Endive;
Hindi	:	Kashini, Kasini.

Distribution : Cultivated in India.

Part Used : Root.

Utilization : Used in dyspepsia, fever and as a tonic, demulcent and warm stimulating.

References : Agarwal (2003), Anonymous (1948), Chopra *et al.* (1956), Kirtikar & Basu (1984), Watt (1889).

Cichorium intybus L. (Pl. 4D)

Syn : *C. divaricatum* Helder. *ex* Nym.

Family : Asteraceae.

English, Hindi & Sanskrit Names:

Eng. : Chicory, Garden Endive, Succory, The Indian or Wild Endive;
Hindi : Hinduba, Kashini, Kasni;
Sans. : Kasani.

Distribution : Cultivated in Bihar, Punjab, H. P., Assam, Maharashtra, Gujarat, Tamil Nadu, Orissa, A. P. & Kerala.

Part Used : Root.

Utilization : Poultice/extr. used in rheumatic pains and high fever. Also known to be prescribed in jaundice, liver enlargement, asthma, abdominal inflammations, gout, rheumatic affections and as a diuretic, emmenagogue, stomachic, febrifuge (one cup decoction twice daily for 10-15 days) and blood purifier.

Active Constituents : These are catechol, tannins, esculetin, 8-deoxylactucin, lacucin, lactupicrin and 11, 13-dihydrolactucopicrin.

Biological Activity : Antimicrobial and hepatoprotective activities confirmed.

References : Anonymous (1948), Asolkar *et al.* (1992), Chatterjee & Pakrashi (1997), Dey (1973), Kaul (1997), Khare (2004), Reddi *et al.* (2005), Said (1997), Sharma (2003), Singh (1996), Sood & Prakash (2007), Watt (1889).

Cimcifuga foetida L.

Syn : *Actea cimcifuga* Linn.; *Cimcifuga frigida* Royle
Family : Ranunculaceae.

English & Hindi Names:

Eng. : Bugbane;
Hindi : Jaunti.

Distribution : Temperate Himalaya (Kashmir-Bhutan): 2,600-4,000 m.

Part Used : Root.

Utilization : Known in India as sudorifice, antiperiodic, emetic, purgative, sedative, and also useful in neuralgia dropsy, rheumatic affections, bronchial diseases, hypochondriasis and high fevers. Hot decoction facilitates quick delivery.

References : Agarwal (2003), Ambasta (1986), Anonymous (1948), Kapur & Nanda (1996), Kaul (1997), Kirtikar & Basu (1935), Singh (1996), Watt (1889).

Cinchona officinalis L.

Syn : *C. condaminea* Humb. & Bonpl.

Family : Rubiaceae.

English, Hindi & Sanskrit Names:

Eng. : Crown Bark;

Sans. : Kunayana.

Distribution : Cultivated in W. Bengal & Nilgiris, Tamil Nadu.

Part Used : Root bark.

Utilization : Root bark - a source of quinine used as febrifuge.

Active Constituents : Contains cinchonine, cinchonidine, quinine, quinamine, and quinidine.

References : Ambasta (1986), Chatterjee & Pakrashi (1997).

Cinnamomum pauciflorum Nees.

Syn : *Laurus recurvata* Roxb.

Family : Lauraceae.

Hindi Name:

Hindi : Dinglapat.

Distribution : Sikkim Himalaya & Assam.

Part Used : Root.

Utilization : Chewed for dysentery and stomachache.

References : Agarwal (2003), Jamir & Rao (1990).

Cirsium arvense (Linn.) Scop.

Syn : *Aplotaxis cirsiodes* DC.; *Breea arvensis* (Linn.) Less.; *Cardus lanatus* Roxb.; *Cnicus arvensis* (Linn.) Roth.; *C. candicans* Wall.; *Sarratula arvensis* Linn.

Family : Asteraceae.

English Names:

Eng. : Canada Thistle, Creeping Thistle.

Distribution : N.W. Himalaya.

Part Used : Root.

Utilization : Poultice relieves headache and its paste used as diuretic. Powdered roots in combination with equal quantities of sugar, fruits of *Foeniculum vulgare* and black pepper cure giddiness. Aq. extr. (1 teaspoon 3 times a day for 7 days) given for the treatment of liver disorder.

Active Constituents : Aplotaxene (Z, Z, Z-1, 8, 11, 14-heptadecatetraene), pentadecane, 1-pentadecane, 1-heptadecane, 8, 9-epoxy-heptadeca-1, 11, 14-triene and 8, 9-dihydroxyheptadeca-1, 11, 14-triene are the principle constituents.

References : Ambasta (1986), Asolkar *et al.* (1992), Rastogi & Mehrotra (1998, 2002), Singh & Rao (2003), Sood & Thakur (2004).

Cirsium sinensis Cl.

Syn : *Cnicus sinensis* Gardener & Champ.

Family : Asteraceae.

Distribution : Marshy Areas on the Khasia Hills.

Part Used : Root.

Utilization : Crushed root used as antiseptic and the juice taken internally for flatulence and externally for ulcers and abscesses.

References : Ambasta (1986), Chopra *et al.* (1956), Lalramnghinglova (1996).

Cirsium verutum (D. Don) Spreng.

Family : Asteraceae.

Part Used : Roots.

Utilization : Used for persistant infectious cuts and sores.

Reference : Negi *et al.* (1993).

Cirsium wallichii (DC.) Cl.

Family : Asteraceae.

Distribution : Himalaya, (Kashmir to Bhutan): 1,800-3,600 m & Nilgiris in S. India.

Part Used : Roots.

Utilization : Used for fever and dyspepsia.

Active Constituents : O-acetyljacoline.

References : Negi *et al.* (1993), Rastogi & Mehrotra (2002).

Cissampelos pareira L. (Pl. 5A)

Syn : *C. caepeda* Linn.; *C. convolvulacea* Willd.; *C. delicatula* Miers.; *C. discolour* Buch.; *C. diversa* Miers.; *C. elata* Miers.; *C. eriantha* Miers.; *C. grallatoria* Miers.; *C. hirsuta* DC.; *C. nepalensis* Rhodes.; *C. obecta* Miers.; *C. orbiculata* Buch.; *C. subpeltata* Thw.

Family : Menispermaceae.

English, Hindi & Sanskrit Names:

Eng. : False Pareira Brava, Ice Vine, Velvet-Leaf Pareira;

Hindi : Akanadi, Akauadi, Harjori, Nirbissi, Pari, Patha;

Sans. : Ambashtha, Avidhakarni, Brihattika, Kucheli, Paatha, Shishira, Tiktapushpa, Trishira, Vara, Venivel, Vidhakarni, Vrittaparni.

Distribution : Throughout India, in Sub-Tropical & Tropical Zones.

Part Used : Root.

Utilization : Recorded to be used in gastric disorders, stomachache, jaundice, malaria, cough, leprosy, inflammation of kidney and bladder, piles, ulcers, dropsy, diabetes (root powder with water once only for 40 days), chest pain, dyspepsia, leucorrhoea, gonorrhoea, toothache, catarrhal affections, acute and chronic cystitis, bowel complaints, diarrhoea and snake bite. Its decoction taken orally for stomachache caused by indigestion, dysentery, enlarged spleen, skin diseases and as diuretic. About 30 ml extr. of fresh root given twice or thrice daily for 3 days to cure fever, cold, bodyache and as an antodote to dog bite. One tablespoon of its infusion given twice a day after breakfast for one month as health tonic, and paste applied to check headache.

Active Constituents : Active ingredients are hayatin, hayatidine, hayatinin, d-quercitol, an essential oil (0.2%), a fixed oil (3.4%), furnished hydrochloride, picrate, chloroplatirate, L-curine, methiocide, methiochloride and d-isochondrodendrine (root), menismine, cissamine, pareine, d-isochondrine, L-berberine, hayatine (root bark).

Biological Activity : Extr. has hypoglycaemic, CNS depressant and muscular relaxant proeprties.

References : Agarwal (2003), Ambasta (1986), Anonymous (1948, 1968), Asolkar *et al.* (1992), Caius (2003), Chatterjee & Pakrashi (1997), Chauhan (1999), Chopra *et al.* (1956), Dash & Mishra (2000), Dastur (1970), Dey (1973), Goel & Rajendran (2000), Islam (1996), Jain (2003), Jain & Singh (1997), Jha & Verma (1996), Khare (2004), Kirtikar & Basu (1935), Kumar & Lal (2003), Maliya & Singh (2003), Masih (2003), Maya *et al.* (2003), Nadeem *et al.* (2001), Painuli & Maheshwari (1996), Parrotta (2001), Rana *et al.* (2000), Rastogi & Mehrotra (1999a, b), Reddi *et al.* (2005), Saxena *et al.* (1997), Sharma (2003), Singh (1999, 2002, 2003), Singh & Kumar (2000a), Singh & Rao (2003), Sinha & Sinha (2001), Sood & Thakur (2004), Srivastava *et al.* (2003), Thakur *et al.* (1989), Varghese (1996), Yoganarasimhan (1981).

Cissus repanda Vahl.

Syn : *Vitis repanda* Wight & Arn.

Family : Vitaceae.

Hindi Name:

Hindi : Pani-Bel.

Distribution : N.W. Himalaya, Assam, Silhet, E. Bengal, W. Peninsula from the Concan Southwards.

Part Used : Root.

Utilization : Paste bandaged over the fractured area.

References : Ambasta (1986), Hooker (1875), Punjani (2004), Varghese (1996).

Cissus repens Lam.

Syn : *Vitis repens* Wight & Arn.

Family : Vitaceae.

Hindi Name:

Hindi : Pureni.

Distribution : Throughout E. Himalaya & W. Peninsula.

Part Used : Root.

Utilization : Paste applied externally for burns, headache, and swollen rheumatic joints. Root powder 5 g given once a day in rheumatism and gastric troubles.

References : Agarwal (2003), Ambasta (1986), Asolkar *et al.* (1992), Brahmam & Saxena (1990).

Citrullus colocynthis (L.) Schrad. *ex* Eckl. & Zeyh. (Pl. 5B)

Syn : *Cucumis colocynthis* Linn.

Family : Cucurbitaceae.

English, Hindi & Sanskrit Names:

Eng. : Bitter Apple, Bitter Cucumber, Colocynth;

Hindi : Indrayan, Tumbi;

Sans. : Aindri, Chitraa, Chitraphala, Gavaakshi, Indra-Varuni, Mrigaani, Mrigaivaaru.

Distribution : Warm, Arid & Sandy Parts Throughout India Upto 1,500 m.

Part Used : Root.

Utilization : Useful in the termination of foetus, enlargement of abdomen, intermittent fever in children, boils, pimples, ascites, biliousness, dropsy, inflammation of breasts, antidote to snake bite, urinary diseases, asthma, cough and as carminative and purgative. Root extr. taken in jaundice and its lukewarm paste applied over the lower abdomen of woman just after delivery to clear the remains of placenta. Pills made of equal quantity of root and *Piper longum* beneficial in rheumatism and piles.

Active Constituents : Elaterin, hentriacontane, saponins, 1, 11-undecanediolmonoacetate, nonyl hexadecanoate and choline are the chief constituents.

References : Agarwal (2003), Ambasta (1986), Anonymous (1948), Chatterjee & Pakrashi (1997), Chhetri (2005), Chopra *et al.* (1956), Dastur (1970), Dwivedi (1993), Khanna & Kumar (2000), Khare (2004), Rastogi & Mehrotra (1998), Reddi *et al.* (2005), Said (1997), Samvatsar & Diwanji (2000), Sarin (1996), Thakur *et al.* (1989), Watt (1889).

Citrus maxima (Burm.) Merrill

Syn : *C. decumana* Watt.; *C. grandis* (Linn.) Osbeck

Family : Rutaceae.

English & Hindi Names:

Eng. : Forbidden Fruit, Pummelo, Shaddock;

Hindi : Chakotra, Mahanibu, Sadaphal.

Distribution : Grown in S. & N. India.

Part Used : Root.

Utilization : Useful as purgative, emetic, anthelmintic and also good for urinary calculus.

Active Constituents : 2′, 2′-dimethyl-(pyrano 5′, 6′:3:4) 1, 5-dihydroxy-6-methoxy-10methylacridone (I) together with 2′2′-

dimethyl (pyrano 5′, 6′:3:4)-1-hydroxy-5-6-dimethoxy-10-methylacridone, xanthyletin and xanthoxyletin have been isolated.

References : Agarwal (2003), Ambasta (19986), Rastogi & Mehrotra (2001).

Citrus medica L.

Syn : C. *aurantifolia* (Christm.) Sw.; *C. medica* var. *medica* Watt.

Family : Rutaceae.

English, Hindi & Sanskrit Names:

Eng. : Citron;
Hindi : Baranimbu, Turang;
Sans. : Beejapura.

Distribution : Khasia & Garo hills; Cultivated Throughout the Warm & Moist Regions of India.

Part Used : Root.

Utilization : Considered as an emetic, anthelmintic, and also employed in constipation, kidney troubles and jaundice. Its decoction with bark of *Saraca asoca* good for stomach ulcer and urinary calculus.

Active Constituents : These are nordentatin, xanthyletin, citflavonone, etrogol, prenylated phenol derivatives and valencic acid.

References : Agarwal (2003), Ambasta (1986), Chatterjee and Pakrashi (1997), Chopra *et al.* (1956), Lalramnghinglova (1996), Saklani & Jain (1994).

Clausena excavata Burm.f.

Family : Rutaceae.

Hindi Names:

Hindi : Aganijal, Ratanjote.

Distribution : Sub-Himalayan Tracts, Assam, Chotanagpur & Orissa.

Part Used : Root.

Utilization : Tribals of Orissa use its decoction for the treatment of malaria. Poultice of pounded roots good for sores.

Active Constituents : Clausenin, clausenidin, clausenidinaric acid, clausarin, nordentatin, xanthoxyletin and heptaphylline have been isolated.

References : Agarwal (2003), Anonymous (1948), Asolkar *et al.* (1992), Prakash & Mehrotra (1991), Rastogi & Mehrotra (1998, 1999a, b).

Cleistanthus collinus Benth. *ex* Hook.f.

Family : Euphorbiaceae.

Part Used : Root.

Utilization : Root poultice applied for boils and wounds, and root infusion for excess menstruation.

Reference : Varghese (1996).

Clematis ladakhiana Grey-Wilson.

Family : Ranunculaceae.

Part Used : Root.

Utilization : One teaspoonful extr. of root and young shoot given to cure stomach pain.

Reference : Chandrasekar & Srivastava (2003).

Clematis smilacifolia Wall.

Family : Ranunculaceae.

Part Used : Root.

Utilization : Decoction said to be useful against carbuncle.

Reference : Yoganarasimhan *et al.* (1981).

Clematis tibetana Kuntze.

Family : Ranunculaceae.

Part Used : Root

Utilization : Root powder applied to kill lice on yak and cattle body.

Reference : Chaurasia *et al.* (2000a).

Clematis triloba Heyne *ex* Roth.

Family : Ranunculaceae.

Hindi & Sanskrit Names:

Hindi : Murhari;
Sans. : Laghuparnika.

Distribution : Throughout S. India.
Part Used : Roots.
Utilization : Paste applied on painful boils.
References : Anonymous (1948), Bhattacharjee (2001).

Cleome gynandra Linn.

Syn : *Cleome pentaphylla* DC.; *Cynandropsis pentaphylla* (L.) DC.
Family : Capparidaceae.

Hindi & Sanskrit Names:

Hindi : Karalia;
Sans. : Ajagantha.

Distribution : Abundant Throughout the Warm Parts of India.
Part Used : Root
Utilization : Decoction used in fever.
References : Hooker (1872), Reddi *et al.* (2005), Subramaniam (2000).

Cleome monophylla L.

Family: Capparidaceae.

Part Used : Root.
Utilization : Root powder applied on lips of unconscious patients.
Reference : Varghese (1996).

Cleome viscosa L. (Pl. 5C)

Syn : *Cleome isocandra* L.; *Polanasia isocandra* (L.) Wight & Arn.; *P. viscosa* (L.) DC.
Family : Capparidaceae.

English, Hindi & Sanskrit Names:

Eng.	:	Sticky Cleome, Wild Mustard;
Hindi	:	Boogra, Hulhul, Jangliharrar, Kamphutia;
Sans.	:	Adityabhakta, Arkakanta, Barbara, Manduki, Raviprita, Ravishta, Sauri, Suryalata, Tilparni, Varada, Vikranta.

Distribution : Throughout Tropical & Sub-Tropical India; Rocky, Sandy Soils, Cultivated Fields & Along Roadsides.

Part Used : Root.

Utilization : 3 drops of the mixture prepared by boiling pounded root with garlic and caster oil added in the ear to treat earache (morning and at bed time for about 3 days). Its powder applied on the lips by Santhals to restore consciousness in persons who have fainted and paste applied to cure headache.

Active Constituents : Main constituents are kaempferide-3-glucuronide, dihydrokaempferide-3-glucuronide, docosanoic acid, 3′, 4′, 5-trihydroxyloflavone-7-O-α-L-rhamnopyranoside, α-amyrin, lupeol, nerigenin4′-galactosidedihydrokaempferol-4′-xyloside, dihydrorobinetin pyranoside, xyloside of OH-yethanoic acid, lupeol, betulin, betulinic acid and β-sitosterol.

References : Asolkar *et al.* (1992), Caius (2003), Kumar & Pulliaiah (1998), Parrotta (2001), Patil & Patil (2002), Rastogi & Mehrotra (1999a, b, 2001), Reddi *et al.* (2005), Saxena *et al.* (2000), Sur & Halder (2002).

Clerodendrum indicum (L.) Kuntze.

Syn : *C. siphonanthus* R.Br.; *Siphonanthus indica* Linn.

Family : Verbenaceae.

Hindi & Sanskrit Names:

Hindi	:	Barangi;
Sans.	:	Bharangi.

Distribution : Common Throughout N.E. Region.

Part Used : Root.

Utilization : Recorded so far for asthma, cough, burns (fresh root paste applied twice a day for 4 days), scrofulous affections, nervous and alimentary disorders (root seedlings given to animals) and as an abortifacient and anthelmintic (5g root powder given at bed time for three days). Crushed roots boiled with seed oil of *Schleichera oleosa* good for massaging the affected parts to cure bodyache. Also mixture comprising its root (5 g) and 21 black peppers (*Piper nigrum*) given in 3 doses at half an hour intervals to cure chest pain and heart palpitation.

References : Ambasta (1986), Brahmam & Saxena (1990), Chatterjee & Pakrashi (1997), Chopra *et al.* (1956), Dhiman (2003), Jha & Verma (1996), Kumar (2002), Tirkey (2004), Varghese (1996).

Clerodendrum phlomoids L.f.

Syn : *C. multiflorum* Burm. *ex* Kuntze.; *C. phlomoids* Willd.; *Volkameria multiflora* Burm.f.

Family : Verbenaceae.

Hindi & Sanskrit Names:

Hindi : Arni, Urni;

Sans. : Agnimantha, Vataghni.

Distribution : Drier Parts of N. & Peninsular India; Often Cultivated as a Hedge Plant.

Part Used : Root.

Utilization : Decoction astringent, stomachic, aromatic, and used in Maharashtra as a demulcent and bitter tonic for children recovering from measles and also for treating gonorrhoea in S. India. These are also chewed to treat sore throat.

Active Constituents : Ceryl alcohol, clerodin, clerosterol and clerodendrin A.

References : Agarwal (2003), Ambasta (1986), Anonymous (1948), Chatterjee & Pakrashi (1997), Chopra *et al.* (1956), Dhiman (2003), Parrotta (2001), Rastogi & Mehrotra (2001), Saini (1996a), Watt (1889).

Clerodendrum serratifolia (L.) Mold

Family : Verbenaceae.

Part Used : Root.

Utilization : Paste of the roots with garlic boiled in water and given orally to cure stomachache caused by flatulence.

Reference : Hosagoudar & Henry (1996b).

Clerodendrum serratum (L.) Moon.

Syn : *C. fortunatum* Burm. *ex* Moldenke; *C. serrulatum* Spreng. *ex* Razi; *Volkameria serrata* L.

Family : Verbenaceae.

English, Hindi & Sanskrit Names:

Eng. : Beetle Killer, Blue-Flowered Glory Tree;
Hindi : Barangi, Bharangi, Duma, Gant-Baharangi;
Sans. : Bharang, Bharngi, Bhargi.

Distribution : S. & E. India, especially in W. Bengal.

Part Used : Root.

Utilization : Known in India for the treatment of malaria, cough, asthma, rheumatism, dyspepsia, fistula, cholera, scrofulous affections, loss of appetite, dropsy, expelling worms, stomachache, paralysis, menstrual disorders, anorexia, colic, anasarca, indigestion, fever, catarrhal affections in malaria, burning sensation, epilepsy, wounds, inflammations, tuberculous glands, diseases of blood, flatulence and snake-bite and also its paste applied externally on goitre, cervical adenitis, swollen legs (2-3 hrs once a day for 1 week), scrotal enlargement and inguinal hernia. Decoction taken orally twice a day for a week's time to give relief from asthma, bronchitis, stomachache, controlling loose motion, lung disorders and nausea (decoction of its root with ginger and coriander).

Active Constituents : Sapogenin, triterpene, starch, resin, glucose and D-(-) mannitol.

Biological Activity : Alc. extr. and saponin isolated from root bark caused release of histamine from lung tissue.

References : Agarwal (2003), Ambasta (1986), Anonymous (1948, 1968), Balasubramanian & Prasad (1996), Chauhan (1999), Islam & Jha (2003), Jain *et al.* (2003), Khare (2004), Kirtikar & Basu (1984), Kothari & Rao (2000), Kumar & Lal (2003), Nadanakunjidam (2003), Parrotta (2001), Rao & Henry (1996), Rastogi & Mehrotra (1999a, b), Reddi *et al.* (2005), Sharma (2003), Singh *et al.* (2003), Tirkey (2004), Varghese (1996), Watt (1889).

Clerodendrum viscosum Vent.

Syn : *C. infortunatum* auct. non L.

Family : Verbenaceae.

Hindi & Sanskrit Names:

Hindi : Bhant, Titabhant;

Sans. : Bhandirah.

Distribution : Throughout India as a Gregarious Shrub.

Parts Used : Root. Root bark.

Utilization : A teaspoonful decoction of root drunk twice a day for three days to cure malarial fever, and its powder with rice water known for excess bleeding or other menstrual problems. Decoction taken orally thrice a day for the treatment of dysentery. Also, paste recorded to be used for blood dysentery, chest pain, menstrual problems, asthma, and as an anthelmintic (prepared in mother's milk) and colic (prepared in butter milk). Root bark given to cure ear infections.

Active Constituents : Active principles are due to lupeol, b-sitosterol, fructose, glucose, galactose, lactose, maltose, raffinose, sucrose, clerodol, clerodolone, clerodone, clerosterol, β-sitosterol, stigmasta-5,25- diene-3-ol and clerosterol, 5,25-stigmastadien-3β-ol, clerodolone as lup-20(30)-en-3β and 19-diol-12-one.

References : Agarwal (2003), Bora (2000), Brahmam & Saxena (1990), Chatterjee & Pakrashi (1997), Jain (1996), Rao

& Henry (1996), Rastogi & Mehrotra (1999a, b), Reddi *et al.* (2005), Saxena *et al.* (1997), Varghese (1996).

Clitorea ternatea L.

Syn : *Terneata vulgaris* H.B.K.

Family : Fabaceae.

English, Hindi & Sanskrit Names:

Eng. : Butterfly Pea, Clitoria, Meczereon, Winged-Leaved Clitoria;
Hindi : Aparajit, Khagia, Vishnukranti;
Sans. : Aparajita, Asphota, Girikarnika, Sankhapushpi.

Distribution : Garden Plant Throughout India; Wild in Assam & S. India.

Parts Used : Root. Root bark.

Utilization : One teaspoonful of crushed fresh root bark taken with a cup of warm milk twice a day for two weeks as diuretic, laxative, demulcent for relieving irritation of bladder and urethra whereas roots prescribed to cure insect bite (root grounded and mixed with milk), gonorrhoea, tumours, swollen joints, bronchitis, leprosy (root paste), eye diseases, corneal ulcers, tuberculous glands, urinary troubles (root juice), elephantiasis, headache, leucoderma, jaundice (root paste mixed with 1 glass of cow's milk), biliousness, inflammations, ulcers, snake bite (5g root powder), and also in ascites, enlargement of abdomen viscera (paste applied on cattle stomach). Poultice prepared from powdered root applied externally for treatment of goitre, leprosy and also as an abortifacient.

Active Constituents : Contains taraxerol and taraxerone.

References : Agarwal (2003), Ambasta (1986), Anonymous (1948, 1968), Asolkar *et al.* (1992), Binu *et al.* (2003), Chatterjee & Pakrashi (1997), Chopra *et al.* (1956), Das & Sharma (2003), Dastur (1970), Dey (1973), Dhiman (2003), Jain *et al.* (1991), Khare (2004), Kirtikar & Basu (1935), Kurian (1995), Pandey & Rout (2006),Parrotta (2001), Reddi *et al.* (2005), Shukla & Verma (1996), Shukla *et al.* (2003), Sood *et al.* (2005c),

Thakur *et al.* (1989), Thomas & Britto (2003), Tirkey (2004).

Coccinia grandis (L.) Voigt.

Syn : *C. cordifolia* Cogn.; *C. indica* W.& A.; *Cephalandra indica* (W.& A.) Naud.
Family : Cucurbitaceae.

English, Hindi & Sanskrit Names:

Eng.	: Ivy Gourd;
Hindi	: Kanduri, Kanduri-Ki-Bel;
Sans.	: Bimbi.

Distribution : Cultivated in Assam, W. Bengal, Bihar, Orissa, Maharashtra, A.P., Tamil Nadu; Wild in Many Parts of India.

Part Used : Root.

Utilization : Cures diabetes (20 ml decoction given orally twice a day), bodyaches, rheumatism, headache and checks vomiting. About 5g of its powder dusted over scorpion sting said to heal affected part of the body.

Active Constituents : Talbutamic β-amtrin, heperol and stigmast-7-ene-3-one are the main active principles.

Biological Activity : Alc. extr. exhibited antidiabetic activity on alloxan diabetic rats.

References : Agarwal (2003), Anonymous (1968), Asolkar *et al.* (1992), Jain *et al.* (1991), Khanna *et al.* (1996), Khare (2004), Rastogi & Mehrotra (1999a, b), Reddi *et al.* (2005), Shukla *et al.* (2003).

Cocculus hirsutus L. Diels.

Syn : *Cocculus villosus* (Lam.) DC.; *Menispermum hirsutum* L.; *M. villosum* L.
Family : Menispermaceae.

English, Hindi & Sanskrit Names:

Eng.	: Broom Creeper, Ink Berry;
Hindi	: Chireta, Dier, Faridburti, Hier, Jalayamani, Jamti-Ki-Bel;

Sans.	:	Dridhakanda, Dridhamani, Garudi, Mahamula, Sauparni, Tiktanga, Vanatiktika, Vasandi.

Distribution : Tropical & Sub-Tropical India, Myanmar, Sri Lanka, W. Asia & Tropical Africa.

Part Used : Root.

Utilization : Generally used as a refrigerant, antipyretic, laxative, alterative in chronic rheumatic, bilious dyspepsia antidote to snake poison, excessive urination, bed wetting, children stomachache, venereal diseases, leucoderma, asthma and strength. Decoction given to enrich blood and in genitourinary disorders, eczema, urticaria, rheumatic and old venereal pains (decoction of fresh roots with a few heads of pepper in goat's milk). One treaspoonful extr. in water given orally once daily for 2-3 days to cure diarrhoea and dysentery.

Active Constituents : Contains D-trilobine, D-coclaurine and magnoflorine.

References : Agarwal (2003), Ambasta (1986), Anonymous (1948, 1968), Bhatt *et al.* (2002), Caius (2003), Chatterjee & Pakrashi (1997), Chopra *et al.* (1956), Khanna (2002b), Kirtikar & Basu (1935), Parrotta (2001), Punjani (2006), Rastogi & Mehrotra (1999a, b), Reddi *et al.* (2005), Srivastava *et al.* (2003), Singh & Pandey (1998), Sinha & Sinha (2001), Watt (1889).

Cocos nucifera Linn.

Family : Arecaceae.

English, Hindi & Sanskrit Names:

Eng.	:	Cocoa Nut, Coconut-Palm, Coconut Tree;
Hindi	:	Nariel, Nariyal;
Sans.	:	Dakshinatriya, Dridanira, Dridhaphala, Garikera, Jataphala, Kaushikaphala, Mriduphala, Nilataru, Sadapushpa, Trinaraja, Tryambakaphala, Uchhataru, Varaphala.

Distribution : Thrives Best Within the Tropic Zone; 20^0 N & 20^0 S on both Sides of Equater.

Part Used : Root.

Utilization : Said to be anthelmintic, diuretic, astringent (gargle in sore throat), and also found useful in uterine diseases, strengthening gums, liver complaints, bronchitis and dysentery.

References : Agarwal (2003), Anonymous (1948), Chopra *et al.* (1956), Dey (1973), Dhiman (2003), Kirtikar & Basu (1984), Kurian (1995).

Codenia procumbens L.

Family : Boraginaceae.

Part Used : Root.

Utilization : Paste applied on head for epilepsy.

Reference : Varghese (1996).

Codonopsis ovata Benth.

Family : Campanulaceae.

Hindi Name:

Hindi : Luduti.

Distribution : W. Himalaya (Kashmir to Garhwal); 2,600-4,000 m.

Part Used : Root.

Utilization : Sweet milky roots eaten raw as a good physical and sexual tonic.

References : Anonymous (1948), Bhattacharjee (2001), Sharma (2003), Sood & Prakash (2007).

Colchicum luteum Baker

Family : Liliaceae.

Hindi & Sanskrit Names:

Hindi : Hirantutiya, Surinjan;

Sans. : Hiranyatutha.

Distribution : N. W. Himalaya: 700-2,800 m (Outskirts of Forests or in Open Grassy Places).

Part Used : Root.

Utilization : Considered carminative, laxative, aphrodisiac and said to heal inflammations, pain, heat of the brain, old piles, wounds, headache, gout, rheumatism, diseases of the liver and spleen.

References : Agarwal (2003), Jain (1968), Kirtikar & Basu (1984).

Colebrookia oppositifolia Sm.

Family : Lamiaceae.

Hindi Name:

Hindi : Binda Pansa.

Distribution : Hilly Parts of India Upto 1, 300 m.

Part Used : Root.

Utilization : Infusion (two teaspoons) given twice daily for 21 days in asthma. Its paste along with black salt given two spoonfuls once a day for the treatment of epilepsy.

Active Constituents : Triacontane, triacontanol, β-sitosterol, palmitic, stearic & oleic acids & 5, 6, 7-trimethoxyflavone and 4′, 5, 6, 7-tetramethoxyflavone have been identified.

References : Agarwal (2003), Ambasta (1986), Chopra *et al.* (1956), Jha & Verma (1996), Rastogi & Mehrotra (2001), Singh & Kumar (2000b), Watt (1889).

Coleus barbatus Benth.

Syn : C. *forskohlii* (Willd.) Briq.

Family : Lamiaceae.

English & Hindi Names:

Eng. : Kaffir Potato;
Hindi : Bhedi, Gurmal.

Distribution : Wild in Dry & Barren Hills of Sub-Tropical Himalaya Ascending to 2,700 m, Deccan Peninsula, Gujarat & Bihar.

Part Used : Root.

Utilization : Recorded in India as an anthelmintic, laxative, tonic (decoction), and for muscular pains, worms and allaying burning festering boils (paste). Grounded roots mixed with mustard oil externally applied to cure eczema and skin diseases.

Active Constituents : These are 1-acetoxycoleosol, coleol, coleonol =forskolin, i.e., 7-β-acetoxy-8, 13-epoxy-1α, 6β, 9α-trihydroxylabd-14-en-11-one, colenols B & C (isomeric 6β-acetoxy-8, 13-epoxy-1, 7, 9α-trihydroxylabd-14-en-11-one), coleonol D, coleonol E (1, 9-dideoxyforskolin), coleonol F, coleonone, coleosol, deoxycoleonol (7α-acetoxy-8, 13-epoxy-1α, 6β-dihydroxylabd-14-en-11-one), 8,8′-diapocarotenedial(crocetin dialdehyde), 1, 9-dideoxy- coleonol B, 8,13-epoxylabd-14-en-11-one and its 6β-hydroxy, 6β, 7β, 9α-trihydroxy and 7β-acetoxy-6β-9α-dihydroxy derivatives, naphthopyrone, α-cis-bergamotene, β-bisabolene, bornylacetate, camphene, α-copaene, cuparene, α-& γ-curcumene, β-cymene, 3-decanone, β-elemene, humulene, α-& β-ionenes, myrcene, 3-nonanone, β-phellandrene, α-& β-pinenes, δ-selinene, α-thujene, crocetin dialdehyde (8, 8′-diapocarotenedial), 2-hydroxymethyl A (1) nor-urs-19α-hydroxy 2 (3), 12 (13)-dien-28-oic acid, 1, 9-dideoxycoleonol B and 1-acetoxycoleosol.

Biological Activity : Bronchodilating, cardiotonic, hypotensive and spasmolytic activities found to be confirmed.

References : Chatterjee & Pakrashi (1997), Negi *et al.* (1993), Rastogi & Mehrotra (1998, 1999a, b, 2001).

Colocasia esculenta L.

Syn : *Arum colocasia* Willd.; *Caladium esculentum* Vent.; *Colocasia antiquorum* Schott.

Family : Araceae.

English, Hindi & Sanskrit Names:

Eng. : Coco, Cocoyam, Dasheen, Eddo, Egyptian Arum, Scratch-Coco;

Hindi : Arvi, Kachalu;
Sans. : Kachu, Kachwi.

Distribution : Moist Parts of India.

Part Used : Root.

Utilization : About 10 ml decoction prepared by boiling roots (50 g) in goat's milk prescribed once daily per dose in diabetes.

References : Agarwal (2003), Ambasta (1986), Dutta & Nath (2000), Sood & Prakash (2007), Watt (1889).

Commelina obliqua Ham.

Syn : *C. communis* Roxb.; *C. paludosa* Blume
Family : Commelinaceae.

Hindi Name:

Hindi : Kanjuna.

Distribution : Low Moist Parts of India, Lower Himalaya (ascending to 1,300 m), Sri Lanka, Myanmar and the Malaya.

Part Used : Root.

Utilization : Useful as refrigerant, antidote to snake bites and laxative; useful in strangury, costiveness, vertigo, fevers and bilious affections. Root powder (20-30g) mixed with a cup of curd and jaggery taken two times a day in leucorrhoea, and as purgative.

References : Ambasta (1986), Caius (2003), Chopra *et al.* (1956), Painuli & Maheshwari (1996), Saklani & Jain (1994), Watt (1889).

Connarus monocarpus Linn.

Family : Connaraceae.

Hindi Name:

Hindi : Sunder.

Distribution : Karnatka to Malabar.

Part Used : Root.

Utilization : Decoction given in syphilis and swellings.

Active Constituents : Rapanone, embelin, bergenin and leucopelargonidin are the principle constituents.

References : Agarwal (2003), Ambasta (1986), Anonymous (1948), Chopra *et al.* (1956), Kirtikar & Basu (1935), Rastogi & Mehrotra (1999a, b).

Coptis teeta Wall.

Family : Ranunculaceae.

English, Hindi & Sanskrit Names:

- **Eng.** : Coptis, Coptis Radix, Gold Thread, Mishmi Teeta;
- **Hindi** : Mamira, Mishmitita, Teeta;
- **Sans.** : Mamira, Mamiran-Chini.

Distribution : Temperate Regions, on Mishmi Hills (Arunachal Pradesh), E. of Assam.

Part Used : Root.

Utilization : So far known to be useful in diarrhoea, gastroenteritis, jaundice, atonic dyspepsia, restoring digestive powers, debility, convalescence after fever, mild form of intermittent fevers, flatulence, conjunctivitis and strengthening immunity, and as bitter tonic and prophylactic for eye diseases.

Active Constituents : Root abounds in a compound of a yellow bitter principle, berberine to the extent of 8.5 % and coptine.

Biological Activity : Alc. extr. showed hypotensive and hypoglycaemic activity in experimental animals.

References : Agarwal (2003), Caius (2003), Jain (1968), Kirtikar & Basu (1935), Thakur *et al.* (1989), Tiwari & Tiwari (1996), Watt (1889).

Corallocarpus epigaea Rottl. & Willd.

Syn : *Archmandra epigaea* Arn. in Hook.; *Bryonia epigaea* Rottler; *B. glabra* Roxb.

Family : Cucurbitaceae.

English, Hindi & Sanskrit Names:

Hindi : Akasgandha, Rakasgaddah;
Sans. : Patalagaruda.

Distribution : Throughout India.

Part Used : Root.

Utilization : Known in India as an aperient, laxative, alterative, and used in chronic dysentery, turnout of plague, diarrhoea, sprue, syphilitic, diabetes, chronic enteritis and painful rheumatism (paste in castor oil and cumin seeds). Decoction given in bronchitis, constipation, toxaemia and worms, and its paste applied on swellings, scorpion stings and snake bite.

Active Constituents : Contains bryonin.

References : Agarwal (2003), Anonymous (1948), Bhatt *et al.* (1999, 2002), Chopra *et al.* (1956), Jain (1991), Mitaliya & Bhatt (2003), Samvatsar & Diwanji (2000), Singh & Pandey (1998), Watt (1889).

Corbichonia decumbens (Forssk.) Excel.

Family : Aizoaceae.

Part Used : Root.

Utilization : Paste given orally thrice a day for three days for the treatment of yellow and white jaundice.

Reference : Ram *et al.* (2002).

Corchorus aestnuans L.

Syn : *C. acutangulus* Lam.; *C. fuscus* Roxb.

Family : Tiliaceae.

Hindi & Sanskrit Names:

Hindi : Titapatta;
Sans. : Cancu, Cuncuh.

Distribution : Found Throughout Tropical & Sub-Tropical India.

Part Used : Root.

Utilization : Mixture comprising its powder (10 g), rhizome powder of *Asparagus* (5 g) and leaves (5 g) of *Ocimum* *tenuiflorum* (50:50) given internally (one teaspoon thrice a day) for about 15 days for asthma.

References : Agarwal (2003), Parrotta (2001), Rao & Pullaiah (2001), Reddi *et al.* (2005), Watt (1889).

Corchorus trilocularis L.

Family : Tiliaceae.

Sanskrit Name:

Sans. : Kaunti.

Distribution : N.W. Provinces, Ambala to the Punjab; Nilgiri Mountains.

Part Used : Root.

Utilization : Root paste known for eczema, scabies and gonorrhoea.

References : Hooker (1874), Reddi *et al.* (2005).

Cordia dichotoma Forst.f.

Syn : *C. latifolia* Roxb.; *C. myxa* Roxb.; *C. obliqua* Willd.

Family : Boraginaceae.

English, Hindi & Sanskrit Names:

Eng. : Sebesten Plum;
Hindi : Lasura;
Sans. : Bahuvarah.

Distribution : Commonly Planted Throughout India.

Part Used : Root.

Utilization : 100 g roots and 50 g 'mishri' (sugar pellets) crushed and mixed with a cup of water for flatulence (twice a day for 2-3 days).

Active Constituents : Contains 5, 7-dimethoxytaxifolin-3-O-α-L-rhamnopyranoside I, β-sitosterol, hesperetin-7-rhamnoside, lup-20(29)-ene-3-O-β-D-maltoside II, and lupa-20 (29)-ene-3-O-α-L-rhamnopyranoside.

References : Agarwal (2003), Rastogi & Mehrotra (1999a, b, 2001, 2002), Reddi *et al*. (2005), Tirkey (2004).

Corydalis govaniana Wall.

Family : Fumariaceae.

Hindi & Sanskrit Names:

Hindi : Bhutkesi, Bhutkis;
Sans. : Bhutakisi.

Distribution : Throughout Himalaya, above 2,500 m.

Part Used : Root.

Utilization : Known in India as tonic, diuretic, carminative, alterative, antiperiodic and also prescribed in syphilic, scrofulous, cutaneous affections, eye infections, muscular and gastric pains.

Active Constituents : Comprised choline, corlumine, bicuculline, isocorydine, protopine and stylopine.

References : Agarwal (2003), Ambasta (1986), Bhattacharjee (2001), Chauhan (1999), Chhetri (2005), Kapur & Nanda (1996), Kaul (1997), Kirtikar & Basu (1935), Rastogi & Mehrotra (1999a, b), Sharma (2003).

Costus speciosus (Koen.) Sm.

Syn : *Banksia speciosa* Koen.;
Costus speciosus var. *nepalensis* (Rosc.) Baker.
Family : Gingiberaceae.

English, Hindi & Sanskrit Names:

Eng. : Spiral Ginger;
Hindi : Kemuk, Keokanda, Keu, Kust;
Sans. : Canda, Kemuka, Kushta.

Distribution : Throughout India Upto 1,300 m.

Part Used : Root.

Utilization : Useful as bitter astringent, stimulant, digestive, anthelmintic, depurative, aphrodisiac, and for catarrhal fever, cold, cough, dyspepsia, worms, lumbago, skin diseases, urinary trouble, tumour, snake bites, headache (root extr.), bronchitis, anaemia, rheumatism, inflammations, leucorrhoea, body pain and postnatal complaints. In Manipur root is used for urinary trouble. One teaspoonful of its paste taken with a cup of milk twice a day for 1 week to treat abdominal pain, and also applied externally on the affected part once a day to cure boils.

Active Constituents : Diosgenin, sitosterol, 5α-stigmast-a (11)-en-3β-ol, 8-hydroxytriacontane-25-one, methyl triacontanoate, β-sitosterol-β-D-glucoside, prosapogenins A and B of dioscin, dioscin, gracillin, 3-O-[α-L-rhamnopyranosyl(1>2)-β-D-glucopyranosyl]-26-O-[β-D-glucopyranosyl]-22α-methoxy-(25R) Furost-5-en-3β-26-diol, protodioscin and methyl protodioscin are the chief ingredients.

References : Agarwal (2003), Anonymous (1948, 1968), Bhattacharjee (2001), Behera & Rout (2003), Chauhan (1999), Chopra *et al.* (1956), Dey (1973), Jain (1968), Jain & Saklani (1992), Kirtikar & Basu (1984), Kumar (2002), Parrotta (2001), Prasad *et al.* (2003), Rastogi & Mehrotra (2001), Reddi *et al.* (2005), Saklani & Jain (1994), Sharma (2004), Sharma *et al.* (2003), Tirkey (2004), Varghese (1996), Watt (1889).

Crataeva magna (Lour.) DC.

Family : Capparidaceae.

English, Hindi & Sanskrit Names :

Eng. : Three Leaved Caper;
Hindi : Barum;
Sans. : Varunah.

Distribution : Upper Godavari Districts of E. Ghats.

Part Used : Root bark.

Utilization : Given to cure gastric trouble.

References : Rao & Henry (1996), Reddi *et al.* (2005).

Crinum asiaticum Linn.

Family : Amaryllidaceae.

Hindi & Sanskrit Names:

Hindi : Kanwal, Pindar;
Sans. : Nagadamani, Nagapatra.

Distribution : Wild or Cultivated Throughout Tropical India.

Part Used : Root.

Utilization : Fresh root emetic, nauseant and diaphoretic.

Active Constituents : Contains narcissine (lycorrine).

References : Anonymous (1948), Chopra *et al.* (1956).

Crotalaria albida Heyne *ex* Roth.

Family : Fabaceae.

Hindi Name:

Hindi : Ban Methi.

Distribution : Hotter Parts of India.

Part Used : Root.

Utilization : Paste of its roots (10 g) and two pieces of pumpkin stock taken orally twice daily for a week in chronic pain of backbone and as purgative.

References : Agarwal (2003), Ambasta (1986), Chopra *et al.* (1956), Sahoo & Bahali (2003).

Crotalaria juncea L.

Syn : *C. tenuifolia* Roxb.
Family : Fabaceae.

English, Hindi & Sanskrit Names:

Eng. : Brown Hemp, False Hemp, Indian Hemp, Jabalpur Hemp, Salsette Hemp, Sunn Hemp, Travancore Flax;
Hindi : San, Sanai, Sunaa, Sunn;
Sans. : Katutikta, Sana, Shanapushpi, Tvaksaara.

Distribution : Throughout the Plains of India, Especially in S. India.

Part Used : Root.

Utilization : Given in intestinal spasm, and as colic and diuretic. 10-20 ml of its decoction given twice a day for 7-10 days to treat measles.

References : Khare (2004), Parrotta (2001), Sharma & Singh (1999).

Crotalaria medicaginea Lamk.

Family : Fabaceae.

Hindi Name:

Hindi : Gulabisun.

Distribution : Scattered Throughout India.

Part Used : Root.

Utilization : 10g paste given in cases of gastritis.

Active Constituents : Chalcone isolated from roots characterised as 3′,5′-di-C-prenyl-4,2′,4′trihydroxychalcone.

References : Agarwal (2003), Kumar & Chauhan (2005), Rastogi & Mehrotra (2002).

Crotalaria nana non Burm.f.

Syn : *C. umbellata* Wight *ex* Wight & Arn.

Family : Fabaceae.

Distribution : W. Peninsular India ascending to 2, 500 m.

Part Used : Root.

Utilization : One teaspoonful of its paste given twice a day in diarrhoea.

References : Asolkar *et al.* (1992), Jain & Patole (2001), Rastogi & Mehrotra (1999a, b).

Crotalaria prostrata Rottl. *ex* Willd.

Family : Fabaceae.

Distribution : Drier Plains of India ascending to 2,000 m.

Part Used : Root.

Utilization : Known in India for stomachache, diarrhoea, toothache and colic.

References : Chopra *et al.* (1996), Nayak *et al.* (2003), Varghese (1996), Watt (1889), Yoganarasimhan (1981).

Croton bonplandianum Baill.

Syn : *Croton sparsiflorus* Morong
Family : Euphorbiaceae.

Hindi Name :

Hindi : Kukka.

Distribution : Moist Parts of India (Roadsides, Fallow Fields, Open Habitats).

Part Used : Root.

Utilization : Used after child birth to stop abnormal bleeding. Its decoction mixed with decoction of stem of 'Arjun' (*Terminalia arjuna* Wt. & Arn.) in 3:2 ratio to prevent cholera (5ml twice daily for 5 days).

References : Agarwal (2003), Chakraborty *et al.* (2003), Parrotta (2001), Sur & Halder (2002).

Croton oblongifolius Roxb.

Syn : *C. roxburghii* Balak.
Family : Euphorbiaceae.

Hindi & Sanskrit Names:

Hindi : Chucka;
Sans. : Bhutamkusam, Bhutankusa, Putrashreni.

Distribution : N.E. India from U.P. (Oudh) to Assam, Chota Nagpur (Bihar), M.P. & Scrub Forests in W. & S. Regions of Peninsular India.

Parts Used : Root. Root bark.

Utilization : So far root used for constipation, gastric disorders, snake bite, sprain, vomiting, rheumatism, antifertility agent and as anthelmintic (juice); also its juice applied externally for body swellings. Decoction of root bark with black pepper, reportedly used to treat diarrhoea, chest pain, flatulence (with lime water), stomach pain, fevers and to relieve inflammations. Also pounded bark with lime juice (2 spoonfuls of juice) administered orally for flatulence, and its poultice applied externally for body swellings.

References : Asolkar *et al.* (1992), Bhandary *et al.* (1996), Jain (1996, 2003), Jha & Verma (1996), Molla & Roy (1996), Parrotta (2001), Rajendran *et al.* (1997), Rao & Henry (1996), Varghese (1996).

Cryptocoryne retrospiralis Fisch.

Family : Araceae.

Distribution : Bengal, Deccan Peninsula & Sri Lanka.

Part Used : Root.

Utilization : Cooked bulbous roots mixed with sugar prescribed to ladies for menorrhagia.

References : Anonymous (1948), Gopal & Chandra (2003).

Cryptolepis buchananii Roem. & Schult.

Syn : *C. reticulata* (Roxb.) Schum.; *Nerium reticulatum* Roxb.

Family : Asclepiadaceae.

English, Hindi & Sanskrit Names:

Eng. : Indian Sarsaparilla (Black Variety);
Hindi : Dudhi, Kalidudhi, Karanta;
Sans. : Gopi, Kaalghatika, Krishnasariva, Shyamalataa.

Distribution : Common Throughout C. & S. India, especially Dry Localities, Deciduous Forests & Hedges.

Parts Used : Root. Root bark.

Utilization : Known for rheumatic pains, reptile bites, joint pains, diarrhoea, jaundice (decoction taken internally for a period of week), fits (root extr.), lactation, antinarcotic and stomachic (extr. recommended by

Kolams-1 to 3 spoonfuls thrice a day). Paste of its root applied over bone fracture wrapped with bamboo pieces and also 5-10 ml of it is taken twice a day for the treatment of abdominal pain and cooling.

Active Constituents : Sarmentogenin, sarmentocymarin, buchanin characterised as sarverogenin-3β-O-α-L-oleandropyranoside, cryptanoside C, germanicol docosanoate, cryptnosides B and C characterised as isosarverogenin-3-O-α-oleandroside and sarverogenin-3-O-[β-D-glucopyranosyl (1→4)-α-(1→4)-α-L-oleandroside] have been isolated.

References : Asolkar *et al.* (1992), Jain & Singh (1997), Jain *et al.* (2003), Khare (2004), Nadanakunjidam (2003), Parrotta (2001), Ravisankar & Henry (1992), Reddi *et al.* (2005), Saxena *et al.* (1997), Sen & Behera (2003), Singh (1996), Singh & Kumar (2000b), Singh & Rao (2003), Sinha & Sinha (2001), Subramaniam (2000), Varghese (1996).

Cucumis callosus (Rottle.) Cogn.

Syn : *Bryomia callosa* Rottl.; *Cucumis eriocarpus* Boiss.; *C. maderaspatanus* Roxb.; *C. pseudo-colocynthis* Royle; *C. pubescens* Wall.; *C. trigonous* Roxb.; *C. turbinatus* Roxb.

Family : Cucurbitaceae.

Hindi Names:

Hindi : Bhakura, Bislambhi.

Distribution : Throughout Drier Upland Parts of India.

Part Used : Root.

Utilization : Decoction taken in indigestion, dropsy, and its paste applied on scorpion sting.

References : Agarwal (2003), Anonymous (1948), Bhatt *et al.* (2002), Watt (1889).

Curculigo orchioides Gaertn.

Syn : *C. malabarica* Wight; *Hypoxis orchioides* Kurz.

Family : Amaryllidaceae.

English, Hindi & Sanskrit Names:

Eng. : Black Muscale;
Hindi : Kali Musli, Mushali, Muslikand, Siyah Musli;
Sans. : Krishna Mushali, Taalpatri, Talamuli.

Distribution : Sub-Tropical Himalaya (Kumaon (U.P.) Eastwards to Assam) W. Regions of Peninsular India from Konkan (Maharashtra) Southwards to 2,400 m.

Parts Used : Root. Root bark.

Utilization : Bitter mucilaginous tuberous root prescribed along with *Mucuna pruriens* seeds, *Pueraria tuberosa* tubers, *Asparagus racemosus* roots as tonic, colic, demulcent, diuretic, antidiabetic (root juice), aphrodisiacs, aromatics, and also given in leucorrhoea, menstrual irregularities, diarrhoea (powdered root stirred in sugary milk and taken in the form of a thick mucilaginous fluid), enlarged spleen (water poured over a piece of red hot iron is mixed with root paste and given to patients three times a day), bone fracture (root paste with 'khair' and 'supari'), jaundice, wounds, snake bite, scorpion bite, asthma, cough, piles, skin diseases, quick ejaculation of sperms during intercourse (root powder given orally with milk and sugar), urinary troubles (root powder), headache (root paste applied on head), bronchitis, ophthalmia, indigestion, vomiting, gonorrhoea, hydrophobia, pains in the joints, itching sensation and skin rashes (root poultice). Its paste mixed with castor oil (*Ricinus communis* L.) rubbed on head in lunacy. Decoction (50 ml) prescribed three times daily for one week in stomach trouble and dysentery. Bark paste applied in case of pimples and also taken with honey once daily for the same.

Active Constituents : Reported to contain glycosides, polysaccharides (hemicellulose & other polysaccharides), starch, resin, tannin, mucilage, fat, calcium oxide, lycorine, sterols including β-sitostrol, yuccagennin, 5,7-dimethoxy myrcetin-3-O-α-L-xylopyranosyl-4-O-β-D-glucopyraniside, 3 methoxy-5-acetyl-31-tritriacotene, 2-methoxy-4-acetyl-5-methyltriacontane, palmitic acid, oleic, linoleic, linolenic, arachidic and behenic acids

Biological Activity : The plant yields a flavone glycoside which is a powerful uterine stimulant in guinea pigs, rats and rabbits.

References : Ambasta (1986), Aminuddin & Girach (1991), Anonymous (1948, 1968), Beg *et al.* (2005) Bhatt *et al.* (2003), Dastur (1970), Dey (1973), Drury (1978), Dwivedi (1993), Gopal & Chandra (2003), Jain *et al.* (2003), Khare (2004), Kirtikar & Basu (1984), Maheshwari *et al.* (1997), Nayak *et al.* (2003), Painuli & Maheshwari (1996), Panda & Das (2000), Parrotta (2001), Rastogi & Mehrotra (2001), Reddi *et al.* (2005), Samant *et al.* (2001), Sarin (1996), Sen & Behera (2003), Singh & Pandey (1996), Sinha & Sinha (2001), Sur & Halder (2002), Thakur *et al.* (1989), Trivedi (2002), Upadhyay (2003), Varghese (1996), Watt (1889).

Curculigo recurvata Dryand

Syn : *C. captulata* Kuntze
Family : Amaryllidaceae.

Part Used : Root.

Utilization : Paste good for cuts and in scabies.

References : Ambasta (1986), Rahman (2000).

Curcuma angustifolia Roxb.

Family : Zingiberaceae.

English, Hindi & Sanskrit Names:

Eng. : East Indian Arrowroot, Narrow Leaved Turmeric, Travancore Starch, Wild Arrowroot;

Hindi : Tavakhira, Tikhur;

Sans. : Gavayodhava, Godhumaja, Payakshira, Pistitika, Talakshira, Yavaja.

Distribution : Outer Himalaya, W. Bihar, N. Bengal, S. India.

Part Used : Root.

Utilization : Root (50-100 g) mixed with milk used three times daily in weakness and rheumatism. Also, useful in consumption, biliousness, leprosy, burning sensation, dyspepsia, loss of taste, bronchitis, asthma, fever, thirst, jaundice, anaemia, leucoderma, stones in the kidney and bladder, strangury, urinary discharges, ulcers, diseases of the blood (Ayurveda), and as tonic, aphrodisiac, demulcent, non-irritating, and nutritive.

References : Agarwal (2003), Ambasta (1986), Kirtikar & Basu (1984), Maheshwari *et al.* (1997), Watt (1889).

Curcuma inodora Blat.

Family : Zingiberaceae.

Part Used : Root.

Utilization : Used for abdominal pain in 'Dangs', (Gujarat, W. India).

Reference : Jain (1995).

Curcuma zedoaria Rosc. (non Roxb.)

Syn : *Amomum zedoaria* Christ.; *Cucuma zerumbent* Roxb.

Family : Zingiberaceae.

English, Hindi & Sanskrit Names:

Eng. : The long & The Round Zedoary, Zedoary;
Hindi : Kachura, Kalihaldi;
Sans. : Dravida, Durlabha, Gandhamulaka, Jatala, Kalpaka, Mukhya, Vedhya.

Distribution : Cultivated Throughout India Upto 1,500 m.

Part Used : Root.

Utilization : Fresh root considered to be refrigerant, diuretic and blood purifier; it checks leucorrhoeal and gonorrhoeal discharges. Also, chewed to correct a sticky taste in mouth and taken by woman to remove weakness after child birth.

References : Agarwal (2003), Ambasta (1986), Kirtikar & Basu (1984), Sood & Prakash (2007), Watt (1889).

Cyananthus lobatus Wall. *ex* Benth.

Family : Campanulaceae.

Distribution : Alpine Himalaya (Garhwal to Sikkim): 3,600-4,600 m.

Part Used : Root.

Utilization : Reported so far as mild purgative, blood purifier and for flatulence and bile related disorders in the body.

References : Hooker (1888), Jain & Goel (2005).

Cyanotis tuberosa Schult.f.

Syn : *C. adscendens* Dals. In Hook.; *Tradescantia tuberosa* Roxb.

Family : Commelinaceae.

Distribution : Chota Nagpur & Deccan Peninsula on the Westside from Konkan to Travancore.

Part Used : Root.

Utilization : Given in long-continued fevers and for strength and worms in cattle. Root paste used in chronic wounds.

References : Caius (2003), Chopra *et al.* (1956), Goud & Pullaiah (1996), Sinha & Sinha (2001), Watt (1889).

Cyathula prostrata (L.) Blume

Family : Amarantaceae.

Distribution : Throughout E. India & Deccan Peninsula.

Part Used : Root.

Utilization : Used as expectorant, demulcent and emetic, and its paste mixed with sugar given for the treatment of tuberculosis. Root extr. used in dysentery and skin diseases.

References : Agarwal (2003), Ambasta (1986), Anonymous (1948), Chaudhary & Neogi (2000), Chopra *et al.* (1956), Maya *et al.* (2003).

Cyclea bicristata Griff.

Family : Menispermaceae.

Distribution : E. Himalaya.

Parts Used : Root. Root bark.

Utilization : The bitter bark chewed with water for dysentery and gas trouble whereas grounded roots taken for stomachache, and also considered a very good remedy for any cancerous tissue in the stomach.

References : Jain *et al.* (1991), Jamir & Rao (1990).

Cyclea peltata (Lam.) Hk.f.

Syn : *C. arnotii* Miers.; *C. barbata* Miers.; *C. burmanii* Miers.

Family : Menispermaceae.

English, Hindi & Sanskrit Names:

Eng. : Pataroot;
Hindi : Parvel, Pathi;
Sans. : Patha.

Distribution : Wild from Konkan Down to Sri Lanka, Assam & Khasia Hills Eastwards.

Part Used : Root.

Utilization : Extr. used for dysentery, jaundice, fever, leprosy, asthma and stomachache. Raw roots chewed to relieve severe stomach pain, and its paste massaged on the back to cure backache by 'Malekudiyas'.

Active Constituents : These are fangchinoline, d-tetradrine, d-isochondrodendrine, D-quercitol, burmannalide $C_{21}H_{23}NO_4$, (-) cuirne, berbamine, isotetrandrine, fangchinoline, cycleanine, cycleanorine, phaeanthine, bisbenzylisoquinoline, (-)-norlimacine, (+) cyclea barbetine, (+) tetrandrine-2'-β-N-oxide, (+) berbamine, (-) repandine, (+) cycleanorine, (+) daphnandrine, (+) coclaurine and (-) N-methyl coclaurine.

Biological Activity : (+) tetrandrine (1.0μm) caused 50% inhibition of thymidine and uridine uptake in He La cells;

tetrandrine (7 mg/kg i.v.) significantly decreased toxic effect of cardiac glycosides, strophanthin and ouabain, in guinea pigs.

References : Ambasta (1986), Balasubramanian & Prasad (1996), Hosagoudar & Henry (1996b), Kshirsagar & Singh (2000a), Maya *et al.* (2003), Rastogi & Mehrotra (1998, 1999a, b, 2001), Reddi *et al.* (2005), Sivakumar *et al.* (2003), Yoganarasimhan (1981).

Cyclista robusta Roxb.

Family : Fabaceae.

Sanskrit Name:

Sans. : Nadinishpava.

Part Used : Root.

Utilization : Prescribed as remedy for dysentery, leucorrhoea and tumours.

Reference : Kirtikar & Basu (1935).

Cyclista scariosa Roxb.

Family : Fabaceae.

Sanskrit Name:

Sans. : Nadinishpara.

Distribution : C., W. & S. India.

Part Used : Root.

Utilization : Prescribed for dysentery, leucorrhoea and also applied along with other drugs as a poultice to reduce tumours.

Active Constituents : Contains tannin.

References : Anonymous (1948), Chopra *et al.* (1956).

Cymbidium aloifolium (L.) Sw.

Syn : *Epidendrum aloifolium* L.

Family : Orchidaceae.

English Name:

Eng. : Boat Orchid.

Distribution : Tropical Himalaya & S. India.

Part Used : Root.

Utilization : Aerial root juice used as ear drops for relieving pain. Its paste plastered tightly on fractured bones.

References : Agarwal (2003), Caius (2003), Ragupathy & Mahadevan (1991), Rao & Henry (1996).

Cynodon dactylon (L.) Pers.

Syn : *Agrostis linearis* Retz.; *Cynodon linearis* Willd.; *C. stellstus* Willd.; *Digitaria dactylon* Scop.; *Panicum dactylon* L.; *Paspalum dactylon* DC.

Family : Poaceae.

English, Hindi & Sanskrit Names:

Eng. : Bahama Grass, Bermuda Grass, Creeping Panic Grass, Couch Grass, Devil's Grass, Dog's Tooth Grass, Doob Grass, Doorwa,

Hindi : Dhoboghas, Dub, Durba, Huryali, Kalighas, Ramghas;

Sans. : Amari, Asitalata, Bahuvirya, Bharghvi, Bhutahantri, Durva, Gauri, Jaya, Mangala, Nanda, Saumya, Shatmula, Shitala, Tiktaparva, Vamini.

Distribution : Throughout India Upto 3,000 m.

Part Used : Root.

Utilization : Roots taken with 'mishri' (sugar pellets) as memory sharpner for students and scholars. Decoction used as diuretic, and for secondary syphilis, fever, urinary problems and leucorrhoea. Cold infusion often stops bleeding from piles. Crushed root mixed with curd taken for chronic gleet and nose bleeding.

References : Agarwal (2003), Anonymous (1948, 1968), Caius (2003), Chauhan (1999), Chopra *et al.* (1956), Dastur (1970), Kapur & Singh (1996), Kirtikar & Basu (1984), Parrotta (2001), Punjani (2006), Reddi *et al.* (2005), Saini (1996b), Upadhyay & Singh (2005), Varghese (1996), Watt (1889).

Cynoglossum lanceolatum Forssk.

Family : Boraginaceae.

Distribution : Throughout the Gangetic Plains.

Part Used : Root.

Utilization : Its paste along with black pepper and cow's urine applied on wounds for quick healing. Pills made from its crushed roots with molasses and 'Hathil' (*Murraya paniculata* (L.) Jacq.) given orally (twice a day for a period of one month) for providing strength, vigour and to cure rheumatic pains.

References : Hooker (1885), Singh & Kumar (2000b), Verma *et al.* (1995).

Cynoglossum wallichii G. Don

Syn : *C. glochidiatum* Wall. *ex* Benth.

Family : Boraginaceae.

Distribution : Mountainous Regions of Sub-Tropical & Temperate Zones.

Part Used : Root.

Utilization : Considered emetic and antiallergic (powdered root applied on swelling caused by allergy). Also root (20g) eaten once a day for 7 days to enhance potency.

References : Anonymous (1948), Brahmam & Saxena (1990), Chopra *et al.* (1956), Kapur & Singh (1996), Tiwari & Tiwari (1996).

Cynoglossum zeylanicum Thunb. *ex* Lehm.

Syn : *C. denticulatum* A.DC. var *zeylanica* (Lehm.) Clarke.; *C. furcatum* Wall. *ex* Roxb.

Family : Boraginaceae.

Distribution : Peninsular India.

Part Used : Root.

Utilization : Paste applied externally as an ointment in case of scabies.

References : Gopal & Chandra (2003), Rastogi & Mehrotra (2002).

Cyperus iria L.

Syn : *C. parviflora* Nees.; *C. umbellata* Roxb.

Family : Cyperaceae.

Hindi Name:

Hindi : Pilipatas.

Distribution : Swampy Places.

Part Used : Root.

Utilization : Root paste tied over body joints to cure rheumatism.

Active Constituents : Cyprene, caryophyllene, γ-cadinene, karacolene, α-cardinol, methyl 3, 7,11-trimethyl-trans-2, trans-6, cis-9, 11-dodecatetraenoate,methyl 3, 7 11- trimethyl trans-2, trans-6, trans-9, 11-dodecatetranoate, methyl 3, 7,11-trimethyl-11-hydroxy-trans-2, trans-6, trans-9-dodecatrienoate and minor components-α-pinene, β-pinene, limonene, p-cymene, linalool, α-capaene, β-elemene, humulene, α-elemene, β-selinene, germacrene, calamine, δ-cadinene, methyl 3, 7, 11-trimethyl-trans-2-trans-6, 10-dodecatrienoate, and cis-3, cis-5-dodecadien-1-ol are the active constituents.

References : Agarwal (2003), Islam & Jha (2003), Rastogi & Mehrotra (1999a, b), Shukla & Verma (1996), Watt (1889).

Cyperus kyllingia Endl.

Syn : *Kyllinga monocephala* Rottb.

Family : Cyperaceae.

English, Hindi & Sanskrit Names:

Eng. : Button Sedge;
Hindi : Nirbishi, Swetgothubi;
Sans. : Musta, Nirvisha.

Part Used : Root.

Utilization : Known so far as tonic, diuretic, demulcent, astringent, febrifuge, refrigerative and antidote to poisons.

References : Caius (2003), Gogoi & Das (2003), Subramaniam (2000).

Cyperus pangorei Rottb.

Family : Cyperaceae.

Part Used : Root.

Utilization : 3-6 ml extr. given to children upto one year of age to cure stomach pain.

Reference : Sharma & Singh (1999).

Cyperus rotundus L.

Syn : *C. hexastachyos* Roxb.; *C. retzii* Nees. non Poir.; *C. tuberosus* sensu Cl. non Roxb.

Family : Cyperaceae.

English, Hindi & Sanskrit Names:

Eng. : Nut Grass;

Hindi : Ghun, Motha, Mustaka, Nagurmotha, Tarmool

Sans. : Abda, Bhadrakshi, Bhadramusta, Gundra, Hima, Kachhola, Kasheru, Kuru, Musta, Valya, Vindakhya.

Distribution : A Weed Throughout India Upto 2,000 m.

Part Used : Root.

Utilization : Regarded stomachic, astringent, tonic, anthelmintic, diuretic, demulcent, stimulant, antipyretic, anti-inflammatory, analgesic, (extr.) and galactagogue (fresh tubers applied to the breasts), given in fevers, dyspepsia, diarrhoea, irritation of the bowels, cholera, urinary calculi, amenorrhoea, wounds, ulcers, pain in the abdomen, scorpion sting, jaundice (root extr.), snake bite (root juice), joint pains, white leucorrhoea, ulcers, dermatitis, and to increase potency. Root paste applied externally to cure boils and alti (reddish small boil-like structure).

Active Constituents : Contains terpenes, oleanic acid, beta-sitosterol, pinene, traces of cineole, mooterpenes, sesquterpenes and patchoulenone.

Biological Activity : Antipyretic, analgesic and anti-inflammatory activities positive.

References : Agarwal (2003), Anonymous (1968), Caius (2003), Chauhan (1999), Dastur (1970), Dey (1973), Gogoi & Das (2003), Khare (2004), Kirtikar & Basu (1984), Parrotta (2001), Rastogi & Mehrotra (1999a, b, 2002), Reddi *et al.* (2005), Saklani & Jain (1994), Sen & Behera (2003), Sharma (2003), Shukla & Verma (1996), Sur & Halder (2002), Sood & Prakash (2007), Thakur *et al.* (1989), Trivadi (2002).

Cyperus scariosus Br.

Syn : *C. pertenuis* Roxb.

Family : Cyperaceae.

Hindi & Sanskrit Names:

Hindi : Nagarmotha;

Sans. : Chakranksha, Charukesara, Kachharuha, Kalapini, Nagarmusta, Shishire, Uchchta.

Distribution : Damp Regions of U.P., Bengal & S. India.

Part Used : Root.

Utilization : Useful in the treatment of biliousness, fever, gonorrhoea, syphilic affections (decoction), dysentery, relieving thirst, fatigue, burning sensation, bad taste in the mouth, and also considered cordial, stomachic, desicant, galactagogue, diaphoretic, diuretic and astringent.

References : Agarwal (2003), Caius (2003), Watt (1889).

Dactylorhiza hatagirea (D. Don) Soo.

Syn : *Orchis latifolia* auct. non L.

Family : Orchidaceae.

Distribution : Trans-N.W.Himalaya (J. & K. & H.P.), N.W. Kumaon & Garhwal Regions of Himalaya: 2,800-4,000 m.

Part Used : Root.

Utilization : Found to cure cuts, bone fracture, fever, urinary troubles, spermatorrhoea (powdered root along with milk), wounds and tumours, kidney complaints, internal bleeding after child birth (5-10g dry root paste given orally to woman with one glass of hot milk thrice a day) and antiageing medicine (dried root powder). Roots also considered as aphrodisiac and serve as a general tonic.

Active Constituents : Known to contain starch, fairly large quantity of mucilage, sugar, albumen, traces of a volatile oil and ash consisting chiefly of potassium and lime.

References : Arya & Prakash (2000), Ballabh & Chaurasia (2006), Bhattacharjee (2001), Chaurasia *et al.* (2000b), Chhetri (2005), Nadeem *et al.* (2001), Negi & Pant (1990), Sharma (2003), Singh & Kumar (2000b), Sood & Prakash (2007), Sood *et al.* (2005c).

Dalbergia pinnata (Lour.) Prain var. *acciaefolia* (Dalz.) Thoth.

Syn : *D. tamarindifolia* Roxb.

Family : Fabaceae.

Hindi Name:

Hindi : Ketikat.

Distribution : Grows on E. Himalaya & W. Peninsula.

Part Used : Root bark.

Utilization : Paste taken with sugar and water as an effective remedy for dysuria and strangury. Also found to possess anthelmintic and masticatory properties.

References : Agarwal (2003), Ambasta (1986), Chopra *et al.* (1956), Lalramnghinglova & Jha (1999).

Dalbergia sissoo Roxb.

Family : Fabaceae.

Hindi & Sanskrit Names:

Hindi : Shisham, Sissu;

Sans. : Aguru, Shinshapa.

Distribution : W. Himalaya Upto 1,500 m from Sikkim to Upper Assam & Extensively Planted Along Roadsides.

Part Used : Root.

Utilization : Known to possess astringent properties.

References : Ambasta (1986), Chopra *et al.* (1956), Dhiman (2003).

Datisca cannabina Linn.

Family : Cactaceae.

Hindi Name:

Hindi : Akalbir

Distribution : Temperate Regions of Himalaya from Kashmir to Nepal.

Part Used : Root.

Utilization : Known to cure rheumatism and carious teeth.

Active Constituents : Datiscin, datiscetin, galangin, isalpinin, 2′3, 5-trihydroxy-7-methoxyflavone (datin), 2′5-dihydroxy-7-methoxyflavone-3-O-β-D-glucopyranosyl(1>6)-α-L-rhamnopyranoside (datinoside), galanginoside, cannabin, chalcone, galangin-3-rutinosides, and datinate-CM (datinate Ca & Mg) are the active principles.

References : Agarwal (2003), Anonymous (1948), Asolkar *et al.* (1992), Chopra *et al.* (1956), Kirtikar & Basu (1984), Rastogi & Mehrotra (1999a, b).

Datura innoxia Miller

Syn : *D. metel* auct. non Linn.

Family : Solanaceae.

English, Hindi & Sanskrit Names:

Eng. : Thorn Apple;

Hindi : Lal Dhatura, Safed Dhatura;

Sans. : Dhattura.

Distribution : Temperate Regions of India & W. Ghats.

Part Used : Root.

Utilization : Juice taken internally to relieve fever and rheumatic pains. Also fed to animals to relieve fever.

Active Constituents : Reported to possess hyoscyamine, hyoscine, meteloidine, 7-hydroxy-3, 6-bis-tigloyioxytropane, (-) 3α, 6β-bistigloyloxytropane, tropine, pseudotropine, 0.3% atropine, 6α propyonyloxy-3α-tigloyloxytropane (I), 3α, 6β-tigloyloxytropane-7β-ol and ψ-tropine.

References : Agarwal (2003), Asolkar *et al.* (2992), Goud *et al.* (2000), Khadel & Jain (2006), Rastogi & Mehrotra (1999a, b), Reddi *et al.* (2005).

Datura metel Linn.

Syn : *D. alba* Nees.; *D. fastuosa* L.;
D. fastuosa var. *alba* (Nees) C.B.Clarke.
Family : Solanaceae.

English, Hindi & Sanskrit Names:

Eng. : Downy Datura, Thorn Apple, Trumpet Flower;
Hindi : Dhattura, Kala Datura, Sadah Dhatura;
Sans. : Dhastura, Kanak, Maheshpriya, Shivapriya, Unmatta.

Distribution : Common Weed in Wastelands of the Country.

Part Used : Root.

Utilization : Powder (one teaspoon 2 times a day for 15 days) given after menstruation for inducing permanent sterility. To prevent miscarriage/abortion, root pulled out in one breath is tied to the waist of woman with the help of a string on Sunday. Its paste said to be good for suppuration of boils and eruptions. Also, considered antidiarrhoeal, antipyretic, antiseptic, and given for insanity, skin diseases, toothache, brushing teeth and ringworm.

Active Constituents : Found to possess hyoscyamine, scopolamine (hyoscine), ditigloyloxytropane, tigloidine, apohyoscine, nor-hyoscine, meteloidine, nor-hyocyamine, cuscohygrine, tropine, atropine, anti-o-lectins; nor-atropine, pseudotropine, 3-α–acetoxytropane, 3α-tigloyloxytropane, 3α, 6β-ditigloyloxytropane, and 3α, 6β-ditigloyloxytropane-7β-ol.

References : Asolkar *et al.* (1992), Chatterjee & Pakrashi (1997), Chopra *et al.* (1956), Dhiman (2003), Dwivedi (1993), Hembrom (1996), Khare (2004), Parrotta (2001), Punjani (2002), Rastogi & Mehrotra (1999a, b), Reddi *et al.* (2005), Sharma (2003), Singh (1996), Singh & Prakash (1994), Thakur *et al.* (1989), Varghese (1996).

Daucus carota Linn. var. *sativa* DC.

Syn : *D. caruta* Linn. (Wild var.)

Family : Apiaceae.

English, Hindi & Sanskrit Names:

Eng. : Bird's Nest, Cultivated Carrot, Queen Anne's Lace, Wild Carrot;

Hindi : Gajar;

Sans. : Garjara, Granjana.

Distribution : Grown in N., S. & C. India.

Part Used : Root.

Utilization : Juice given orally at the dosage of 2 teaspoonfuls twice daily to cure night blindness. Known as deobstruent, diuretic, blood purifier, antiseptic, refrigerant, stimulant; also used in leprosy, piles, cuts, bruises, tumours, biliousness, jaundice, palpitation and burning micturition.

Active Constituents : Cultivated carrot root contains sugar, pectin, carotene, vitamins, minerals, asparadine, ã-bisabolene, α-bergamotene, bornyl acetate, carophyllene, β-farnesene, falcarinol, falcaindiol, α-humulene, myristictin, sabinene, α -terpineol, ã-terpinene, terpinene-4-ol, terpinolene;α-β-ã-carotenes, 2-nonenal, octanal, citric, isocitric, fumaric, malic, succinic & tartaric acids; anthocyanins, 3 methyl-6-methoxy-8-hydroxy-3,4-dihydroxyisocaumarin.umbelliferose-(α-D-galatopyranosyl-2-α-D-glucopyranosyl-2β-D-fructofuranoside), n-alkanes(C13-58) with n-nonacosne, n-heptacosane, N-hentriacontane, β-sitosterol, leucine, valine, tryptophan, praline, alanine, glycine, glucose, fructose, α- & β-carotenes and lutein.

Biological Activity : β-carotene protects both the insides and outsides of the cells against free radical damage.

References : Anonymous (1948), Asolkar *et al.* (1992), Bhattacharjee (2001), Chatterjee & Pakrashi (1997), Chopra *et al.* (1956), Khare (2004), Rastogi & Mehrotra (1998, 1999a, b, 2001, 2002), Sen & Batra (1997), Sharma (2003).

Delphinium cachemirianum Royle

Family : Ranunculaceae.

Distribution : W. Himalaya, Kashmir & Tibet, at 3,300-5,300 m.

Part Used : Root.

Utilization : Relieves cough, cold and fever.

References : Kaul (1997), Watt (1889).

Delphinium denudatum Wall. *ex* Roylc

Syn : *D. pauciflorum* Royle

Family : Ranunculaceae.

English, Hindi & Sanskrit Names:

Eng. : Larkspur;

Hindi : Juswar, Nirbishi, Nirvisi;

Sans. : Aparvisha, Avisha, Nirvisha, Vishabhava, Vivisha, Vishavairini.

Distribution : Outer Range of Temperate Himalaya (Kashmir to Nepal): 1,500-2,700 m.

Part Used : Root.

Utilization : Considered stimulant, tonic, anthelmintic, alterative, astringent, adulterant of aconites and for respiratory complaints, ulcers, removing lice and ticks in domestic animals, cough and cold, high fever (root extr.), digestive disorders, sluggish liver, abdominal affections, and rheumatic pains. Dried roots chewed in toothache and its decoction taken in case of gall bladder and kidney stones.

Active Constituents : Chief constituents are denudatine, denudatidine, condephine, telatizidine, iso- talatizidine, stigmasterol, sitosterol, avenasterol and condelphine

Biological Activity : Aq. extr. exhibited a protective effect against hepatocellular damage in rats; Et OH (50%) extr. showed anticonvulsant activity.

References : Agarwal (2003), Ambasta (1986), Anonymous (1948), Asolkar *et al.* (1992), Aswal (1996), Caius (2003), Chhetri (2005), Chopra *et al.* (1956), Jain & Goel (2005), Kapur & Singh (1996), Kaul (1997), Khare (2004), Kirtikar & Basu (1935), Negi *et al.* (1999), Rana *et al.* (1996), Rastogi & Mehrotra (1999a, b), Samant *et al.* (2001), Singh (1996), Sinha & Sinha (2001), Verma & Chauhan (2006), Watt (1889).

Delphinium roylei Munz.

Family : Ranunculaceae.

Part Used : Root.

Utilization : Useful in rheumatic pain, tooth cavity pain and liver infection in cattle. Fresh root extr. used for high fever.

References : Kaul (1997), Singh (1996).

Delphinium vestitum Wall. *ex* royle

Family : Ranunculaceae.

Distribution : W. & C. Temperate Himalaya: 2,200-4,000 m (Nepal to Kunawer).

Part Used : Root.

Utilization : Useful for cuts, wounds, rheumatism, cooling, cough, diseases of blood, toothache, piles and as stimulant in debility and brain tonic. The powdered root along with oil applied externally on boils, blisters, cuts and wounds.

References : Hooker (1872), Kapur & Nanda (1996), Negi & Pant (1990), Singh & Kumar (2000b).

Dendrocnide sinuata (Bl.) Chew.

Syn : *Laportea crenulata* Wedd.; *L. gigantea* Gaud.; *L. latifolia* Gaud.; *Urtica crenulata* Roxb.

Family : Urticaceae.

Distribution : Tropical Himalaya; Sikkim Eastwards and Assam.

Part Used : Root.

Utilization : Used as an abortifacient and its poultice applied to swellings and blind abscesses. Its decoction beneficial for jaundice.

References : Asolkar *et al.* (1992), Gogoi *et al.* (2003a), Hooker (1985).

Derris robusta (Roxb. *ex* DC.) Benth.

Syn : *Brachypterum robustum* Dalz.& Gibs.; *Dalbergia krowee* Roxb.; *D. robusta* Roxb.

Family : Fabaceae.

Distribution : Outer Himalaya; Ganges Eastwards to Assam, E. Bengal.

Part Used : Root.

Utilization : Warm root stick chewed in toothache.

Active Constituents : Found to contain robustenin, 4-isoflavonesderrustone, derubone, robustone, 5-coumarines, derrusnin, robustic acid and their Me esters.

References : Anonymous (1948), Asolkar *et al.* (1992), Prakash & Mehrotra (1991), Watt (1889).

Desmodium gangeticum (L.) Dc.

Syn : *D. gangeticum* var. *maculatum* (L.) Baker.; *D. maculatum* (L.) DC.; *Hedysarum gangeticum* L.; *H. maculatum* L.; *Meibomia gangetica* (L.) Kuntze.

Family : Fabaceae.

English, Hindi & Sanskrit Names:

Eng. : Salpan;

Hindi : Salpan, Salpani, Salwan, Sarivan;

Sans. : Amshumati, Guhaa, Shalaparni, Sthiraa, Somyaa.

Distribution : Forests of India Upto 1,500 m.

Part Used : Root.

Utilization : Reported to be beneficial for asthma, diarrhoea, jaundice (root chewed), typhoid, biliousness, whooping cough (1 spoonful twice a day, till cure) and also as diuretic, aphrodisiac, antiinflammatory, analgesic, carminative, astringent, anticatarrhal, febrifuge and antidiarrhoeal (decoction). Paste given orally to the victim of snake bite once or twice. Two tablespoons of its juice given in dysentery, fevers, biliousness, cough and to check vomiting

Active Constituents : Roots afford pterocarpanoids, gangetin, gangetnin, desmodin, aptrocarpan and several alkaloids.

Biological Activity : Aq. extr. exhibited anti-inflammatory, antibacterial, antifungal and diuretic activities.

References : Agarwal (2003), Ambasta (1986), Asolkar *et al.* (1992), Chopra *et al.*(1956), Jha & Verma (1996), Khare (2004), Kirtikar & Basu (1935), Maya *et al.* (2003), Mukherjee & Namhata (1990), Parrotta (2001), Patil (2000), Rao & Henry (1996), Reddi *et al.* (2005), Saini (1996b), Saren *et al.* (2000), Sharma (2003), Thakur *et al.* (1989), Varghese (1996).

Desmodium laxiflorum DC.

Syn : *D. diffusum* DC. non (Willd.) DC.; *D. recurvatum* (Roxb.) Graham *ex* Wight & Arn.; *Hedysarum diffusum* Roxb.; *H. recurvatum* Roxb.

Family : Fabaceae.

Sanskrit Name:

Sans. : Prshnniparni.

Distribution : Peninsular India (900 m) N. India (1,300 m) Valleys & Shaded Forest Habitats.

Part Used : Root.

Utilization : 50 ml decoction of 50-100 g of its root and powder of three fruits of *Piper nigrum* L. taken daily for 10-15 days to cure jaundice.

References : Pandey *et al.* (1996), Parrotta (2001).

Desmodium microphyllum Miq.

Syn : *D. parvifolium* DC.; *D. scoparium* Wall.; *D. stipulaceum* Hassk.

Family : Fabaceae.

Distribution : Plains, Ascending to Temperate Region, Reaching 2,700 m in Kumaon.

Part Used : Root.

Utilization : Main root used as abortive stick. Its paste with curd prescribed in empty stomach early in the morning for 3 to 5 days for patients of jaundice.

References : Ambasta (1986), Hooker (1879), Saren *et al.* (2000).

Desmodium tiliaefolium G. Don

Syn : *D. argenteum* Wall.; *D. nutans* Wall.; *Hedysarum tiliaefolium* Don

Family : Fabaceae.

Hindi Names:

Hindi : Chamkat, Chara, Chamyat, Gurshagal, Laber, Murt, Samru.

Distribution : N. India, Temperate & Tropical regions.

Part Used : Root.

Utilization : Recorded as bitter, diuretic, carminative and tonic; and useful to improve appetite, enrich blood, urethral discharges, piles, ophthalmia, lumbago, amenorrhoea and bilious complaints.

References : Agarwal (2003), Ambasta (1986), Chopra *et al.* (1956), Kapur & Nanda (1996), Kirtikar & Basu (1935), Watt (1889).

Desmodium triangulare (Retz.) Merr.

Syn : *D. cephalotes* Wall.

Family : Fabaceae.

Distribution : Throughout India in Plains. Himalaya. (Nepal Eastwards ascending to 1,200 m).

Part Used : Root.

Utilization : Extr. (30-40 ml) given as antidote for snake bite.

Active Constituents : Phenethylamine, salsolidine, hordenine, tyramine, candicine and choline are the principle constituents.

References : Ambasta (1986), Asolkar *et al.* (1992), Sharma & Singh (1999).

Desmodium triflorum (L.) DC.

Syn : *Hedysarum triflorum* Linn.

Family : Fabaceae.

Hindi Name:

Hindi : Kudaliya.

Distribution : Throughout the Plains of India & in the Himalaya Upto 2,300 m.

Part Used : Root.

Utilization : Used for cough, asthma and applied to wounds and abscesses.

Active Constituents : Phenethylamines, indole-3-alkylamines, stachydrine, trigonelline, choline and betaine have been isolated from roots.

References : Chopra *et al.* (1956), Rastogi & Mehrotra (1999a, b).

Desmodium trifolium (L.) DC.

Family : Fabaceae.

Hindi Name:

Hindi : Kudaliya.

Part Used : Root.

Utilization : Known as carminative, tonic and diuretic.

References : Ambasta (1986), Maya *et al.* (2003).

Desmodium triquetrum (L.) DC.

Family : Fabaceae.

Hindi Name:

Hindi : Adkhapanal.

Distribution : Occurs from C. & E. Himalaya & S. India.

Part Used : Root.

Utilization : 3g powdered drug prepared from its roots along with *Terminalia bellerica* stem bark and 5 black peppers given to children for rickets.

References : Agarwal (2003), Aminuddin & Girach (1996).

Desmodium velutinum (Willd.) DC.

Syn : *D. lasiocarpum* DC.; *D. latifolium* (Roxb. *ex* Ker.) DC.; *Hedysarum lasiocarpum* Beau.; *H. volutinum* Willd.

Family : Fabaceae.

Sanskrit Name:

Sans. : Prshparni.

Distribution : Throughout Most of India Upto 1,200 m (Open Habitats or in Deciduous Forests.

Part Used : Root.

Utilization : Crushed root reportedly given to infants as a remedy for inability to digest mother's milk among the 'Kathodas' of Rajasthan. The roots mixed with small hot pepper used to cure blood in urine.

References : Aminuddin & Girach (1996), Parrotta (2001), Yoganarasimhan *et al.* (1981).

Desmos dumosus (Roxb.) Saff.

Syn : *Unona dumosa* Roxb.

Family : Annonaceae.

Part Used : Root.

Utilization : Decoction used for strangury and genitourinary disorders.

References : Ambasta (1986), Lalramnghinglova & Jha (1997).

Desmostachya bipinnata (L.) Stapf.

Syn : *Eragrostis cynosuroids* Beauv.

Family : Poaceae.

English, Hindi & Sanskrit Names:

Eng. : Sacrificial Grass (Smaller Variety);
Hindi : Dab, Kusha;
Sans. : Kusha, Suuchyagra, Yajnabhuushana.

Distribution : Plains of India in Dry & Hot Areas & in Sandy Deserts.

Part Used : Root.

Utilization : Decoction of roots (20 g) with little honey given thrice a day for 3 days to stop abdominal and colic pain and for dysuria. Also known as bitter, sweet, cooling, diuretic, emollient, galactagogue, astringent, and used in asthma, jaundice, vitiated conditions of (pitta) and hyperdypsia.

References : Agarwal (2003), Ambasta (1986), Dhiman (2003), Khare (2004), Satapathy & Brahmam (2000).

Dichroa febrifuga Lour.

Syn : *Adamia cyanea* Wall.

Family : Saxifragaceae.

Hindi Names:

Hindi : Aseru, Basak.

Distribution : Forest of the E. Himalaya Ascending between 1,500-2,400 m.

Part Used : Root bark.

Utilization : Decoction efficacious as febrifuge and tonic.

Active Constituents : Contains dichrin A and dichrin B and two alkaloids dichroine A, dichroine B, febrifugine and isofebrifugine.

Biological Activity : Dichroine isolated from roots used as antimalarial drug. Febrifugine approximately 100 times as active as quinine against *Plasmodium lophusae* in ducks.

References : Agarwal (2003), Anonymous (1948), Chopra *et al.* (1956), Dhiman (2003), Rastogi & Mehrotra (2001).

Dichrostachys cinerea (L.) Wight & Arn.

Syn : *Acacia cinerea* (L.) Spreng.; *A. dalea* Desv.; *Caelliea cinerea* Macbride; *Desmanthus cinereus* (L.) Willd.; *Mimosa cinerea* L.

Family : Fabaceae.

Hindi & Sanskrit Names:

Hindi : Bilatri, Kheri, Murmiar, Vurutuli;

Sans. : Bahuvaraka, Dirghamula, Krichhari, Mahakapitha, Sandnaka, Vellantara, Viradru.

Distribution : Dry Scrub Forests & Arid Hills of N.W. & Peninsular India.

Parts Used : Root. Root bark.

Utilization : Roots useful for rheumatism, white and red leucorrhoea, paralysis, fractured bone, nephrocalcinosis, uretral stones, strangury, urinary calculi and renal troubles, diseases of vagina and uterus, retention of urine, chicken pox and also as appetizer and astringent. Extr. of root taken orally by 'Ḳolams' for menstrual disorders (2-3 spoonfuls daily in the morning for three days) whereas extr. of bark crushed with black pepper and garlic given for healing bone fracture (2 spoonfuls twice daily till cure) and/or plastered over fracture.

Active Constituents : These are n-octacosanol, β-amyrin, friedelan-3-one, friedelan-3β-ol and β-sitosterol.

References : Agarwal (2003), Ambasta (1986), Anonymous (1948), Chopra *et al.* (1956), Kirtikar & Basu (1935), Parrotta (2001), Rastogi & Mehrotra (1999a, b), Rao & Henry (1996), Ravisankar & Henry (1992), Reddi *et al.* (2005), Samvatsar & Diwanji (2000).

Dictamnus albus Linn.

Family : Rutaceae.

English Names:

Eng. : Bastard Dittany, Burning Bush, Fraxinella, Gas Plant Dittany.

Distribution : Temperate W. Himalaya (Kashmir to Kunawr): 2,000-2,600 m.

Parts Used : Root. Root bark.

Utilization : Root bark aromatic, given in intermittent fevers (1-2 teaspoons extr. twice a day), nervous diseases, amenorrhoea, hysteria and joint pain (poultice). Decoction of roots given for scabies and other skin diseases, and its paste good against rheumatic pains and skin eruptions.

Active Constituents : Active principles are due to obalactone, fraxinellone, dictamine, γ-fagarine, isodictamnine and isomaculosidine.

References : Agarwal (2003), Ambasta (1986), Anonymous (1948), Asolkar *et al.* (1992), Chopra *et al.* (1956), Kapur & Srivastava (1996), Kaul (1997), Kirtikar & Basu (1935), Rastogi & Mehrotra (1999a, b), Singh (1996).

Dillenia pentagyna Roxb.

Syn : *Colbertia coromandelina* DC.; *C. augusta* Wall.

Family : Dilleniaceae.

English, Hindi & Sanskrit Names:

Eng. : Dog Teak;

Hindi : Aggai, Kallai, Karmal;
Sans. : Aksikiphala.

Distribution : Deciduous Tree of Oudh, Bengal, Assam, C., S. & W. India.

Part Used : Root.

Utilization : Decoction of fresh roots taken once a day in early morning for one month to treat high blood pressure and body pain (30 ml three times daily).

References : Ambasta (1986), Anonymous (1948), Maheshwari *et al.* (1997), Reddi *et al.* (2005), Tirkey (2004), Watt (1889).

Dioscorea bulbifera L. (Pl. 5D)

Syn : *D. crispata* Roxb.; *D. pulchella* Roxb.; *D. sativa* auct. non L.; *D. versicolor* Buch. Ham. *ex* Wall.

Family : Dioscoreaceae.

English, Hindi & Sanskrit Names:

Eng. : Air Potato, Potato Yam;
Hindi : Pitalu, Ratauloo, Suaralu;
Sans. : Yarachi.

Distribution : Common Throughout India Upto 1,400 m.

Part Used : Root.

Utilization : Used against snake bite.

Active Constituents : Known to possess norditerpenes and diosbulbins.

References : Agarwal (2003), Anonymous (1948), Deshmukh & Rothe (2003), Rastogi & Mehrotra (1999a, b), Sharma (2003), Singh (1999), Sood & Prakash (2007).

Dioscorea hispida Dennst.

Syn : *D. daemona* Roxb.; *D. hirsuta* Dennst.; *D. triphylla* Linn.

Family : Dioscoreaceae.

Hindi & Sanskrit Names:

Hindi : Baichandi, Karukanda;
Sans. : Hastyaluka.

Distribution : Humid Parts of the Country, Sikkim, Khasia Hills (Meghalaya), Peninsular India Upto 300 m in the E. & W. Ghats.

Part Used : Root.

Utilization : Roasted powder (5-10g) mixed with a cup of milk and given to children two-three times a day in ricketssiae and weakness.

References : Painuli & Maheshwari (1996), Parrotta (2001).

Dioscorea oppositifolia L.

Family : Dioscoreaceae.

Hindi Name:

Hindi : Kanta-Alu.

Distribution : E. & W. Coasts of S. India, extending to Khandalla; Frequent Occurrence in the Sub-Alpine Forests.

Part Used : Root.

Utilization : Paste applied on bitten area in case of scorpion bite. Ground root heated and applied as poultice to reduce swellings and for curing snake bite and scorpion sting.

References : Sood & Prakash (2007), Venkataratnam & Raju (2004), Watt (1889).

Diospyros melanoxylon Roxb.

Syn : *D. dubia* Wall. *ex* A. Dc.; *D. exsculpa* Bedd.; *D. melanoxylon* var. *beddomei* Clarke; *D. tupru* Buch-Ham.; *D. wightiana* A.DC.

Family : Ebenaceae.

English, Hindi & Sanskrit Names:

Eng. : Coromandel Ebony, Ebony, Malabar Ebony, Persimmon Ebony;

Hindi : Abnus, Kendu, Tendu, Timburni;

Sans. : Dirghapatraka, Kenduka.

Distribution : Deciduous Forests & Dry Rocky Hills in Peninsular India From Bihar & M.P. Southwards to Karnataka & Tamil Nadu.

Part Used : Root.

Utilization : Mixture of its powder (10 g) with pepper (5 g) and turmeric (5 g), given (one spoon twice a day) for malaria till cure and also for syphilis.

References : Parrotta (2001), Rao & Pullaiah (2001), Reddi *et al.* (2005).

Dipsacus inermis Wall.

Family : Dipsacaceae.

Distribution : Temperate Himalaya: 2,000-4,000 m.

Part Used : Root.

Utilization : Pounded root taken with warm water to cure dysentery.

Reference : Kumar (2002).

Dipteracanthus patulus (Jacq.) Nees.

Syn : *D. erectus* Nees.; *Petalidium patulum* Dalz & Gibs.; *Ruellia patula* Jacq.

Family : Acanthaceae.

Distribution : Deccan Peninsula, Extending to Scinde, Rajpootana & Bundelkund.

Part Used : Root.

Utilization : Decoction given in fever.

References : Bhatt *et al.* (2002), Hooker (1885).

Dipteracanthus suffruticosus Voigt

Syn : *Ruellia suffruticosa* Roxb.

Family : Acanthaceae.

Distribution : Gangetic Plains, M.P., Bihar & Orissa.

Part Used : Root.

Utilization : Paste known in India for renal infections, gonorrhoea, eye sores, syphilis, and abortion.

References : Agarwal (2003), Ambasta (1986), Anonymous (1952), Chopra *et al.* (1956).

Distemon indicum Wedd.

Syn : *D. grossum* Wedd.; *Urtica grossa* Wall.

Family : Urticaceae.

Distribution : Assam.

Part Used : Root bark.

Utilization : Poultice good against dog or snake bite.

References : Hooker (1890), Jamir (1997).

Dolichos lablab L. var. *typicus*

Family : Fabaceae.

English & Hindi Names:

Eng. : Bonavist Bean, Hyacinth Bean, Indian Butter Bean, Lablab Bean;

Hindi : Sem.

Distribution : Cultivated Throughout India.

Part Used : Root.

Utilization : 2g powder given orally twice a day with milk in leucorrhoea.

References : Agarwal (2003), Ambasta (1986), Shukla *et al.* (2003).

Doronicum roylei DC.

Syn : *Fullaromia kumaonensis* DC.

Family : Asteraceae.

Distribution : W. Himalaya from Kashmir to Garhwal, Upto 3,300 m.

Part Used : Root.

Utilization : Considered bitter, pectoral, carminative, cardiotonic, stomachic, alexiteric; and used in cough, chest pain, inflammations, headache, and giddiness on ascending heights.

References : Ambasta (1986), Caius (2003), Chopra *et al.* (1956), Kirtikar & Basu (1984), Watt (1889).

Dracaena elliptica Thunb.

Family : Liliaceae.

Part Used : Root.

Utilization : One cup full decoction given at bed time to check excessive bleeding during menstruation.

Reference : Prakash & Mehrotra (1991).

Dracaena petiolata Hk.L.

Family : Liliaceae.

Distribution : Upper Assam.

Part Used : Root.

Utilization : Juice good against stomachache and vomiting.

Reference : Kumar (2002).

Dracaena spicata Roxb.

Family : Liliaceae.

Part Used : Root.

Utilization : Juice swallowed for stomachache.

Reference : Lalramnghinglova & Jha (1999).

Dregea volubilis (L.f.) Benth. in B. & H.

Syn : *Wattakaka volubilis* (L.) Stapf.

Family : Asclepiadaceae.

Part Used : Root.

Utilization : Paste applied for pain due to external injury.

Reference : Varghese (1996).

Drosera burmanni Vahl.

Family : Droseraceae.

Part Used : Root.

Utilization : Good against wounds of cattle.

Reference : Varghese (1996).

Ebermaiera staurogyne Nees.

Family : Acanthaceae.

Distribution : Khasia Hills Upto 1,000 m & N. Silhet.

Part Used : Root.

Utilization : Crushed roots made into pills and given for the treatment of typhoid.

References : Chaudhury & Neogi (2000), Hooker (1885).

Ecbolium viride (Forssk.) Alston var. *dentata* (Klein. *ex* Link.) Rizada & Santapau

Syn : *E. linnaeanum* Kurz. var. *dentata* (Klein *ex* Link.) Cl.; *Justicia dentata* Klein *ex* Link.

Family : Acanthaceae.

English & Hindi Names:

Eng. : Blue Fox Tail;

Hindi : Udajati.

Distribution : Hedges, Wasteplaces, Under Shade in Assam, W. Bengal, Orissa & Deccan Peninsula.

Part Used : Root.

Utilization : Used against jaundice, menorrhagia, rheumatism and tumours. Its decoction fed to cure headache.

Active Constituents : Posseses isoorientin, orientin, isovitexin and vitexin.

References : Agarwal (2003), Asolkar *et al.* (1992), Chatterjee & Pakrashi (1997), Chopra *et al.* (1956), Rastogi & Mehrotra (1999a, b), Sur & Halder (2002).

Echinops echinatus Roxb.

Family : Asteraceae.

Hindi & Sanskrit Names:

Hindi : Gokhru, Utakanta;

Sans. : Brahma Dandi, Kantalu, Kan Taphala, Shrigala, Shunakashana, Utati, Unthkaantaka, Vrittaguchha.

Distribution : Almost Throughout India Upto 1,500 m; Common in Fields, Grasslands, Open Forests & Roadsides in Peninsular India.

Part Used : Root.

Utilization : Powder used against seminal diseases, vomiting, wounds in cattle to destroy maggots, lice (mixed with acacia gum and applied to the hair), eczema (paste applied externally two times a day for 15 days), pneumonia (a teaspoonful powder taken for three days with milk), diarrhoea and dysentery (Kolams-2 to 3 spoonfuls thrice a day for 3 days). Its decoction placed in the navel of woman for easy and quick delivery by the Garasia tribe in Pali District. 'Bhils' take orally the extr. of roots (5 ml twice daily with luke warm water) to cure whooping cough and cold. The tribals in Dungarpur district apply the root paste on the body in fever and also to cure scorpion sting, gout and wounds. A small piece of root tied to the hair of the expectant mother at the time of delivery to induce labour pain, and removed immediately after delivery to be kept in a box. Root pieces collected on Sunday used as a necklace to cure hoarse cough in children.

Active Constituents : Allophanic acid and w-methylallophanic acid.

References : Chopra *et al.* (1956), Deshmukh & Rothe (2003), Dwivedi (2003), Jadhav (2006), Katewa *et al.* (2001), Kumar & Pullaiah (2000), Parrotta (2001), Patil (2000), Punjani (2002), Rastogi & Mehrotra (2002), Ravisankar & Henry (1992), Reddi *et al.* (2005), Sharma (2002), Singh & Rao (2003), Singh & Pandey (1998).

Echinops niveus Wall.

Family : Asteraceae.

Distribution : Himalaya (Kashmir to Garhwal): 1,200-2,700 m.

Part Used : Root bark.

Utilization : Powdered bark mixed with honey taken internally (2-3g dose) to cure cough and asthma.

References : Chauhan (1999), Rastogi & Mehrotra (2002).

Eclipta prostrata L. (Pl. 6A)

Syn : *Cotula alba* L.; *Eclipta alba* (L.) Hassk.; *E. erecta* Linn.; *Verbesina alba* L.; *V. prostrata* L.

Family : Asteraceae.

English, Hindi & Sanskrit Names:

Eng. : Trailing Eclipta Plant;

Hindi : Babri, Bengra, Bhangra, Mochkand;

Sans. : Ajagara, Angaraka, Bringraja, Kesharaja, Markara, Nilapushpa, Patanga, Shyamala.

Distribution : Throughout India Upto 2,000 m on the Hill.

Part Used : Root.

Utilization : Considered emetic, purgative, tonic, antiseptic, alterative, applied externally on ulcers and wounds in cattle, excessive bleeding of wounds, cataract, and diarrhoea. One teaspoonful of its paste given orally and a portion applied on affected victim bitten by snake or scorpion. It is one of the most favourite drugs used for jaundice in S. India.

Active Constituents : Roots afforded 5′-isovalerylloxymethylene-2-(4-iso-valeroxybut-3-ynyl) dithiophene, heptacosanol, hentriacontanol, stigmasterol, and 5′-tigloyloxy-methylene-2-(4-isovaleryloxybut-3-ynyl) dithiophene (III).

References : Agarwal (2003), Ambasta (1986), Anonymous (1952), Caius (2003), Chatterjee & Pakrashi (1997), Chauhan (1999), Chopra *et al.* (1956), Dwivedi (1993), Kumar (2002), Nadeem *et al.* (2001), Panda & Das (2000), Parrotta (2001), Rastogi & Mehrotra (2001, 2002), Reddi *et al.* (2005), Saren *et al.* (2000), Singh & Pandey (1998), Subramaniam (2000), Sur & Halder (2002).

Ehretia aspera Willd

Syn : *E. laevis* Roxb. var. *aspera* C.B.Clarke.; *E. obtusifolia* Hochst.
Family : Boraginaceae.

Distribution : Deccan Peninsula Extending Northwards to U.P. & Punjab.

Part Used : Root.

Utilization : Decoction of fresh roots given in venereal diseases.

References : Ambasta (1986), Chopra *et al.* (1956), Watt (1889).

Ehretia buxifolia Roxb.

Syn : *Carmona microphylla* Lam.;
C. retusa (Vahl.) Masamune; *Ehretia microphylla* Lam.
Family : Boraginaceae.

Distribution : Jungles of the Deccan Peninsula.

Part Used : Root.

Utilization : Sweet root prescribed as an alterative in syphilis and debility.

References : Chopra *et al.* (1956), Watt (1889).

Eichhornia crassipes (Mart.) Solms.

Syn : *E. speciosa* Kunth.
Family : Pontederiaceae.

English & Hindi Names:

Eng. : Water Hyacinth;
Hindi : Jalkumbhi.

Distribution : Ponds in Tropical Regions of India.

Part Used : Root.

Utilization : Paste applied on teeth as dentifice and for the treatment of goitre.

Active Constituents : Contains bound auxin.

References : Agarwal (2003), Asolkar *et al.* (1992), Chakraborty *et al.* (2003).

Elaeagnus caudata Scht. *ex* Mom.

Syn : *E. conferta* Roxb.; *E. kolaga* Scht.; *E. latifolia* auct. non Linn.; *E. simoni* Hort.

Family : Elaeagnaceae.

English & Hindi Names:

Eng. : Bastard Oleaster;
Hindi : Gaura, Ghiwain.

Distribution : Sub-Tropical & Temperate Himalaya (600-2,300 m) to Sikkim (1,600-2,700 m).

Part Used : Root.

Utilization : 100 ml decoction efficacious against retained placenta (twice daily), fever. Also, its paste applied all over the body for checking fever.

References : Agarwal (2003), Hooker (1890), Hosagoudar & Henry (1996b), Lalramnghinglova & Jha (1997), Parrotta (2001).

Elaeagnus pyriformis HK.f.

Family : Elaeagnaceae.

Distribution : Upper Assam.

Part Used : Root.

Utilization : One cup full of its paste drunk for appendix ailment.

References : Hooker (1885), Lalramnghinglova & Jha (1997).

Elephantopus scaber L.

Family : Asteraceae.

English, Hindi & Sanskrit Names:

Eng. : Prickly Leaved Elephant's Foot;
Hindi : Gobhi, Gojilata, Karipadam;
Sans. : Gojihra, Gojivha, Kari Padam.

Distribution : Entire N.E. Region: Wasteland & Moist Grassland.

Part Used : Root.

Utilization : So far reported for vomiting, toothache, healing wounds, snake bite, filaria, colic pain, tympanites, diabetes (1 teaspoon of root powder and 'mishri' (sugar pellets) taken orally 2 times a day for 10-15 days), rheumatism, menstrual disorders, amoebic dysentery, scabies (10 g powdered roots mixed with slightly warmed mustard oil), cornea ulcer (roots rubbed on hand palm), stomach pain, leprosy, eczema, acidity, urinary troubles, liver, fever, gonorrhoea, swellings, diarrhoea, dysentery, cattle wounds (root paste), and also given to patients with heart, liver trouble and malaria. Decoction given orally as an anthelmintic. Pieces of root tied with thread and hung in the sleeping area of suckling mother for better lactation. Its paste applied in headache, tonsillitis and to cure pimples in children. Also chewed raw or its paste applied on forehead as antinarcotics to subside influence of liquor. For curing pyorrhoea, mixture of 5 g powdered root and three black peppers is kept in dental cavity at bed time for 2-3 days.

References : Agarwal (2003), Ambasta (1986), Aminuddin & Girach (1991), Anonymous (1952), Asolkar *et al.* (1992), Bondya & Sharma (2004), Brahmam & Saxena (1990), Chatterjee & Pakrashi (1997), Chopra *et al.* (1956), Dwivedi (1993), Girach (1992), Nadanakunjidam (2003), Jain & Patole (2001), Jain *et al.* (1991, 1995), Kharkongar & Joseph (1997), Kumar (2002), Kumar & Lal (2003), Khanna (2002a, b), Maliya (2004), Panda & Das (2000), Pandey & Verma (2002), Patil (2000), Rao & Henry (1996), Reddi *et al.* (2005), Saren *et al.* (2000), Singh (2000), Sur & Halder (2002), Thomas & Britto (2003), Varghese (1996), Watt (1889).

Eleusine indica (L.) Gaertn.

Syn : *Cynosurus indicus* Linn.

Family : Poaceae.

English & Hindi Names:

Eng. : Crow Foot Grass, Crab Grass;

Hindi : Malankuri, Mandla.

Distribution : Wet Plains & Low Hills of India.

Part Used : Root.

Utilization : Infusion used for diarrhoea. Its juice good for jaundice (2-3 teaspoons twice a day given for a week), and considered sudorifice and febrifuge.

References : Agarwal (2003), Chhetri (1994), Jain & Sikarwar (1998), Srivastava & Singh (2003), Watt (1889).

Elytraria acualis (Linn.f.) Lindau.

Family : Acanthaceae.

Hindi Names:

Hindi : Dasmori.

Distribution : Deccan Peninsular India to Gangetic Plains & E. Himalaya.

Part Used : Root.

Utilization : Useful against eczema, maggot infested sores, skin diseases, wounds, rotten teeth, snàke bite, anasarca, mammary tumours, abscesses, rickets (root along with mother's milk), maramus, pneumonia, infantile diarrhoea. Decoction administered to subside boils and body swellings.

References : Agarwal (2003), Asolkar *et al.* (1992), Ranjan (1996), Reddi *et al.* (2005), Shah & Singh (1990).

Embelia ribes Burm.f.

Family : Myrsinaceae.

Hindi & Sanskrit Names:

Hindi : Barberang, Wawrung;

Sans. : Vidanga, Vrishanasana.

Distribution : Throughout India from the C. Himalaya to Peninsular India Ascending to 1,700 m.

Parts Used : Root. Root bark.

Utilization : Bark paste applied externally to chest in pneumonia and its powder considered a good remedy for toothache. Root infusion beneficial in cough, diarrhoea and as an antifertility drug.

Biological Activity : Powder showed 100 % antifertility activity in female albino rats at a dose of 100 mg/kg.

References : Agarwal (2003), Ambasta (2003), Asolkar *et al.* (1992), Chatterjee & Pakrashi (1997), Chopra *et al.* (1956), Sharma (2003).

Embelia tsjeriam-cottam DC.

Family : Myrsinaceae.

Part Used : Root.

Utilization : Considered antidote of snake poison.

Reference : Kothari & Moorthy (1996).

Emblica officinalis Gaertn.

Syn : *Cicca emblica* Kurtz.; *Phyllanthus emblica* L.; *P. glomeratus* Roxb.; *P. taxifolius* Don

Family : Euphorbiaceae.

English, Hindi & Sanskrit Names:

Eng. : Indian Goseberry;
Hindi : Amala, Ambolati, Amlika, Aonla, Yeonlah;
Sans. : Adiphala, Amalaka, Amlika, Dhatri.

Distribution : Wild or Cultivated Throughout Tropical Part of India.

Part Used : Root.

Utilization : Known so far as an astringent and remedy for jaundice.

Active Constituents : Roots contain ellagic acid.

References : Ambasta (1986), Asolkar *et al.* (1992), Dhiman (2003), Watt (1889).

Emilia sonchifolia (L.) DC.

Syn : *Cacalia sonchifolia* L.
Family : Asteraceae.

Hindi & Sanskrit Names:

Hindi : Hirankhur, Hirankhuri;
Sans. : Sasarutih, Sasasruti.

Distribution : Common in Cultivated Fields & Open, Disturbed Habitats Throughout Most of India Upto 1,200 m.

Part Used : Root.

Utilization : Known in India to treat diarrhoea, bowel complaints, mouth dryness, cataract and redness of the eyes (root juice mixed with salt and water).

References : Agarwal (2003), Anonymous (1952), Asolkar *et al.* (1992), Chopra *et al.* (1956), Kumar & Lal (2003), Parrotta (2001), Reddi *et al.* (2005).

Enicostema hyssopifolium (Willd.) Verdoorn

Family : Gentianaceae.

Hindi Name:

Hindi : Chota-Chirayata.

Part Used : Root.

Utilization : Root paste or decoction administered twice for three days to cure malaria and general fever (not to be taken on empty stomach).

References : Asolkar *et al.* (1992), Sharma (2002).

Ephedra gerardiana Wallich *ex* Stapf. *emend* Florin

Syn : *E. major* Host.; *E. nebrodensis* Tineo.; *E. vulgaris* Rich.
Family : Gnetaceae.

English, Hindi & Sanskrit Names:

Eng. : Ephedra;
Hindi : Asmania, Khanda, Phok;
Sans. : Ephedra.

Distribution : Drier Regions of Higher Himalayan Ranges: 2,000-4,000 m.

Part Used : Root.

Utilization : Ash applied on injuries and wounds; and also as astringent, alterative, stomachic and tonic.

Active Constituents : Contains ephedrine.

Biological Activity : Liquid extr. of ephedrine has the property for controlling asthmatic paroxysms.

References : Anonymous (1968), Jain (1968), Pandey & Pande (2000), Reddi *et al.* (2005).

Eranthemum purpurascens Nees in Wall.

Family : Acanthaceae.

Part Used : Root.

Utilization : Decoction employed for bodyache, pain, swelling and venereal diseases.

Reference : Varghese (1996).

Eranthemum roseum R.Br.

Syn : *Daedalacanthus roseus* T.anders.

Family : Acanthaceae.

Hindi Name:

Hindi : Gulsham.

Distribution : Moist, Shaded Forest Habitats in the W. & S. Regions of Peninsular India from N. Maharashtra to Karnataka & Tamil Nadu.

Part Used : Root.

Utilization : Root boiled in milk is a popular remedy for leucorrhoea. Also given to cattle to promote growth of the foetus.

References : Agarwal (2003), Ambasta (1986), Anonymous (1952), Chopra *et al.* (1956), Parrotta (2001).

Ervatamia divaricata (L.) Burkill

Syn : *E. coronaria* Stapf.; *Nerium divaricatum* Linn.; *Tabernaemontana coronaria* R.Br.; *T. divaricata* Blume.

Family : Apocynaceae.

English, Hindi & Sanskrit Names:

Eng.	: East Indian Rosebay;
Hindi	: Chandni, Tagar;
Sans.	: Nandivriksha, Tagara.

Distribution : Cultivated Throughout India; N.W. Himalaya in Kumaon, Eastwards & Southwards.

Part Used : Root.

Utilization : Used in blood diseases, paralysis, eye troubles, and employed as a local anodyne and also chewed for relief from toothache.

Active Constituents : Tabernemontanine, dregamine, vobasine, vocamine, voacangine, 5-hydroxy-6-oxocoronaridine, 5-oxocoronaridine, 6-oxocoronaridine and 19 hydroxy coronaridine are the active ingredients.

References : Ambasta (1986), Hooker (1890), Nayak *et al.* (2003), Rastogi & Mehrotra (2001).

Eryngium caeruleum Bieb.

Syn : *Eryngium planum* Lindl.

Family : Apiaceae.

Hindi Name:

Hindi	: Dudhali.

Distribution : W. Himalaya at 2,000 m.

Part Used : Root.

Utilization : Reported in India as nervine tonic, aphrodisiac, expectorant, diuretic, diaphoretic and haematinic, and also its powder rubbed over weak limbs to give strength and vigour.

References : Agarwal (2003), Anonymous (1952, 1968), Chopra *et al.* (1956), Singh (1996), Sinha & Sinha (2001), Sood *et al.* (2005b), Watt (1889).

Erythrina resupinata Roxb.

Family : Fabaceae.

Distribution : Upper Gangetic Plains.

Part Used : Root.

Utilization : Mixture of its root (10 g) grounded with 1 g each of 'pipeli' (*Piper longum*), clove, ginger, black pepper, 'kala-zira' (*Nigella sativa*) and chillies given for fever, leprosy and rheumatism.

References : Brahmam & Saxena (1990), Rastogi & Mehrotra (1999a, b).

Erythrina stricta Roxb.

Syn : *E. corallodendron* Linn.; *E. indica* Lam.; *E. variegata* L.; *E. variegata*var *orientalis* (L.) Merr.

Family : Fabaceae.

English, Hindi & Sanskrit Names:

Eng. : Indian Coral Tree;
Hindi : Ferrud, Mandar;
Sans. : Mandaara, Mura.

Distribution : Common in Bengal & S. India.

Part Used : Root bark.

Utilization : For threadworms, about one teaspoonful of root bark juice given twice daily for five consecutive days.

References : Ambasta (1986), Anonymous (1968), Borthakur (1993), Rastogi & Mehrotra (1999a, b).

Erythroxylum monogynum Roxb.

Syn : *E. indicum* (DC.) Bedd.; *Sethia indica* DC.

Family : Erythroxylaceae.

English, Hindi & Sanskrit Names:

Eng. : Bastard Sandal, Red Cedar;
Hindi : Goranti;
Sans. : Devataru.

Distribution : Dry Evergreen Forests of Peninsular India.

Part Used : Root.

Utilization : Powdered root (10 g) applied twice a day for 10 days for acute skin diseases.

Active Constituents : These are 1αH, 5αH-tropan-3α-yl-3′, 4′, 5′-trimethoxybenzoate (I) & 1αH, 5H-tropan-3α-yl-3′,4′,5′-trimethoxycinnamate (II), bases-1αH, 5αH-tropan-3α, 6β-diol-3-(3′, 4′, 5′-trimethoxycinnamate)-6-benzoate (III) and 1αH, 5βH-tropan-3α, 6β, 7β-triol-3-(3′,4′,5′-trimethoxybenzoate) (IV).

References : Agarwal (2003), Ambasta (1986), Asolkar *et al.* (1992), Parrotta (2001), Rao & Pullaiah (2001), Rastogi & Mehrotra (1999a, b), Reddi *et al.* (2005).

Eulophia dabia (D. Don) Hochr.

Family : Orchidaceae.

Distribution : C. & N.W. Himalaya from Kumaon to Sikkim Upto 2,000 m.

Parts Used : Root.

Utilization : Known as purgative and aphrodisiac.

References : Negi *et al.* (1999), Sood & Prakash (2007).

Eupatorium ayapana (L.) Br.

Syn : *E. triplinerve* Vahl.

Family : Asteraceae.

Hindi Name:

Hindi : Ayapana.

Distribution : Ornamental Plant in Indian Gardens.

Part Used : Root.

Utilization : Its paste (25g) given orally, at an interval of 2 hrs as antidote for snake venom.

References : Ambasta (1986), Anonymous (1952), Maiti & Mishra (2000).

Euphorbia acaulis Roxb.

Syn : *E. fusiformis* Buch-Ham *ex* D. Don
Family : Euphorbiaceae.

Distribution : Tropical Himalaya, Kumaon, Nepal, Oudh, Bengal & Konkan.

Parts Used : Rootstock. Root.

Utilization : Powdered root (3-5g) given twice a day to improve digestion, and also prescribed to nursing mother having deficient secretion of milk. Paste of its rootstocks together with mustard oil applied externally in paralysis for a long period of time.

References : Chopra *et al.* (1956), Girach & Aminuddin (1995), Singh & Maheshwari (1992).

Euphorbia caducifolia Roxb.

Syn : *E. caudiculosa* Boiss.
Family : Euphorbiaceae.

Distribution : Dry Regions of India.

Part Used : Root.

Utilization : As an abortifacient.

References : Agarwal (2003), Singh & Pandey (1998), Trivedi (2002).

Euphorbia hirta L.

Syn : *Chamaesvce hirta* (L.) Millsp.; *Euphorbia pilulifera* L.
Family : Euphorbiaceae.

English, Hindi & Sanskrit Names:

Eng. : Asthma Herb, Asthma Weed, Australian Asthma Herb, Pill-Bearing Spurge;

Hindi : Bahidudhi, Dudhi, Laldudhi;

Sans. : Dugdhika, Nagarjuni, Pusitoa.

Distribution : Roadsides & in Open Fields Throughout Tropical & Sub-Tropical India.

Part Used : Root.

Utilization : Extr. used as febrifuge and emetic. Also said to be useful in case of excessive urination and snake bites.

References : Barua *et al.* (2000), Caius (2003), Kurian (1997), Parrotta (2001), Reddi *et al.* (2005).

Euphorbia nerifolia L.

Syn : *E. ligularia* Roxb.

Family : Euphorbiaceae.

English, Hindi & Sanskrit Names:

Eng. : Holy Milk Hedge;

Hindi : Pattonisend, Sehund, Thohar, Thuar;

Sans. : Gudaa, Samantdugdhaa, Sehunda, Snuhi, Sudhaa, Vajjri, Vajra.

Distribution : Wild in C. India, Orissa & S. India. Cultivated in W. Bengal.

Part Used : Root.

Utilization : Paste applied to the wounds and its extr. given orally every 3 hrs as antidote for snake bite and as antispasmodic. Decoction reportedly taken internally as an abortifacient among the 'Bhils'.

Active Constituents : Affords cycloartenol, euphorbol, ingenol triacetate, 12-deoxyphorbol-13, 20-diacetate, delphinidin-3, 5-diglucoside, 24-methylenecycloartenol, tulipanin-3, 5-diglucoside, 3, 4, 3'-tri-o-methylellagic acid-4'-rutinoside and 3, 4-di, o-methylellagic acid-4'-rutinoside.

Biological Activity : Exhibits antiseptic and emetic properties.

References : Chatterjee & Pakrashi (1997), Chopra *et al.* (1956), Kadel & Jain (2006), Khare (2004), Maiti & Mishra (2000), Parrotta (2001), Rastogi & Mehrotra (1998, 2001), Watt (1889).

Euphorbia nivula Buch-Ham.

Family: Euphorbiaceae.

English, Hindi & Sanskrit Names:

Eng.	:	Dog's Toungue;
Hindi	:	Katalhohar, Thor;
Sans.	:	Patrasnuhi, Snuhi, Vajri.

Distribution : Throughout Dry Rocky Regions in India.

Part Used : Root. Root bark.

Utilization : Root given to cure paralysis and root bark for dropsy.

Active Constituents : Root contains 24-Me-enecycloartenol, cycloartenol, ingenol-tri-OAC, euphorbol and deoxyphorbol-OAC.

References : Agarwal (2003), Ambasta (1986), Anonymous (1952), Asolkar *et al.* (1992), Chopra *et al.* (1956), Reddi *et al.* (2005).

Evolvulus alsinoids (L.) L.

Family : Convolvulaceae.

English, Hindi & Sanskrit Names:

Eng.	:	Sankapushpi;
Hindi	:	Shyama Kantha, Vishnugandhi;
Sans.	:	Vishnugandhi, Vishnukrantha.

Distribution : Common Weed Throughout India.

Part Used : Root.

Utilization : Beneficial for fever, hiccups, jaundice, whitlow, epilepsy and as aphrodisiac.

References : Agarwal (2003), Ambasta (1986), Asolkar *et al.* (1992), Reddi *et al.* (2005), Watt (1889).

Fagopyrum esculentum Moench

Family : Polygonaceae.

English & Hindi Names:

Eng.	:	Brank, Buck Wheat;
Hindi	:	Kotu, Phapra.

Distribution : Himalaya from Kashmir to Khasia Hills.

Part Used : Root.

Utilization : Decoction used against rheumatic pains, lung diseases, typhoid while juice useful in urinary disorders.

References : Agarwal (2003), Ambasta (1986), Asolkar *et al.* (1992), Chauhan (1999), Jain *et al.* (1991).

Feronia elephantum Corr.

Syn : *F. limonia* (L.) Swingle

Family : Rutaceae.

English, Hindi & Sanskrit Names:

Eng. : Elephant Apple, Wood Apple;
Hindi : Bilin, Kaithbel, Kavitha;
Sans. : Kapitha.

Distribution : Throughout Dry Plains of India.

Part Used : Root.

Utilization : Root chewed slowly after 8 hrs as antidote for snake venom.

Active Constituents : Yields geranyl umbelliferone, osthol, bergapton, isopimpinellin, xanthotoxin, marmesin, marmin, fernolin and aurapton.

References : Agarwal (2003), Maiti & Mishra (2000), Rastogi & Mehrotra (1999a, 2001, 2002).

Ferula assafoetida L.

Syn : *F. alliaceae* Boiss.; *F. foetida* Regel.

Family : Apiaceae.

English, Hindi & Sanskrit Names:

Eng. : Asafoetida;
Hindi : Hing;
Sans. : Balhika, Hingu.

Distribution : N.W. Parts of India.

Parts Used : Root. Gum -Resin (exuded from incisions in living root).

Utilization : Root antipyretic. Gum-resin known as antispasmodic, anthelmintic, aphrodisiac, carminative, diuretic, emmenagogue, expectorant, mild laxative and nervine tonic, and useful remedy for asthma, bronchitis, flatulence, colic pain, spasmodic movement of the bowels, infantile convulsions, diarrhoea, habitual abortion, indigestion, liver troubles and ringworm.

Active Constituents : Foetidin (sesquiterpenoid coumarin) isolated from root.

References : Ambasta (1986), Chatterjee & Pakrashi (1997), Rastogi & Mehrotra (2002).

Ferula jaeschkeana Vatke.

Family : Apiaceae.

Distribution : Pakistan to H.P.; 2,400-3,600 m.

Part Used : Root.

Utilization : Poultice considered useful in articular rheumatism.

Active Constituents : These are jaeskeanin, jaeskeanidin, ferutinone, ferutinianin and angeloylferutinianin.

References : Aswal & Mehrotra (1994), Ballabh & Chaurasia (2006), Rastogi & Mehrotra (1998, 2002), Sood *et al.* (2001).

Ficus benghalensis L.

Syn : *F. banyana* Oken; *F. cotonaefilia* Vahl.; *F. indica* L.; *Urostigma bengalensis* (L.) Gasp.

Family : Moraceae.

English, Hindi & Sanskrit Names:

Eng. : Banyan Tree;

Hindi : Bar, Bargad, Bargat, Bor;

Sans. : Bahupada, Nyagrodhah, Vata.

Distribution : Sub-Himalayan Hills in the North Through Peninsular India & Sri Lanka.

Parts Used : Root. Root bark.

Utilization : Decoction of crushed tips of adventitious roots cow's milk served hot in piles. Root poultice applied on syphilitic lesins and bleeding piles. Warm paste of aerial roots together with egg paste plastered with the help of *Dendrocalamus strictus* for 3 days for treating bone fracture. Juice from aerial roots administered orally to children in dysentery. Root bark efficacious in the treatment of diabetes, gonorrhoea, rheumatism, bruises, toothache, fever, arthritis, ulcers and as aphrodisiac. Tender roots useful for female sterility and leucorrrhoea. Paste of root applied on the skull as hair tonic and for menorrhagia. Prop root considerd good for abortion, while the young tips of hanging roots considered good for obstinate vomiting.

References : Anonymous (1956), Asolkar *et al.* (1992), Bhatt *et al.* (2003), Chopra *et al.* (1956), Dhiman (2003), Dwivedi (1993, 2003), Girach (1992), Jain & Singh (1997), Mohanty & Padhy (1996), Mukherjee & Namhata (1990), Parrotta (2001), Reddi *et al.* (2005), Saklani & Jain (1994), Singh & Prakash (1994), Subramaniam (2000), Tewari *et al.* (2001).

Ficus cunia Buch-Ham.

Syn : *F. conglomerata* Roxb.

Family : Moraceae.

English & Hindi Names:

Eng. : Ghwi, Kassa, Khewnau, Khurhur;

Hindi : Jahraphali, Khain, Khenan.

Distribution : Sub-Himalayan Tract, Chenab Eastwards to Bengal & Myanmar; Ascending to 1,300 m.

Part Used : Root.

Active Constituents : Juice given in bladder complaints and for visceral obstructions.

References : Anonymous (1956), Chopra *et al.* (1956), Watt (1889).

Ficus dalhousie Miq.

Family : Moraceae.

Hindi & Sanskrit Names:

Hindi : Kallal;
Sans. : Somavalkhom.

Distribution : Grows on Nilgiris.

Part Used : Root bark.

Utilization : Paste of grounded root bark with salt brushed on the aching tooth.

References : Agarwal (2003), Ambasta (1986), Hosagoudar & Henry (1996b).

Ficus fulva Reinw.

Syn : *F. apiculata* Miq.; *F. chlorocarpa* Miq.; *F. suborbicularis* Miq.

Family : Moraceae.

Distribution : Myanmar, Andaman Islands & Malay Peninsula.

Part Used : Root.

Utilization : Consumed to neutralise poison in food.

References : Hooker (1890), Jamir & Rao (1990).

Ficus hispida L.f.

Syn : *F. daemona* Koen. *ex* Vahl.; *F. daemonum* Koen.; *F. oppositifolia* Roxb.

Family : Moraceae.

English, Hindi & Sanskrit Names:

Eng. : Devil Fig;
Hindi : Daduri, Dagarin, Gobla, Kagsha, Katguilaria, Konea-Dumbar;
Sans. : Kakodumaru, Malapu, Phalgu.

Distribution : Outer Himalaya, C. & S. India & Andaman Islands.

Part Used : Root.

Utilization : So far considered useful in case of dog bite, diarrhoea, diabetes and intrinsic haemorrhage (root pounded with rice water). Aerial roots given to treat paralysis (paste applied to the affected parts twice daily for 15-21 days).

References : Beg *et al.* (2005), Khare (2004), Parrotta (2001), Reddi *et al.* (2005), Tigga *et al.* (2003).

Ficus microcarpa L.f.

Family : Moraceae.

Hindi & Sanskrit Names:

Hindi : Kamarupa;

Sans. : Plaksah.

Distribution : Common in A. P.

Part Used : Root.

Utilization : Mixture of root powder with latex of the plant and a pinch of salt taken orally for the treatment of dental caries and pyorrhoea. Also given to cure wounds.

References : Rao & Henry (1996), Rao *et al.* (1996), Reddi *et al.* (2005).

Ficus racemosa L.

Syn : *Covellia glomerata* Miq.; *F. glomerata* Roxb.; *F. goolereea* Roxb.; *F. chittagonga* Miq.; *F. mollis* Miq. (non Vahl.); *F. semicostata* Bailey; *F. vesca* Muell. *ex* Miq.

Family : Moraceae.

English, Hindi & Sanskrit Names:

Eng. : Cluster Fig, Gular Fig;

Hindi : Dimere, Gular, Lelka, Paroa, Tue, Umrai;

Sans. : Hemdugdhak, Jantuphala, Udumbarah, Yagyaanga.

Distribution : Throughout India.

Part Used : Root.

Utilization : Useful for pain during urination, hydrophobia, dysentery, diabetes (root paste) and abortion.

References : Agarwal (2003), Beg *et al.* (2005), Chopra *et al.* (1956), Dhiman (2003), Hosagoudar & Henry

(1996b), Jha & Kumar (2003), Khare (2004), Kirtikar & Basu (1984), Kothari & Rao (2000), Kumar & Lal (2003), Parrotta (2001), Singh *et al.* (2003), Watt (1889).

Ficus religiosa L.

Syn : *Urostigma religiosum* (L.) Gasparrini
Family : Moraceae.

English, Hindi & Sanskrit Names:

Eng. : Bot Tree, Peepal Tree, Sacred Bo Tree, Sacred Fig;
Hindi : Pipal;
Sans. : Asvatthah, Bodhidru, Bodhivriksha, Gajabhaksha, Kashiradruma, Peeppal, Pippala.

Distribution : Sub-Himalayan Tracts. W. Bengal, C. & S. India, Planted Throughout India.

Parts Used : Root. Root bark.

Utilization : Root used for gout, malaria and as an antidote to snake poison while root bark used to cure stomatitis, vomiting, ulcers, leucorrhoea, lumbago and as aphrodisiac and astringent. Also its extr. with butter milk administered for paralysis (2 spoonfuls twice a day for 30 days).

Active Constituent : Phytosterolins.

Biological Activity : Phytosterolin from bark exhibited powerful stimulant and hypoglycaemic activities.

References : Asolkar *et al.* (1992), Khare (2004), Kirtikar & Basu (1984), Masih (2003), Parrotta (2001), Rao & Henry (1996), Reddi *et al.* (2005).

Ficus tinctoria Forst.f.

Syn : *Ficus gibbosa* Bl.
Family : Moraceae.

Hindi & Sanskrit Names:

Hindi : Chanchri;
Sans. : Udumber.

Distribution : Practically Throughout India.

Parts Used : Root. Root bark.

Utilization : Root aperient while its bark considered aperiant and stomachic.

References : Agarwal (2003), Ambasta (1986), Kirtikar & Basu (1984), Subramaniam (2000), Watt (1889).

Flacourtia indica (Burm.f.) Merr.

Syn : *F. ramontchii* L'Herit.; *F. ramontchii* L'Herit var. *sapida* (Roxb.) Hook.f. & Thoms.; *F. ramontchii* L' Herit var. *latifolia* Hook.f. & Thoms. var. *occidentalis* Hook.f.& Thoms.; *F. sepiaria* Roxb.

Family : Flacourtiaceae.

English, Hindi & Sanskrit Names:

Eng. : Madgascar Plum;
Hindi : Baichi;
Sans. : Svadukantaka.

Distribution : Sub-Himalayan Tract, Upper Gangetic Plains & Peninsula.

Part Used : Root.

Utilization : Known so far as an expectorant, and also for healing body pains, bronchial allergy, blood pressure, skin affections, dysentery, stomach heat, excessive body heat and snake bite.

References : Jain *et al.* (1991), Rastogi & Mehrotra (2002), Reddi *et al.* (2005), Varghese (1996).

Flemingia bracteata (Roxb.) Wt.

Syn: *F. strobilifera* R.Br. var. *bracteata* Baker; *Mogania bracteata* Hui-Linn.

Family: Fabaceae.

Distribution : Hilly Areas of A. P.

Part Used : Root.

Utilization : These are kept under the bed of pregnant woman for easy delivery. Juice administered orally for rheumatism and body pain. Also reported in India for epilepsy, hysteria, stomach pain, headache, rheumatism, body pain and inducing sleep.

References : Ambasta (1986), Jain & Patole (2001), Rao & Henry (1996), Upadhyay (2003), Varghese (1996).

Flemingia chappar Ham. *ex* Benth.

Syn : *Moghania chappar* Buch-Ham.
Family : Fabaceae.

Hindi Name:

Hindi : Salpan.

Distribution : Grows Throughout S. India.

Part Used : Root.

Utilization : Considered useful in hysteria, epilepsy, inducing sleep, pain and body swelling (root paste). Also about 20 ml of its decoction given with cow milk at bed time in spermatorrhoea and body swelling (root juice, 2 times a day for 3 days). Root pounded together with 5g *Smilax* sp. (Atkir) and given twice a day to treat spermatorrhoea.

Active Constituents : Flemichapparin C, 7-O-methylmedicagol, flemichapparin-β, β-sitosterol and 7-OH-flavanone characterised in roots.

References : Ambasta (1986), Asolkar *et al.* (1992), Girach & Aminuddin (1995), Rastogi & Mehrotra (1999a, b), Shukla & Verma (1996), Singh & Prakash (1994), Varghese (1996).

Flemingia macrophylla Blume *ex* Miq.

Syn : *Crotalaria macrophylla* Willd.; *F. congesta* Roxb. *ex* Aiton; *F. stricta* Ait.; *Moghania macrophylla* Kuntze.; *Rhynchosia crotalaria* DC.
Family : Fabaceae.

Hindi Names:

Hindi : Bara-Salpan, Bhalia, Kusunt, Supta;

Distribution : Very Common Throughout the Hotter Parts of India.

Part Used : Root.

Utilization : Paste recommended for bodyache, muscular pain, blindness, cholera, dysentery, fever, fistula, spleen complications, ulcers, swellings mainly of the neck and as aphrodisiac.

Active Constituents : Main constituents are flemichin A, flemichins B, C, D, narigenin, genistein, 5, 7, 2', 4'-tetra-hydroxyisoflanone, genistein and β-sitosterol glucoside.

References : Agarwal (2003), Ambasta (1986), Anonymous (1956), Chopra *et al.* (1956), Kumar (2002), Rajendran *et al.* (1997), Rastogi & Mehrotra (1999a, b, 2001), Saklani & Jain (1994), Sood & Thakur (2004).

Flemingia strobilifera L. Ait.

Syn : *Hedysarum bracteatum* Roxb.; *H. strobilifera* L.; *Moghania strobilifera* J.St. Hilaire.

Family : Fabaceae.

Hindi Names:

Hindi : Kanphuti, Kusrunt.

Distribution : Distributed Throughout India.

Part Used : Root.

Utilization : Powdered root (5g) given twice a day to treat constipation. Its decoction (one teaspoonful every night) given as cooling tonic in case of diarrhoea and blood dysentery. One teaspoonful of root paste given twice daily for 3-5 days for the treatment of epilepsy, hysteria, gastric disorder, fever, pain in stomach and inducing sleep and its paste applied locally in scabies.

Active Constituents : Found to contain chalcone, N-triacontane and β-sitosterol.

References : Agarwal (2003), Asolkar *et al.* (1992), Aminuddin & Girach (1996), Anonymous (1956), Chopra *et al.* (1956), Dutta & Nath (2000), Girach & Aminuddin

(1995), Goel & Rajendran (2000), Kirtikar & Basu (1935), Kumar & Lal (2003), Parrotta (2001), Reddi *et al.* (2005), Saren *et al.* (2000).

Flemingia wightiana Grah.

Syn : *F. ferruginea* Wall.

Family : Fabaceae.

Distribution : Nilghiris.

Part Used : Root.

Utilization : One teaspoonful of its paste taken with a glass of milk in early morning for one month to cure weakness.

References : Hooker (1879), Tirkey (2004).

Foeniculum vulgare Gaertn.

Syn : *F. cappilaceum* Gilib; *F. officinale* Ait.

Family : Apiaceae.

English, Hindi & Sanskrit Names:

Eng. : Fennel;
Hindi : Bari Saunf, Saunf;
Sans. : Madhurica, Misreya.

Distribution : Cultivated Throughout India Ascending Upto 2,000m; Often Wild.

Part Used : Root.

Utilization : Known as diuretic and purgative.

Active Constituents : Contains β-myrcene, myristicin, petersilienapiol, α- & β-terpinene, umbelliferone, ascorbic acid, niacin, riboflavin, α-β-& γ-tocopherol, α-& β-tocotrienol, choline, trigonelline, p-cymene and bergapten.

References : Ambasta (1986), Asolkar *et al.* (1992), Chatterjee & Pakrashi (1997), Chopra *et al.* (1956), Dey (1973).

Fumaria indica Pugs.

Syn : *F. parviflora* auct. non Lamk.
Family : Fumariaceae.

Hindi & Sanskrit Names:

Hindi : Pit Papra, Shahterah, Turu;
Sans. : Araka, Kalapanga, Nakra.

Distribution : Greater Parts of India & Himalaya Upto 2,500 m.

Part Used : Root.

Utilization : 5g powder given orally twice a day in fever.

Active Constituents : Protopine, tetrahydrocoptisine, â-sitosterol and octacosanol characterised in roots.

References : Agarwal (2003), Ambasta (1986), Rastogi & Mehrotra (2001), Shukla *et al.* (2003).

Gardenia florida Linn.

Syn : *Gardenia angusta* (L.) Merr.;
G. jasminoides Ellis; *Verneria angusta* L.
Family : Rubiaceae.

English, Hindi & Sanskrit Names:

Eng. : Cape Jasmine;
Hindi : Gandharaj;
Sans. : Gandharaja.

Distribution : Cultivated in Indian Gardens in Tropical Parts.

Part Used : Root.

Utilization : Prescribed in dyspepsia and nervous disorders. In Konkan, its paste applied on the top of the head as a remedy for headache during pregnancy; and also given internally for hysteria either alone or in combination.

References : Ambasta (1986), Chatterjee & Pakrashi (1997), Chopra *et al.* (1956), Watt (1889).

Gardenia turgida Roxb.

Syn : *Ceriscoids turgida* (Roxb.)
Tirvengadum; *Gardenia montana* Roxb.
Family : Rubiaceae.

Hindi & Sanskrit Names:

Hindi : Ghurgia, Karamba, Karhar, Khuriari, Thanella;

Sans. : Bharangi, Mahapindi.

Distribution : Dry Deciduous Forests of the Lower Foot Hills of the Himalaya (Garhwal to Bhutan) & Throughout Peninsular India from Bihar & M. P. Southwards to Tamil Nadu.

Part Used : Root.

Utilization : Reported to be known as abortifacient, emmenagogue, and for spermatorrhoea, abdominal pain (4-5 teaspoonfuls freshly prepared root extr.), headache (crushed roots with water and applied on top of the head), dyspepsia and indigestion (root infusion).

Active Constituents : Active principles are due to D mannitol, hexabenzoate, gardnin A, B, E, oleanolic acid, α-amyrin and β-sitosterol.

References : Agarwal (2003), Anonymous (1956), Bhattacharjee (2001), Chopra *et al.* (1956), Dhiman (2003), Jain & Singh (1997), Kumar & Goel (1998), Maheshwari *et al.* (1997), Parrotta (2001), Rastogi & Mehrotra (1999a, b), Reddi *et al.* (2005), Singh (1996), Singh & Prakash (1994), Trivedi (2002), Watt (1889).

Gentiana carinata Griseb.

Family: Gentianaceae.

Distribution : Kashmir & N.W. Himalaya: 2,700-4,000 m.

Part Used : Root.

Utilization : Useful in the treatment of stomach complaints.

References : Chaurasia *et al.* (2000b), Hooker (1885).

Gentiana kurroo Royle

Syn: *Pneumonanthe kurroo* D. Don
Family: Gentianaceae.

English, Hindi & Sanskrit Names:

Eng. : Indian Gentian;
Hindi : Kamalphul, Karu, Kutki, Nilkanth, Trayaman;
Sans. : Anujaa, Balbhra, Girijaa, Traayanti, Trayamana.

Distribution : N.W. Himalaya (1,500-3,500 m), Garhwal, Kashmir & Adjoining Hills.

Part Used : Root.

Utilization : Known as a bitter tonic, stomachic, febrifuge, anthelmintic, blood purifier, carminative, diaphoretic, digestive; and also in hepatic disorders, urinary affections and abdomonal pains. Its extr. beneficial in spleenic disorders, skin diseases and haemorrhage (mixed with honey and sugar).

Active Constituents : Gentiopicrin, gentianic acid, apectin, catapol, 6′-cinnamoyl, 6-O-cinnamoyl, 6-O-vanilloyl, 6-O-feruloyl derivatives and aucubin are the chief ingredients.

References : Bhattacharjee (2001), Chopra *et al.* (1956), Jain (1968), Kaul (1997), Khare (2004), Rastogi & Mehrotra (1998), Reddi *et al.* (2005), Sharma (2003), Watt (1889).

Geodorum densiflorum (Lam.) Schltr.

Syn : *G. dilatatum* R.Br.; *G. purpureum* R.Br.
Family : Orchidaceae.

Distribution : Grows in Low Altitudes in N. E. & S. India.

Parts Used : Rootstock. Root.

Utilization : Juice and paste of rootstock applied on rabbit and dog bites. Roots given to goats in diarrhoea, and its extr. applied to keep away flies in cattle.

References : Agarwal (2003), Ambasta (1986), Brahmam & Saxena (1990).

Geranium nepalense Sw.

Family : Geraniaceae.

English & Hindi Names:

Eng. : Nepal Geranium, Nepalese Cranesbill;
Hindi : Bhanda.

Distribution : Kashmir to Sikkim, Nilgiris, Pulney Hills & Manipur.

Part Used : Root.

Utilization : Prescribed in the treatment of diseases related to lungs, nerves, veins, arteries, liver disorders, swellings in limbs, toothache, ulcer, wounds and stomach complaints.

Active Constituents : Contains red colouring matter, tannins, gallic acid, quercetin and citric acid.

References : Agarwal (2003), Ambasta (1986), Jain & Goel (2005), Sharma (2003).

Geranium pratense Linn.

Family : Geraniaceae.

English & Hindi Names:

Eng. : Crow Foot, Grace of God, Greatia Dei, Meadow Crane's Bill;
Hindi : Banda.

Distribution : W. Temperate Himalaya (3,000-4,500 m).

Part Used : Root.

Utilization : Popularly applied as a poultice to bruises and for stomach trouble.

Active Constituents : Therapeutic properties are due to iodotannin, isokempferid, geranin and hexahydroflavone.

References : Aswal & Mehrotra (1994), Chopra *et al.* (1956), Sood *et al.* (2001).

Geranium walichianum D. Don

Family : Geraniaceae.

English & Hindi Names:

Eng. : Crane's Bill, Wallich Cranesbill;
Hindi : Laljhari.

Distribution : Himalaya & Nepal: 2,000-4,000 m.

Part Used : Root.

Utilization : Decoction taken in cold and cough, rheumatism (herbal tea prepared from fresh or dried roots), fever, bodyache, dysentery, abdominal pain. Root juice antiseptic, and taken in gonorrhoea, phthalmia, and toothache and its paste given in the stomach disorders of infants and also applied on the cut bruises. Root juice given in gonorrhoea and ophthalmia. Mustard oil with dried root soaked overnight massaged at the site of bone fracture of livestock.

Active Constituents : Possesses tannins, 25-32 % and non-tannins, 18 %.

References : Agarwal (2003), Ambasta (1986), Anonymous (1956), Bhattacharjee (2001), Chhetri (2005), Kumar & Rao (2001), Negi *et al.* (1993), Sharma (2003), Singh (1996), Sood & Prakash (2007), Tiwari & Pande (2006).

Geum elatum Wall.

Family : Rosaceae.

English & Hindi Names:

Eng. : Avens;
Hindi : Gogjimul.

Distribution : Alpine Himalaya from Kashmir to Sikkim Upto 4,500 m.

Part Used : Root.

Utilization : Extr. used in dysentery.

References : Agarwal (2003), Kaul (1997).

Gillenia trifolia (L.)

Family : Rosaceae.

English Name:

Eng. : Moench Indian Psysic.

Distribution : Native to N. America.

Parts Used : Root. Root bark.

Utilization : Root bark reported so far to treat intestinal worm, rheumatism, dyspepsia, chronic diarrhoea and fever. Dried roots considered as laxative.

Reference : Bhattacharjee (2001)

Girardinia diversifolia (Link.)Friis.

Syn : *G. heterophylla* Decne.; *G. heterophylla* var. *palmata* Hook.f.; *G. heterophylla* var. *zeylanica* Hook.f.; *Urtica diversifolia* Link.; *U. heterophylla* Vahl.; *U. palmata* Forssk.

Family : Urticaceae.

English, Hindi & Sanskrit Names:

Eng. : Nilghiri Nettle;
Hindi : Alla, Awa, Bichua, Chichr;
Sans. : Laghukacchu, Lomasa, Romalu, Santativardhini, Vatavidhvamsini.

Distribution : Himalaya, W. Ghats, W., C. & S. India.

Part Used : Root.

Utilization : Paste applied on sprains, suppuration of boils, hydrocele, oedema and bone fracture.

References : Agarwal (2003), Maity *et al.* (2003), Parrotta (2001), Sharma & Rana (2000), Singh & Kumar (2000b).

Girardinia palmata (Forssk.) Goud.

Syn : *G. leschenaultiana* Decne.; *Urtica palmata* Leschen.

Family : Urticaceae.

Distribution : Nilghiri and Sri Lanka : 1,600-2,000 m.

Part Used : Root.

Utilization : Mixture of its roots crushed with that of *Amaranthus viridis* L. given for the treatment of urinary disorders, and kidney troubles.

References : Ambasta (1986), Chhetri (2005), Hooker (1890), Negi *et al.* (1999).

Globa marantina L.

Family : Zingiberaceae.

Distribution : Khasia Hills, Meghalaya.

Part Used : Root.

Utilization : One teaspoonful (2g) of the paste made out of its root (10g), 7 black peppers, 'Pipli' (*Piper longum*) and 1g, 'Ajowan' (*Trachyspermum ammi*) given once daily against asthma, cough, rheumatism, bile troubles and reeling.

References : Asolkar *et al.* (1992), Brahmam & Saxena (1990).

Glochidion oblatum Hk.f.

Family : Euphorbiaceae.

Distribution : Sikkim Himalaya.

Part Used : Root.

Utilization : Extr. taken for dysentery.

References : Hooker (1890), Mao (1993).

Gloriosa superba L.

Family : Liliaceae.

English, Hindi & Sanskrit Names:

Eng. : Climbing Lily, Malabar Glory Lilly, Superb Lily, Tiger's Claw;
Hindi : Kalihari, Karkari, Langali, Ravigani;
Sans. : Agnimukhi, Agnishikhaa, Ailni, Anantaa, Garbhanut, Halini, Indrapushpi, Langaliki, Sirikramaa, Vahni-Amukhi.

Distribution : Throughout the Plains & Low Hills India Upto 2,350 m.

Part Used : Root.

Utilization : Considered useful in piles, ulcers, neuralgic pain, scabies, colic pain, periodic fever, pirhi disease, leucoderma gonorrhoea (starch obtained after detoxification by repeated washing of the paste of root), gout, insect bite, intermittent fever, tumour, bowel complaints, rheumatic fevers, and as laxative, purgative, antidiabetic, anthelmintic, cooling agent, and antibiotic. For expulsion of the placenta after delivery, root paste applied on palms and soles and also given internally with *Nigella sativa*, long pepper (*Piper longum*) and alcohol. Root powder rubbed on infected part of the body to cure leprosy, snake bite and scorpion sting. 5g mixture of fresh tuberous root with jaggery and hot water given orally early in the morning to induce abortion whereas oil extr. applied externally in obstinate skin diseases. To facilitate delivery, about 5 cm long piece of root is tied to the waist of woman and removed immediately after delivery.

Active Constituents : Possesses colchicines; 0.23-0.3 %.

References : Agarwal (2003), Anonymous (1956), Asolkar *et al.* (1992), Bhatt *et al.* (2003), Chopra *et al.* (1956), Dey (1973), Gogoi & Das (2003), Jain & Patole (2001), Jain & Singh (1997), Jain *et al.* (2003), Khare (2004), Kirtikar & Basu (1984), Kothari & Rao (2000), Kumar & Lal (2003), Kumar & Pullaiah (1998), Parrotta (2001), Reddi *et al.* (2005), Samant *et al.* (2001), Sharma (2003), Thakur *et al.* (1989), Varghese (1996), Watt (1889).

Glossogyne bidens (Retz.) Alston

Syn : *Bidens rigida* Hort.Calc.; *B. pinnatifolia* Heyne; *Glossogyne pinnatifolia* DC.; *Zinnia bidens* Retz.

Family : Asteraceae.

Distribution : Plains of India from Jammu & Garhwal to W. Bengal, Bihar & Southwards to Madras.

Part Used : Root.

Utilization : Useful in toothache, snake bite and scorpion sting. Also mixture comprising its root and that of *Peucedodanum dhana* Ham. (Tejraj) and edible oil applied as massage on affected parts to cure rheumatic joint pain.

References : Ambasta (1986), Aminuddin & Girach (1993), Anonymous (1956), Varghese (1996), Watt (1889).

Glycosmis arborea (Roxb.) DC.

Syn : *G. chylocarpa* W & A.; *G. pentaphylla* auct. Mult non DC.; *G. retzii* Roem.; *Limonia arborea* Roxb.; *L. pentaphylla* Retz.; *Myxospermum chylocarpum* Roem.

Family : Rutaceae.

Hindi & Sanskrit Names:

Hindi : Ban-Nimbu, Girgiti;

Sans. : Asyashakhota, Kupiluh, Vadadru.

Distribution : Wasteplaces, Roadsides Throughout Tropical & Sub-Tropical Parts of India.

Part Used : Root.

Utilization : Efficacious for facial inflammations, rheumatism, jaundice, anaemia, stomach pain (root juice) and slow fever (root decoction).

Active Constituents : Pharmacologically active constituents are dictamine, g-fagarine, skimmianine, b-sitosterol, glycosolone, glycozolidal, glycozolinol, glycozolinine, de-N-methylacromycine, de-N-methyl-noracronycine and nor-acronycine.

References : Agarwal (2003), Anonymous (1968), Asolkar *et al.* (1992), Balasubramanian & Prasad (1996), Binu *et al.* (2003), Chatterjee & Pakrashi (1997), Rastogi & Mehrotra (1999a), Reddi *et al.* (2005), Watt (1889).

Glycosmis mauritiana (Lam.) Tanaka

Syn : *G. pentaphylla* DC.

Family : Rutaceae.

Distribution : Throughout the Greater Part of India.

Part Used : Root.

Utilization : Pounded roots with black pepper applied on aching tooth. Its paste good for suppuration of boils, swellings of toes and in low grade fever.

References : Asolkar *et al.* (1992), Chopra *et al.* (1956), Hosagoudar & Henry (1996b), Kshirsagar & Singh (2000a), Rastogi & Mehrotra (1998).

Glycyrrhiza glabra Linn. (Pl. 6B)

Family : Fabaceae.

English, Hindi & Sanskrit Names:

Eng. : Liquorice, Sweet Wood;
Hindi : Jethimadh, Maithilakdi, Mulhatti;
Sans. : Jalayashti, Klitaka, Madhuka, Madhuyashti, Sthalayashti, Yashti-Madhu.

Distribution : Cultivated in J. & K., Punjab, Sub-Himalayan Tracts.

Part Used : Dried Root.

Utilization : Useful in India as tonic, expectorant, demulcent, diuretic emmenagogue, spasmolytic, anti-inflammatory, adrenal agent, mild laxative, and also used to treat bronchial catarrh, bronchitis, chronic gastritis, skin diseases, nervous weakness, cough, hoarseness, peptic ulcer, colic, genitourinary disorders, sore throat, asthma, headache, eye troubles, rheumatoid arthritis, dermatitis and in scorpion sting. Its powder mixed with citrus juice efficacious in catarrhal, and with honey in jaundice. In combination with ginger and milk, acts as a good tonic during convalescence.

Active Constituents : Possesses asparagines, fat, resins, mannitol, glycyrrhizin caumarin, liquiritin (flavanone), isoliquiritin (chalkone), rhamnoglucoside, 4', 7-dihydroxyxylavanone(liquiritigenin), 7-hydroxy-4'-methoxyisoflavone (formonctin), liquorice acid, glycyrrhizic acid, glabrol, 7-acetoxy-2-methylisoflavone (I), 7-methoxy-2-methylisoflavone

(II), 7-hydroxy-2-methylisoflavone (III), liquiritic, 18-α-glycyrrhetic, 24-hydroxyglycyrrhetic, 11-deoxyglycyrrhetic, 21, 24-dihydroxy-11-deoxyglycyrrhetic acids, glabranins A & B. glabrolide, deoxoglabrolide, isoglabrolide, glycyrrhetol, 18 α-hydroxyglycyrrhetic acid, liquiritic acid and olean-12-en-3β-ol-30-oic acid.

References : Agarwal (2003), Anonymous (1968), Asolkar *et al.* (1992), Bhattacharjee (2001), Chatterjee & Pakrashi (1997), Chopra *et al.* (1956), Dastur (1970), Dey (1973), Greenish (1999), Jain (1968), Karnick (1994), Kirtikar & Basu (1935), Rastogi & Mehrotra (1999a, 2001, 2002), Said (1997), Sarin (1996), Watt (1889).

Gmelina arborea Roxb.

Syn : *G. indica* Burm.; *G. rheedii* Hook.; *G. tomentosa* Wall.; *Premna arborea* Roth.; *P. tomentosa* Miq.

Family : Verbenaceae.

English, Hindi & Sanskrit Names:

Eng. : Candahar Tree, White Teak;

Hindi : Gamari, Gambhar, Gamhar, Kamhar, Khamer, Shiwan;

Sans. : Bhadra, Bhadraparni, Gamhari, Hiraa, Kaasmari, Madhuparnikaa, Modini, Pitarohini, Sadaabhadraa, Vatah, Vidari.

Distribution : Throughout India Upto 1,700 m on the Hills & Andaman Islands.

Parts Used : Root. Root bark.

Utilization : So far known as an acrid, bitter, sweet, heating, indigestible, galactagogue, stomachic, laxative, tonic, anthelmintic, appetizer, antidote for snake bite and insect stings and useful in hallucinations, thirst, piles, abdominal pains, diarrhoea, burning sensations, gout, fevers, 'tridosha', urinary discharges, malaria, blood purification and anasarca. Root bark used as tonic in Orissa and also said to improve thirst and relieves abdominal pain.

Active Constituents : Known to contain apiosylskimmin, a coumarin characterised as umbelliferone-7-apiosyl (1→6) glucoside, gmelofuran, â-sitosterol, ceryl alcohol, hentriacontanol-1, gmelinol and octacosanol.

References : Agarwal (2003), Anonymous (1956, 1968), Asolkar *et al.* (1992), Chatterjee & Pakrashi (1997), Chhetri (1994), Dey (1973), Dhiman (2003), Dwivedi (1993), Khare (2004), Kirtikar & Basu (1984), Kumar & Lal (2003), Mehrotra *et al.* (1994), Parrotta (2001), Rastogi & Mehrotra (1999a, b), Reddi *et al.* (2005), Sarin (1996), Singh & Pandey (1998), Varghese (1996), Watt (1889).

Gmelina asiatica L.

Syn: *G. coromandelica* Burm.f.; *G. inermis* W. & A.; *G. parviflora* Roxb.; *Michelia spinosa* Amman; *Premna parviflora* Roth.

Family : Verbenaceae.

Hindi & Sanskrit Names:

Hindi : Badhara, Nag-Phul;

Sans. : Biddari, Gopabhadra, Kasmari, Vikarini.

Distribution : Peninsular India (Scrub Forests, Mainly in E. & S. Regions), Planted Elsewhere in India in Gardens, Particularly as a Hedge Plant.

Part Used : Root.

Utilization : Its paste with cloves used twice a day for 7 days to allay toothache. Also considered as an aromatic, alterative, astringent, demulcent (to treat catarrah of the bladder), blood purifier and for gonorrhoea, rheumatism, and yaws.

References : Agarwal (2003), Ambasta (1986), Anonymous (1956), Chopra *et al.* (1956), Drury (1978), Parrotta (2001), Rao *et al.* (1996), Watt (1889).

Gnaphalium stewartii Clarke

Family: Asteraceae.

Distribution : W. Himalaya, Kashmir Upto 3,300 m; Karakoram, 4,300 m.

Part Used : Root.

Utilization : Used as antipyretic.

References : Hooker (1887), Kaul (1997).

Gossypium arboreum L.

Syn: *G. cernuum* Tod.; *G. indicum* Tod.; *G. intermedium* Tod.; *G. nanking* Meyen; *G. neglectum* Tod.; *G. obtusifolium* Roxb. in part.; *G. sanguineum* Hassk.

Family: Malvaceae.

English, Hindi & Sanskrit Names:

Eng. : Cotton, Tree Cotton;

Hindi : Deokapas, Kapas, Nurma;

Sans. : Karpaasee, Karpasamu.

Distribution : All Over India.

Parts Used : Root. Root bark.

Utilization : Root bark used for amenorrhoea, dysmenorrhoea while root known to cure fever.

References : Agarwal (2003), Anonymous (1968), Caius (2003), Chopra *et al.* (1956), Dwivedi (1993), Kirtikar & Basu (1935), Sikarwar (2002).

Gossypium herbaceum L.

Syn: *G. obtusifoium* Roxb. (in part); *G. wightiana* Tod.

Family: Malvaceae.

English, Hindi & Sanskrit Names:

Eng. : Asiatic Cotton, Common Cotton, Indian Cotton, Levent Cotton;

Hindi : Deokapas, Kupasnarma, Rui-Kapas;

Sans. : Anagnika, Badara, Chavya, Kalakanta, Karpasasarini, Maghani, Patada, Patahena, Pichu, Samudranta, Sutrapuspa, Tula, Tundikeri.

Distribution : Cultivated in Punjab, Maharashtra, Assam, Karnatka, A.P. & Tamil Nadu.

Parts Used : Root. Root bark.

Utilization : Root bark prescribed as a substitute for ergot as an abortifacient, emmenagogue and in uterine disorders. Decoction of root prescribed in strangury, urinary poblems, abortion, dysuria, fever and gravel. Root with rice water said to good for leucorrhoea.

Active Constituents : Cotton root bark contains gossypol (a sesquiterpene).

Biological Activity : Root bark stimulates uterine contractions and hastens a difficult labour. Also it encourages the blood to clot. Gossypol from root considered safe antifertility drug for males.

References : Agarwal (2003), Asolkar *et al.* (1992), Chatterjee & Pakrashi (1997), Caius (2003), Chopra *et al.* (1956), Dastur (1970), Dhiman (2003), Jain *et al.* (1995), Khare (2004), Kurian (1995), Reddi *et al.* (2005).

Grangea maderaspatana Poir.

Syn : *Artemisia maderaspatana* Roxb.; *Cotula maderaspatana* Willd.; *C. sphaeranthus* Link.; *Grangea sphaeranthus* Koch.

Family : Asteraceae.

English, Hindi & Sanskrit Names:

Eng. : Madras Warm Weed;
Hindi : Mukhatari, Mustaru;
Sans. : Mustaru.

Distribution : Throughout India.

Part Used : Root.

Utilization : So far known as galactagogue and appetiser.

References : Agarwal (2003), Reddi *et al.* (2005), Watt (1889).

Grewia asiatica Linn.

Syn : *G. asiatica* var. *tiliaefolia* Brandis; *G. tiliaefolia* Vahl.; *G. subinaequalis* DC.

Family : Tiliaceae.

English, Hindi & Sanskrit Names:

Eng. : Asiatic Grewia;
Hindi : Dhamin, Karra, Parusha, Phalsa, Shukri;

Sans. : Alpasthi, Dhanuivriksha, Dharmana, Mriduphala, Nilacharma, Nilamandala, Paravata, Roshana.

Distribution : Deciduous Forests in the Sub-Himalayan Region & Peninsular India.

Part Used : Root bark.

Utilization : Good for rheumatism and to relieve cow-itch. Grounded with water applied as plaster on bone fracture for relief and also given with whey or curd and sugar to the children in cases of acute dysentery.

References : Agarwal (2003), Caius (2003), Chopra *et al.* (1956), Khare (2004), Parrotta (2001), Singh (2006), Watt (1889).

Grewia aspera Roxb.

Syn : *G. asiatica* Wall.; G. *obtusifolia* Juss.; *G. macrophyll* G. Don;
Family : Tiliaceae.

Distribution : W. Peninsula, from the Circas & Kurg to Malabar.

Part Used : Root.

Utilization : 10-15 ml of extr. given twice a day for 15-20 days to cure stomachache.

References : Hooker (1872), Sharma & Singh (1999).

Grewia hirsuta Vahl.

Syn : *G. polygama* Mast.
Family : Tiliaceae.

Hindi & Sanskrit Names:

Hindi : Gursukri, Kakarundah, Kukurbicha, Sri-Gur-Sukri;
Sans. : Gudasarkara.

Distribution : Sub-Himalayan Tract Upto 1,400 m & Peninsular India from Bihar Southwards.

Part Used : Root.

Utilization : One teaspoonful root powder given with 100g fresh cow-milk thrice a day for two or three weeks in acute leucorrhoea. 5g paste chewed or applied on bitten area in snake bite. Pounded root also prescribed for diarrhoea, hastening suppuration, dressing for wounds, cholera, hydrophobia, kidney pain, piles, gonorrhoea, chest pain, venereal diseases, inflammation of testicles, blisters, cuts, rheumatic pains, amoebic dysentery (15 g in 200 ml water twice daily for 7 days), and as astringent and anthelmintic.

References : Agarwal (2003), Asolkar *et al.*(1992), Caius (2003), Chopra *et al.* (1956), Hosagoudar & Henry (1996b), Jain (2003), Kirtikar & Basu (1935), Maliya (2004), Parrotta (2001), Rao & Henry (1996), Varghese (1996), Venkataratnam & Raju (2004).

Grewia rothii DC.

Family : Tiliaceae.

Distribution : E. Himalaya.

Parts Used : Root. Root bark.

Utilization : Root paste (50g) applied (thrice a day) for gonorrhoea until cure, and the extr. of its bark with black pepper (*Piper nigrum*) prescribed for venereal diseases (2-3 spoonful twice a day till cure).

Active Constituents : Root bark contains lupeol and leucoanthocynidins.

References : Asolkar *et al.* (1992), Rao & Henry (1996), Rao & Pullaiah (2001).

Grewia sapida Roxb.

Family : Tiliaceae.

Distribution : Tropical Himalaya from Garhwal to Assam & in Kheri, U.P.

Part Used : Root.

Utilization : Paste applied on boils.

Biological Activity : Alc. extr. spermicidal, CNS depressant and anticancer.

References : Aminuddin & Girach (1993), Asolkar *et al.* (1992).

Grewia sclerophylla Roxb. Hort. Beng.

Syn : *G. carrea* Ham.; *G. obliqua* Roxb. not of Juss.; *G. pilosa* Wall. (in part); *G. scabrophylla* Roxb.; *G. sulcata* Wall.

Family : Tiliaceae.

Hindi Name:

Hindi : Pharsia.

Distribution : Sub-Himalayan Tract & Outer Hills of Kumaon Upto 1,200 m, Sikkim, Assam, and Chittagong.

Part Used : Root.

Utilization : Prescribed in cough, irritable conditions of the intestine and bladder and its decoction used as an emollient enema.

References : Ambasta (1986), Anonymous (1956), Chopra *et al.* (1956), Kirtikar & Basu (1935), Watt (1889).

Grewia serrulata DC.

Syn : *G. disperma* Rottl.; *G. multiflora* Mast in part.

Family : Tiliaceae.

Distribution : N.E. India & W. Ghats.

Part Used : Root.

Utilization : 30 ml of extr. taken twice a day for 15-20 days in case of bone fracture.

References : Ambasta (1986), Asolkar *et al.* (1992), Sharma & Singh (1999).

Grewia tenax (Forssk.) Fiori

Syn : *G. populifolia* Vahl.

Family : Tiliaceae.

Hindi Name:

Hindi : Gango.

Distribution : N., W. & S. India.

Part Used : Root.

Utilization : Infusion of dry roots soaked in water for 2 hrs given orally in snake bite. Saharia tribals eat young roots to cure diarrhoea.

Active Constituents : Contains sterol, terpenoids and alkaloids.

References : Agarwal (2003), Ambasta (1986), Asolkar *et al.* (1992), Ram *et al.* (2002), Reddy *et al.*(1996), Singh & Pandey (1998).

Gymnema sylvestre (Retz.) R.Br. & Schult.

Syn : *Asclepias geminata* Roxb.; *Periploca sylvestris* Willd.

Family : Asclepiadaceae.

English, Hindi & Sanskrit Names:

Eng. : Periploca of the Woods;

Hindi : Chhota-Duddhilata, Gur-Mar, Merasingi, Meshasingi;

Sans. : Madhunashini, Meshasringi.

Distribution : C. & S. India, on the W. Ghats & in Goa.

Part Used : Root.

Utilization : Considered astringent, stomachic, galactagogue, tonic, refrigerant, emetic, expectorant, antidiabetic (decoction), and also prescribed to cure eye diseases, snake bite and scorpion sting (root extr.). 1teaspoonful of root and stem powder given in piles, bone fever and cataract.

Active Constituents : Gymnemic acid.

References : Agarwal (2003), Anonymous (1968), Chatterjee & Pakrashi (1997), Chopra *et al.* (1956), Dwivedi (1993), Jain & Patole (2001), Rao & Henry (1996), Reddi *et al.* (2005), Watt (1889).

Gymnosporia spinosa

Family : Celastraceae.

Distribution : Drier Parts of India, Punjab, Rajasthan, Gujarat, Maharashtra, M. P., Orissa, Bihar & A.P.

Part Used : Root.

Utilization : Good for gastro-intestinal troubles especially dysentery.

Reference : Dwivedi (1993).

Gynandropsis gynandra Merrill.

Syn : *Cleome gynandra* Sp.; *C. pentaphylla* Linn.; *Gynandropsis pentaphylla* DC.,

Family : Capparidaceae.

Hindi & Sanskrit Names:

Hindi : Churota, Gandhuli, Hulhul, Karalia, Lalhulhul, Safed Hulhul;

Sans. : Ajagandha, Arkapushpika, Avigandha, Kabari, Karnaspota, Sugandhika, Tilaparni, Ugragandha.

Distribution : Weed in All Tropical Countries.

Part Used : Root.

Utilization : Regarded stomachic and good in ascites, tumours, ulcers, earache, spleen enlargement and bilious fevers. Its decoction also said to be a mild febrifuge.

References : Agarwal (2003), Ambasta (1986), Chopra *et al.* (1956), Gogoi & Das (2003), Kirtikar & Basu (1935), Watt (1889).

Gynura pseudo-china DC.

Family : Asteraceae.

English & Hindi Names:

Eng. : Velvet Plant;

Hindi : Malhmalipandha.

Distribution : E. Himalaya & Kerala.

Part Used : Root.

Utilization : Powder used in blood coagulation, internal injury and leprosy. Mixed with tea it is given as a cooling medicine to parturient women.

References : Agarwal (2003), Anonymous (1956).

Habenaria plantaginea Lindl.

Family : Orchidaceae.

Distribution : Hilly Areas of A.P.

Part Used : Root.

Utilization : Paste applied on affected parts twice a day for 1 week in treatment of piles, and thrice a day for 4 days to treat wounds.

References : Rao & Henry (1996), Tirkey (2004).

Hedychium acuminatum Rosc.

Syn: *H. spicatum* Buch.-Ham. *ex* Smith

Family: Zingiberaceae.

English, Hindi & Sanskrit Names:

Eng. : Lesser Galangal, Spiked Ginger Lily;

Hindi : Gandhapalashi, Kapurakachari, Sit-Ruti;

Sans. : Amlaharidra, Durva, Gandhaashati, Kachhora, Karpura, Palashasathi, Samudra, Savata, Tuni.

Distribution : C. Himalaya: 1,100-2,500 m, E. India & Hills of S. India.

Part Used : Root.

Utilization : Paste given in internal problems to strengthen the muscles, and applied to heal wound; and also along with wheat flour (100g once a day for 2-4 days) given orally against cough in cattle. Also known to be used in diarrhoea, inflammation, liver complaints, asthma, vomiting, gastric, bronchitis, snake bite, gonorrhoea, and as tonic, blood purifier, stomachic, carminative, expectorant, and anthelmintic.

References : Aswal (2001), Jain (1995), Kirtikar & Basu (1984), Khare (2004), Nadeem *et al.* (2001), Samant *et al.* (2001), Sharma *et al.* (2003), Singh & Kumar (2000b), Sood & Thakur (2004).

Hedychium coccineum Ham.

Family : Zingiberaceae.

Part Used : Root.

Utilization : Used for headache in Meghalaya in N.E. India.

Reference : Jain (1995).

Hedychium coronarium Koenig.

Syn : *H. flavum* Roxb.

Family : Zingiberaceae.

English & Hindi Names:

Eng. : Common Ginger Lily, Garland Flower;
Hindi : Karchura.

Distribution : Commonly Grown in Gardens as an Ornamental Throughout India.

Part Used : Root.

Utilization : Extr. used as eye drops to check redness, pain, excessive excretion and watering.

References : Agarwal (2003), Parrotta (2001), Verma *et al.* (1995).

Hedychium gracile Roxb.

Family : Zingiberaceae.

Part Used : Root.

Utilization : Prescribed for chest pain in Meghalaya.

Reference : Jain (1995).

Hedyotis scandens Roxb.

Syn: *Oldenlandia scandens* K.Schum.

Family: Rubiaceae.

Hindi Name:

Hindi : Bhedelilata.

Distribution : C. & E. Himalaya.

Part Used : Root.

Utilization : Known for abdominal pain, diarrhoea (Arunachal), abortion, parturition pain (Darjeeling), intestinal worms, jaundice (Meghalaya), dysentery, sprains, expelling tape worms (W. Sikkim).

References : Agarwal (2003), Ambasta (1986), Jain & Saklani (1992), Jain *et al.* (1991).

Hedyotis uricenella Hk. & Arn.

Family : Rubiaceae.

Distribution : Meghalaya.

Part Used : Root.

Utilization : Juice used in cough.

Reference : Kumar (2002).

Helenium hoopesii Gray.

Family : Asteraceae.

Distribution : Native to N. America.

Part Used : Root.

Utilization : A tea prepared from its roots good for diarrhoea, intestinal worms and stomachache. By rubbing the dried roots, pains due to pulmonary diseases and rheumatism can be relieved.

Reference : Bhattacharjee (2001).

Helianthus annus L.

Family : Asteraceae.

English, Hindi & Sanskrit Names:

Eng. : Sunflower;
Hindi : Suraj Mukhi;
Sans. : Surya-Mukhi.

Distribution : Widely Grown Throughput India.

Part Used : Root.

Utilization : Good against tooth and gum problems.

Biological Activity : Root extr. at 10.0 mg inhibited germination of mustard, amaranthus, wild oats by 52, 72, 60 % and at 50.0 mg by 90, 100, 100%, respectively.

References : Agarwal (2003), Pandey & Verma (2002), Rastogi & Mehrotra (2002).

Helianthus tuberosus Linn.

Family : Asteraceae.

English & Hindi Names:

Eng. : Artichoke, Girasole, Jerusalem Artichoke, Topinambur;
Hindi : Artichok, Hatichuk.

Distribution : Cultivated in Garhwal Himalaya, Assam, Bengal & W. India.

Part Used : Root.

Utilization : Tubers eaten raw or as vegetable and these are said to be good to diabetic patients.

References : Agarwal (2003), Ambasta (1986), Sharma & Rana (2000).

Helicteres isora L.

Family : Sterculiaceae.

English, Hindi & Sanskrit Names:

Eng. : East Indian Screw Tree;
Hindi : Avarttani, Bhendu, Javitri, Jonkaphal, Kapasi, Maraphal, Merosi, Muradsing, Mringashinga;
Sans. : Aavartaki, Avaratam, Aavartphala, Mrigashinga.

Distribution : Dry Forests Throughout India.

Parts Used : Root. Root bark.

Utilization : Recorded so far to cure amoebiasis, diabetes, flatulence, cuts, snake bite, dysentery, dog bite, chest pain, cholera, asthma, cough, sciatica, wounds, fractures, boils and scabies (root paste), empynema and stomach affections (root juice), and also as astringent, demulcent, diuretic and antigalactagogue. About 4 teaspoons of root juice mixed in hot water given twice a day for about a week to treat rabies. And its extr. mixed with that of *Cocculus hirsutus* given orally to treat post natal pains. Decoction of 1 glassful of 50 g root bark and 50 g stem bark taken on empty stomach once a day for three days for terminating two/three months old pregnancy.

References : Agarwal (2003), Ambasta (1986), Anonymous (1956), Balasubramanian & Prasad (1996), Bhattacharjee (2001), Caius (2003), Chatterjee & Pakrashi (1997), Chauhan (1999), Chopra *et al.* (1956), Dastur (1970), Dhiman (2003), Dwivedi (1993), Hosagoudar & Henry (1996a), Kirtikar & Basu (1935), Khare (2004), Kumar & Lal (2003), Kumar & Pullaiah (1998), Parrotta (2001), Punjani (2006), Rastogi & Mehrotra (2002), Reddi *et al.* (2005), Singh (1996), Singh & Pandey (1998), Thomas & Britto (2003), Tirkey (2004), Varghese (1996), Venkataratnam & Raju (2004).

Heliotropium ellipticum Led.

Syn : *H. eichwaldi* Steud. *ex* DC.

Family : Boraginaceae.

Distribution : Punjab, Rajputana Desert, Sind & Baluchistan, Ascending in Kashmir Upto 1,700 m.

Part Used : Root.

Utilization : Chewed to cure hypertension.

References : Chopra *et al.* (1956), Saini (1996b).

Heliotropium indicum L.

Syn: *H. anisophyllum* Beaur.; *Heliophytum indicum* DC.; *H. velutinum* DC.; *Tiaridium anisophyllum* G. Don; *T. indicum* Lehm.

Family : Boraginaceae.

English, Hindi & Sanskrit Names:

Eng. : Indian Heliotrope;
Hindi : Hatta-Juri, Hatta-Sura, Hatti Sura, Siriyari;
Sans. : Bhurundi, Hastasunda, Hastishundi, Vrischikali.

Distribution : Wasteplaces Throughout India.

Part Used : Root.

Utilization : Considered bitter, astringent, febrifuge, ophthalmic, aphrodisiac, and also useful against ringworm, rheumatism and cough. Paste of freshly collected and crushed roots taken orally or applied on scorpion and snake bites. Roots chewed to cure weakness.

Active Constituents : Contains estradiol.

References : Ambasta (1986), Anonymous (1956), Chatterjee & Pakrashi (1997), Jadeja (1999), Jain & Sharma (2000), Kumar & Goel (1998), Nayak *et al.* (2003), Patil (2001), Reddi *et al.* (2005), Saini (1996a), Watt (1889).

Hemidesmus indicus (Linn.) Schultes. R.Br.

Syn : *Asclepias pseudosarsa* Roxb.;
Hemidesmus wallichi Miq.; *Periploca indica*L.
Family : Asclepiadaceae.

English, Hindi & Sanskrit Names:

Eng. : False Sarsaparilla, Indian Sarsaparilla;
Hindi : Annantmul, Hindi-Salsa, Magrabu, Salsa, Shwet-Sariva.
Sans. : Ananta, Anantamula, Gopakanya, Gopasuta, Gopimulam, Naga-Jihva, Sariba, Sugandhi.

Distribution : Upper Gangetic Plains. Eastwards to W. Bengal. M.P.to S. India.

Part Used : Root.

Utilization : Known in India for cough, cold, dyspepsia (juice mixed with a glass of lukewarm water and given twice daily for 10 days), loss of appetite, stomach ulcers, itching, arthritis, leprosy, eczema, erysipelas, chronic fevers, rheumatism, intrinsic haemorrhage, cholera, diarrhoea, excess menstruation, menorrhagia, foot cracks, biliousness, urinary disorders, toothache, debility, diabetes, wounds, leucorrhoea, body pains, stomachache, headache, skin diseases, spermatorrhoea (root powder mixed with cow's milk), respiratory disorders, and as alterative, cooling drink, blood purifier, anti-inflammatory, tonic, emetic and styptic. Grinded root mixed with milk given for leucorrhoea and tuberculosis, and paste (2 teaspoonfuls given once a day for 5-7 days) given to the patients of epilepsy and hysteria. Also paste made from its root and aerial root of *Ficus bengalensis* Linn. applied externally for pimples. Mixture comprising 10g of its root and 3g stem bark of 'Jamun' (*Syzygium cumini*) given with water on empty stomach within 3 days of delivery to increase lactation. Root made into powder with *Cryptolepis buchanani* (10 g) is given twice a day to treat malarial fever. Decoction of 250 g roots boiled in 1 litre water taken with honey in 14 equal doses to heal wounds caused by syphilis. One teaspoonful root decoction given once a day for 3 days to check reddish urine. Paste of grounded root tuber with fennel (*Foeniculum vulgare* Gaertn.) taken with molasses twice a day for three days to check diarrhoea and dysentery.

Active Constituents : Therapeutic properties attributed to hentriacontane, lupeol octacosanoate, β-amyrin acetate, lupeol acetate, α-& β-amyrin, lupeol, β-sitosterol, 2-hydroxy-4-methoxybenzaldehyde, hexatriacontane and three coumarinolignoids (hemidesmine, hemidesmin-1 and hemidesmin-2).

Biological Activity : Paramethoxy salicyclic aldehyde in the crude aq. extr. of roots exhibited bacteriostatic activity in mice infected with *Mycobacterium leparae*.

References : Agarwal (2003), Ambasta (1986), Aminuddin & Girach (1991), Bhattacharjee (2001), Borthakur *et al.* (2004), Chatterjee & Pakrashi (1997), Chopra *et al.* (1956), Dash & Mishra (2000), Dastur (1970), Dwivedi (1993), Girach (1992), Girach & Aminuddin (1995), Gogoi *et al.* (2003b), Jain (1968, 2003), Jain &

Singh (1997), Jain *et al.* (2003), Khare (2004), Khare & Khare (2000), Kirtikar & Basu (1984), Kumar & Lal (2003), Kumar & Rao (2001), Kurian (1995), Masih (2003), Mitaliya & Bhatt (2003), Mohanty & Padhy (1996), Nayak *et al.* (2003), Parrotta (2001), Prasad *et al.* (2002), Pullaiah & Kumar (1996), Rana *et al.* (2000), Rao & Henry (1996), Rastogi & Mehrotra (1998, 1999b), Reddi *et al.* (2005), Reddy *et al.* (1996), Said (1997), Saini (1996b), Sarangi & Sahu (2004), Saren *et al.* (2000), Sha & Singh (1990), Sharma (2003), Singh & Kumar (2000a), Singh & Rao (2003), Subramanian & Goraya (2003), Sur & Halder (2002), Tewari *et al.* (2001), Thakur *et al.* (1989), Thomas & Britto (2003), Trivedi (2002), Varghese (1996), Viswanathan (1997), Watt(1972).

Hemionitis arifolia (Burm.) Moore.

Family : Polypodiaceae.

English Name:

Eng. : Male Fern.

Distribution : Plains & Mountains of S. India Upto 1,200 m, W. Bengal, Bihar & Orissa.

Part Used : Root.

Utilization : Paste externally applied to treat polio in children.

References : Ambasta (1986), Asolkar *et al.* (1992), Balasubramanian & Prasad (1996).

Heracleum lanatum Michx.

Syn : *H. candicans* Wall. *ex* Dc.; *H. nepalense* D. Don

Family : Apiaceae.

English Names:

Eng. : Cow Parsnip, Health Root, Hogweed, Master Wort.

Distribution : Kashmir to Kumaon, 2,000-4,000 m.

Part Used : Root.

Utilization : Efficacious in the treatment of leucoderma, psoriasis, leucorrhoea and menstrual complaints.

Active Constituents : These are xanthotoxin, coumarin, bergapton, furocoumarin, heraclenin, heraclenol, xanthotoxol, spondin, cnadicanin, α-pinene(12.6), β-pinene (30.5), carveol acetate (17.1), dihydrocarveol (19.6), phenylpropinaldehyde (2.53) and phenyl ethyl acetate (2.9), heraclenin, imperatorin, 8-geranyloxypsoralen, candicopimaric acid, a new coumarin glucoside-tert-O-β-glucosyl-heraclenol, pimpinellin, isopimpinellin, sphendin, β-sitosterol, isobergapten,vaginidiol, sphondin, scopoletin, umbelliferone, ferulic acid, p-coumaric acid, apterin, angelicine and stearic acid.

Biological Activity : Scopoletin and umbelliferon highly inhibitory to growth of Chinese cabbage seedlings; bergapten and sphondin to growth of cucumber cuttings.

References : Chauhan (1999), Rastogi & Mehrotra (1998, 1999a, b, 2001), Samant *et al.* (2001), Sharma & Sood (1997), Singh (1999), Sood *et al.* (2001).

Heracleum maximum Bartr.

Family: Apiaceae.

Part Used : Root.

Utilization : Prescribed for asthma, indigestion, diarrhoea, colic, gas, rheumatic pains, gum trouble and heart palpitations.

Reference : Bhattacharjee (2001).

Herminium monorchis (Linn.) R.Br.

Family : Orchidaceae.

Part Used : Root.

Utilization : Paste of tuberous roots good against kidney troubles.

Reference : Chaurasia *et al.* (2000b).

Heteropogon contortus (L.) Beauv. *ex* R.Br.

Family : Poaceae.

English Names:

Eng. : Bellary Grass, Spear Grass.

Distribution : Throughout India, ascending in Himalaya Upto 1,700 m.

Part Used : Root.

Utilization : Paste applied for snake bite, and also considered stimulant and diuretic.

References : Dwivedi (1993), Kothari & Moorthy (1996), Varghese (1996).

Heterostemma decanense (Talb.) Swarup. & Mangaly

Family : Asclepiadaceae.

Part Used : Root.

Utilization : Useful for kidney troubles and stomachache.

Reference : Reddi *et al.* (2005).

Heterostemma piperfolium King & Gamble

Family : Asclepiadaceae.

Part Used : Root.

Utilization : Decoction of root (5 g) given for fever in children.

Reference : Brahmam & Saxena (1990).

Hibiscus lobatus (Murr.) Kuntz.

Family : Malvaceae.

Distribution : Shivpuri, M.P.

Part Used : Root.

Utilization : Juice with turmeric given for curing stomachache and dysentery.

References : Asolkar *et al.* (1992), Hosagoudar & Henry (1996b).

Hibiscus rosa-sinensis Linn.

Family : Malvaceae.

English, Hindi & Sanskrit Names:

Eng. : Chinese Hibiscus, Scarlet Rose-Mallow, Shoe Flower Plant;
Hindi : Jasum, Jasut, Lal Ambari;
Sans. : Arkapriya, Aruṇa, Aundrapushpa, Harivallabha, Japapushpa, Joba, Odhrapushpa, Pratika, Raktapushpi, Rudrapushpa, Trisandhya.

Distribution : Gardens Throughout India.

Part Used : Root.

Utilization : Considered good for fever and venereal diseases of cattle and as substitute to *Althaea officinalis.* Its powder given in menorrhagia and the juice of fresh root good against gonorrhoea. Also, considered demulcent and valuable in cough and colds.

Active Constituents : Methyl (R) 2-hydroxysterculate, methyl sterculate, methylmalvalate, methyl (E) 11-methoxy- 9-oxo-10-nonadecenoate, 10-methoxy-8-oxo-9-octadecenoate, and methyl (R) Z-hydroxysterculate.

Biological Activity : Antifungal.

References : Agarwal (2003), Anonymous (1959), Caius (2003), Chatterjee & Pakrashi (1997), Chopra *et al.* (1956), Dhiman (2003), Khare (2004), Kirtikar & Basu (1935), Kurian (1995), Rastogi & Mehrotra (1998), Reddi *et al.* (2005).

Hibiscus syriacus Linn.

Family : Malvaceae.

English Names:

Eng. : Rose of Sharon, Shruby Althea.

Distribution : Grown in Indian Gardens.

Part Used : Root.

Utilization : Root mucilaginous, demulcent and used in diarrhoea, dysentery and dysmenorrhoea.

Reference : Ambasta (1986).

Hibiscus talbotii (Rakshit) Paul & Nayar

Family : Malvaceae.

Part Used : Root.

Utilization : 10-30 ml of root given twice a day for treating excessive bleeding during menstruation.

Reference : Sharma & Singh (1999).

Hibiscus tiliaceous L.

Family : Malvaceae.

English, Hindi & Sanskrit Names:

Eng. : Corkwood, Lime-Tree-Leaved Hibiscus;
Hindi : Bola, Chelwa;
Sans. : Bala.

Distribution : E. & W. Coasts of India; Common in Andamans & Sunderbans.

Part Used : Root.

Utilization : Reported in India as an emollient, febrifuge, aperitive, sudorific, diuretic, laxative, and for treating lumbago and rheumatism.

Active Constituents : Gossypol, mansonons D and F characterised in roots.

References : Agarwal (2003), Ambasta (1986), Caius (2003), Jain *et al.* (1995), Chopra *et al.* (1956), Rastogi & Mehrotra (2001).

Holarrhena antidysentrica (Roxb. *ex* Flem.) Wall.

Syn : *Chonemorpha antidysentrica* (Roth.) G. Don; *Echites antidysentrica* Roxb.; *E. pubescens* Buch.-Ham.; *Holarrhena codaga* G. Don; *H. pubescens* Wall.; *Wrightia antidysentrica* Grah.

Family : Apocynaceae.

English, Hindi & Sanskrit Names:

Eng. : Conessi Tree, Easter Tree, Kurchi Tree, Tellicherry Tree;
Hindi : Dudhi, Indrajav, Indrayava, Kura, Kurchi, Kutja;
Sans. : Girimallika, Indra, Indrayan, Kalinga, Pandura, Vatsata.

Distribution : Tropical Himalaya Upto 1,000 m & Drier Parts of India.

Parts Used : Root. Root bark.

Utilization : Reported so far to cure acute stomach pain, leucorrhoea, ulcers, stomach disorders, rheumatism, gout, paralysis, stomach ulcers, easy delivery, expelling worms (root paste applied on anus of children), snake bite, jaundice (root decoction given early in the morning in empty stomach once daily for 7 days), colic pain, spleen complaints, piles and haemorrhage in child birth, nose bleeding, toothache (root paste with water filled into tooth cavities). Decoction of root and stem bark given twice daily for 2-3 days to check dysentery. To cure cough and cold, decoction of root mixed with equal quantity of 'Harada' (*Terminalia Chebula*), 'Bahada' (*Terminalia bellirica*) and 'Bhejri' fruits (*Solanum surattense*) fruits recommended in a dose of two teaspoonful twice daily for almost a week.

Active Constituents : Holacetine isolated from root bark characterised as (20 S) - acetamido-5-pregnan-3β-ol.

References : Agarwal (2003), Aminuddin & Girach (1991, 1996), Brahmam & Saxena (1990), Jadeja (1999), Khare (2004), Kumar & Lal (2003), Parrotta (2001), Rao & Henry (1996), Rastogi & Mehrotra (1999a, b), Reddi *et al.* (2005), Saren *et al.* (2000), Sood & Thakur (2004), Subramanian & Goraya (2003), Thakur *et al.* (1989), Tirkey (2004), Varghese (1996), Verma & Chauhan (2006), Watt (1889).

Holmskioldia sanguinea Retz.

Syn : *Hastingia cocinea* Smith.; *H. scandens* Roxb.; *Holamskioldia rubra* Pers.; *Platunium rubrum* Juss.

Family : Verbenaceae.

English & Hindi Names:

Eng. : Chinese-Hat-Plant, Parasol Flower;
Hindi : Kapni.

Distribution : All Over N.E. Region at the Outskirts of Forests.

Part Used : Root.

Utilization : Juice taken to relieve fever and headache.

References : Agarwal (2003), Ambasta (1986), Kharkongor & Joseph (1997), Kumar (2002), Watt (1889).

Holostemma ada-kodien Schult.

Syn : *Asclepias annularis* Roxb.; *H. annularis* K.Schum.; *H. brunonianum* Royle; *H. rheedei* Wall.; *Sacrostemma annulare* Roth.

Family : Asclepiadaceae.

Hindi & Sanskrit Names:

Hindi : Chirvel;
Sans. : Arkapushpi, Jivanti.

Distribution : Most Parts of India, Ascending Upto 2, 000 m.

Part Used : Root.

Utilization : Two teaspoons of root juice given twice daily for 10 days to treat spleen enlargement and as health tonic once daily. Also, used as refrigerant, ophthalmic, emollient, alterant, stimulant, aphrodisiac and in cough, fever, diabetes, stomachalgia, burning sensation, and remedy for scalding in gonorrhoea.

Active Constituents : α-amyrin, lupeol, β-sitosterol, alanine, aspartic acid, glycine, serine, threonine and valine are the chief constituents.

References : Anonymous (1959), Chopra *et al.* (1956), Jha & Verma (1996), Patil & Patil (2002), Rastogi & Mehrotra (2001), Reddi *et al.* (2005), Watt (1889).

Homonia riparia Lour.

Syn : *Adelia nerifolia* Roth.; *Croton salicilifolius* Geisel.; *Hematospermum nerifolium* Wall.; *H. riparium* Wall.; *H. salicinum* Baill.; *Ricinus salicinus* Hassk.; *Spathiostemon javense* Blume; *S. salicinus* Hassk.

Family : Euphorbiaceae.

Hindi & Sanskrit Names:

Hindi : Kshudrapashanbeda;
Sans. : Kshudrapashanabheda, Pashanabhedaka.

Distribution : Practically Throughout India.

Part Used : Root.

Utilization : Known to be beneficial as laxative, emetic, diuretic, and for ulcers, chest pain and toothache. Its decoction given in piles, stone in the bladder, gonorrhoea and syphilis and paste given for a week's time for urinary infection.

Active Constituents : Contains unsaponifiable matter, α-spinasteryl-OAC.

References : Agarwal (2003), Ambasta (1986), Anonymous (1959), Asolkar *et al.* (1992), Caius (2003), Chopra *et al.* (1956), Kirtikar & Basu (1984), Kothari & Moorthy (1996), Kumar & Lal (2003), Nadanakunjidam (2003), Watt (1889).

Houttuynia cordata Thumb.

Family : Saururaceae.

English Name:

Eng. : Tsi.

Distribution : Entire N.E. Region Upto 2,000 m.

Part Used : Root.

Utilization : Considered diuretic; and its extr. good for stomach ulcers.

Active Constituents : Diuretic action attributed to the presence of quercitin and inorganic salts like potassium, chloride, potassium sulphate.

References : Agarwal (2003), Kumar (2002).

Humboldtia unijuga Bedd.

Family : Fabaceae.

Distribution : Travancore Mountains, 1,000-1,500 m.

Part Used : Root Galls.

Utilization : Paste ('Kashanti veru') prescribed orally for asthma.

Reference : Radhakrishnan *et al.* (1996).

Hybanthus enneaspermus (L.) F.Muell.

Syn : *Ionidium enneaspermum* (L.) Vent.; *I. suffruticosum* (L.) R. & S.; *Viola enneasperma* L.; *V. suffruticosa* L.

Family : Violaceae.

Hindi & Sanskrit Names:

Hindi : Ratanpurus;

Sans. : Amburuha, Charati.

Distribution : Warmer Parts of India, from U.P. southwards to the Deccan Peninsula.

Part Used : Root.

Utilization : Useful in bowel complaints of children, and its powder used by 'Sahariyas' for urinary troubles and gonorrhoea (infusion).

References : Agarwal (2003), Ambasta (1986), Anonymous (1959), Asolkar *et al.* (1992), Chatterjee & Pakrashi (1997), Chopra *et al.* (1956), Rajendran & Sikarwar (2003).

Hydrangea arborescens L.

Family : Hydrangeaceae.

English Name:

Eng. : Hydrangea.

Part Used : Root.

Utilization : Regarded as tonic, diuretic, laxative, and also believed to treat kidney stones.

Reference : Bhattacharjee (2001).

Hygrophylla auriculata (Schum.) Heine.

Syn : *Asteracantha longifolia* Nees.; *Barleria auriculata* Schumach.; *Hygrophylla schulli* (Ham.) M.R. & S.M.Almeida; *H. spinosa* T.Anders.

Family : Acanthaceae.

English, Hindi & Sanskrit Names:

Eng. : Long-Leaved Barleria;
Hindi : Gokhalkanta, Gokshura, Juliakanta, Kulia Kanta, Talmakhana, Untkataru;
Sans. : Kokilaksah, Ikshugandha.

Distribution : Throughout the Plains of India, Marshy Margins of Canals & Rice Fields.

Part Used : Root.

Utilization : Decoction employed as tonic, refrigerant, diuretic, used in rheumatism, gonorrhoea, diseases of genito-urinary tract, dropsy, jaundice, anasarca, calculi, fever, swelling of face and entire body.

References : Agarwal (2003), Bhatt *et al.* (2002), Jain (1968, 1991), Parrotta (2001), Reddi *et al.* (2005), Varghese (1996), Watt (1889).

Hymenodictyon excelsum (Roxb.) Wall.

Syn : *Cinchona excelsa* Roxb.; *C. thyrsiflora* Roxb.; *Hymenodictyon horsfieldianum* Miq.; *H. obovatum* Wight.; *H. thyrsiflorum* Wall.; *H. utile* Wight.

Family : Rubiaceae.

Hindi & Sanskrit Names:

Hindi : Bandaru, Bhalena, Bhaulan, Bhurkus, Dhauli, Kukurkat Phargur, Potur;
Sans. : Bhramarchhallika, Ugragandha.

Distribution : Base of Himalaya & W. India.

Part Used : Root.

Utilization : Extr. (1 teaspoon once a day, given for 3 days) given after menstruation to check conception. Its paste used as an application on the joints in case of gout.

Active Constituents : Attributed to rubiadin, lucidin, nordamnacanthal, damnacanthal, 2-benzylxanthopurpurin, anthragallol, soranjidol and morindone.

References : Agarwal (2003), Ambasta (1986), Rastogi & Mehrotra (1999b), Saxena *et al.* (1997), Singh & Prakash (1994), Watt (1889).

Hyptis suaveolens (L.) Poit.

Family: Lamiaceae.

English, Hindi & Sanskrit Names:

Eng.	:	Wild Spikenard, Hyptis;
Hindi	:	Ganga Tulsi, Wilayati-Tulsi;
Sans.	:	Bhustrna.

Distribution : Wasteplaces, Agricultural Fields. Roadsides. Railway Lines Throughout S. India, N.E. India & Andaman & Nicobar Islands.

Part Used : Root.

Utilization : 3-4 teaspoonfuls of root extr. given for relief in haematuria. Paste used to cure fever and decoction prescribed as appetizer and stomachic.

Active Constituents : Contains β-sitosterol, α- & β-amyrin; 3β-hydroxylup-12-en-28-oic, oleanolic and urs-12-en-3β-ol-27-oic (α-peltoboykinolic) acids, hentriacontane, hentriacontanone, campesterol, friedelin and lupeol.

References : Agarwal (2003), Ambasta (1986), Anonymous (1959), Chatterjee & Pakrashi (1997), Jain & Sharma (2000), Kumar & Goel (1998), Rao & Henry (1996), Rastogi & Mehrotra (2001), Reddi *et al.* (2005).

Ichnocarpus frutescens (Linn.) R.Br.

Syn: *Apocynum frutescens* L.; *Echites frutescens* (L.) Roxb.; *Ichnocarpus dasycalyx* Miq.; *I. radicans* Wall.

Family: Apocynaceae.

English, Hindi & Sanskrit Names:

Eng.	:	Black Creeper;
Hindi	:	Dudhilata, Kalidudhi, Siama-Lata;
Sans.	:	Paravalli, Sariva, Shariva, Syamalata.

Distribution : Almost Throughout India to an altitude of 1,200 m from W. Himalaya Southwards to Tamil Nadu & Kerala to Sri Lanka.

Part Used : Root.

Utilization : Useful in fever, epilepsy, dysentery, dyspepsia, skin diseases, anorexia, leucorrhoea, rheumatism, syphilis, urinary calculi, body pain and on hard and paining breasts; and also as aphrodisiac, demulcent, alterative, tonic, diaphoretic and diuretic. The root powder with milk administered for diabetes, gall bladder stones and as blood purifier. About 5g of its powder given for 21 days in filaria.

References : Agarwal (20030, Ambasta (1986), Anonymous (1959), Chatterjee & Pakrashi (1997), Jain (1991), Jha & verma (1996), Kumar (2000), Kumar & Lal (2003), Nayak *et al.* (2003), Parrotta (2001), Rao & Henry (1996), Reddi *et al.* (12005), Shukla & Verma (1996), Upadhyay & Chauhan (2000), Varghese (1996), Watt (1889).

Impatiens angustifolia Bl.

Family: Balsaminaceae.

Distribution : Arunachal Pradesh, Khasia Mountains: 1,000-2,000 m.

Part Used : Root.

Utilization : Powder given in urinary complaints.

Reference : Kumar (2002).

Impatiens tripetala Roxb.

Syn: *I. multiflora* Wall.
Family: Balsaminaceae.

Hindi Name:

Hindi : Damdoka.

Distribution : Tropical Himalaya Throughout E. India.

Part Used : Root.

Utilization : Root juice given in haematuria (One tola of juice is mixed with one tola of milk).

References : Agarwal (2003), Ambasta (1986), Chhetri (2005), Chopra *et al.* (1956), Kirtikar & Basu (1935).

Indigofera astragalina DC.

Syn: *I. hirsuta* auct. (non L.)
Family: Fabaceae.

English Name:

Eng. : Hairy Indigo.

Distribution : Throughout Plains, Ascending to 1,200 m in Kumaon Hills.

Part Used : Root.

Utilization : Decoction used for cough. Its powder valued for chest pain, tooth troubles and fever (root crushed and given 2-3 spoonfuls twice a day till cure).

References : Anonymous (1989), Bhalla & Bhalla (2002), Deshmukh *et al.* (2000), Rao & Henry (1996), Rastogi & Mehrotra (2001).

Indigofera cassioides Rottl. *ex* DC.

Syn: *I. arborea* Roxb.; *I. elliptica* Roxb.; *I. pulchella* auct. (non Roxb.); *I. purpurascens* Roxb.; *I. violacea Roxb.*
Family: Fabaceae.

Hindi Names:

Hindi : Nil.

Distribution : Moist Evergreen Forests Throughout the Himalayan Tract & Hills Upto 1,500 m on Peninsular India.

Part Used : Root.

Utilization : Extr. efficacious in stomachache, asthma (extr. of root crushed with *Piper nigrum*), preventing conception, and haematuria. Its decoction used by the 'Santhalis' to treat cough, and powder applied externally to relieve chest pains. Root of this plant and that of *Andrographis paniculata* (Bhuinim) and *Smilax* spp. (Atkir) grounded in equal proportions (10 g) and prescribed twice daily for three days in cases of haematuria.

Biological Activity : Extr. is antiviral and spasmolytic.

References : Agarwal (2003), Asolkar *et al.* (1992), Girach & Aminudin (1995), Islam & Jha (2003), Kirtikar & Basu (1935), Parrotta (2001), Rao & Henry (1996), Reddi *et al.* (2005), Sood & Thakur (2004), Yoganarasimhan (1981).

Indigofera linifolia (L. f.) Retz.

Syn: *Hedysarum linifolium* L.f.; *Sphaeridiophora linifolium* (L.f.) Desv.; *S. abyssinicum* Jaub.
Family: Fabaceae.

Hindi Names:

Hindi : Torki.

Distribution : Distributed Throughout the Plains of India.

Part Used : Root.

Utilization : Paste applied on scorpion bite.

References : Anonymous (1959), Jha & Verma (1996), Parrotta (2001), Sood & Thakur (2004).

Indigofera spicata Forssk.

Syn: *I. enneaphylla* Linn.; *I. caespilosa* Wight; *Hedysarum prostratum* Linn.
Family: Fabaceae.

English Name:

Eng. : Creeping or Trailing Indigo.

Distribution : Plains of India, Himalaya, Sri Lanka & Myanmar.

Part Used : Root.

Utilization : Juice administered orally to relieve from stomach pain.

References : Asolkar *et al.* (1992), Balasubramanian & Prasad (1996), Hooker (1879).

Indigofera tinctoria L.

Family: Fabaceae.

English, Hindi & Sanskrit Names:

Eng. : Indian Indigo;
Hindi : Gouli, Lil, Nil;
Sans. : Aklika, Bhadra, Dola, Gramya, Nilaa, Nilapushpa, Nili, Nilini, Ranjani, Shaaradi, Shodhini, Tuni, Tutthaa.

Distribution : Cultivated in Many Parts of India.

Part Used : Root.

Utilization : Given to treat jaundice, inflammation of the liver, scorpion sting, urinary complaints, bronchitis, antidote to arsenic poisoning, hepatitis, difficult micturition, snake bite, caries of the teeth, fever, and as nervine tonic. Root pounded with goat's milk prescribed in dysuria. For asthma, root (50 g) crushed with black pepper (5 g) taken (one spoon twice a day) for 21 days.

References : Agarwal (2003), Ambasta (1986), Chatterjee & Pakrashi (1997), Chopra *et al.* (1956), Deshmukh *et al.* (2000), Khare 2004), Kirtikar & Basu (1935), Patil & Patil (2002), Rao & Pullaiah (2001), Reddi *et al.* (2005), Varghese (1996).

Inula cappa (Buch.-Ham. *ex* D. Don) DC.

Family: Asteraceae.

Distribution : Himalaya from H.P. to Bhutan, (1,200-3,000 m), Khasia Hills (1,200-1,500 m).

Part Used : Root.

Utilization : Decoction (30-40 ml) given in typhoid for 1 week two or three times a day.

References : Panthi & Chaudhary (2003), Rastogi & Mehrotra (2001).

Inula racemosa Hook.f.

Syn: *I. helenium* Linn.
Family: Asteraceae.

English, Hindi & Sanskrit Names:

Eng. : Elecampane, Scabwort;

Hindi	:	Pohakarmul, Poshkar;
Sans.	:	Padmapatra, Pauskaram, Pushkar Mool.
Distribution	:	W. Himalaya Upto 3,500 m.
Part Used	:	Root.
Utilization	:	So far useful in cough, rheumatism and various gastro-intestinal complaints in Ladakh, asthma, chest pain, skin diseases, flatulence, cardiac debility, amenorrhoea; and as febrifuge, aphrodisiac, anti-inflammatory, antipyretic, antiseptic, antiasthmatic, anthelmintic, expectorant, diuretic, emmenagogue, stimulant, deodorant and anodyne. Extr. of fresh root taken twice a day for joint pain and its decoction good for cough and liver problem.
Active Constituents	:	Active principles are due to alantolactone, β-sitosterol, inulin, iso-alantolacton, dihydro-isoalantolactone, dihydroalantolactone, eudesma-5, 7(11)-diene-8 β, 12-olide & eudesma -4 (15), 7(11) - dien -8 β, 12-olide, 1-deoxy-8 epiivangustin, germacrene-D lactone, isohumulene, trideca 1-en-3, 5, 7, 9, 11-pentyne, (-) dammar-20-24-dien-3 β-yl acetate and alloalantolactone.
Biological Activity	:	Antiseptic, diuretic, anthelmintic, antiallergic, anthelmintic, antiasthmatic properties strongly positive.
References	:	Agarwal (2003), Ambasta (1986), Asolkar *et al.* (1992), Ballabh & Chaurasia (2006), Chauhan (1999), Chopra *et al.* (1956), Jain (1991), Kapur & Singh (1996), Kaul (1997), Khare (2004), Mehrotra *et al.* (1995), Rastogi & Mehrotra (1999a, b, 2001), Reddi *et al.* (2005), Sharma (2003), Singh (1996), Sood *et al.* (2000).

Inula rhizocephala Schrenk. var. *rhizocephaloids* (Clarke) Kitam.

Family: Asteraceae.

Distribution	:	W. Tibet (2,300-3,000 m) & Ladakh, (4,000-4,700 m).

Part Used : Root.

Utilization : Beneficial in the treatment of cold, cough and chest complaints.

References : Chaurasia *et al.* (2002b), Hooker (1887).

Ipomoea digitata L.

Syn : *Convolvulus paniculatus* L.; *I. gossypifolia* Willd.; *I. mauritiana* Jacq.; *I. paniculata* (L.) R.Br. non Burm.f.

Family : Convolvulaceae.

English, Hindi & Sanskrit Names:

Eng. : Sweet Potato;

Hindi : Belaikanda, Mithaalu;

Sans. : Bhumi Kushmanda, Kandagranthi, Pichhila, Pindalu, Pinditaka, Ramashakandaka, Svadukandaka, Vidaree.

Distribution : Grassy Grounds & Scrub-Jungles of Various Parts of India.

Part Used : Roots.

Utilization : Used in inflammation, liver trouble and treatment of scorpion sting. Known to be considered sweet, cooling, laxative, alterative, cholagogue, demulcent, mucilaginous, lactagogue, purgative, diuretic, tonic and aphrodisiac; and said to be useful in strangury, burning sensation, thirst, urinary discharges, diarrhoea, emaciation in children, liver and spleen disorder, inflammation, scorpion sting, menorrhagia, debility, fat accumulation, consumption (decoction), fevers, and bronchitis. In combination with coriander and fenugreek seeds, it is recommended for spermatorrhoea.

Active Constituents : Paniculatin, scopoletin, scoparone, β-sitosterol, its 3-O-glucoside, taraxerol acetate, umbelliferone, 1-o-ethyl-glucopyranoside & mandelamide characterised in roots

References : Agarwal (2003), Ambasta (1986), Anonymous (1959), Chatterjee & Pakrashi (1997), Chopra *et al.* (1956), Dey (1973), Dwivedi (1993), Islam (1996), Watt (1889).

Iris ensata Thunb.

Syn: *I. moorcroftiana* Wall.
Family: Iridaceae.

Hindi Names:

Hindi : Irisa, Sosun.

Distribution : Occurs in W. Himalaya Upto 3,000 m.

Parts Used : Root.

Utilization : Used so far as antiallergic, alterative, blood purifier, and for venereal infection, dropsy and liver complaints.

Active Constituents : Ceryl alcohol.

References : Agarwal (2003), Ambasta (1986), Anonymous (1959), Chopra *et al.* (1956), Kaul (1997), Rastogi & Mehrotra (1999a, b), Watt (1889).

Iris kumaoensis D. Don *ex* Royle

Syn : *I. longifolia* Royle
Family: Iridaceae.

Hindi Name:

Hindi : Karka.

Distribution : Temperate zones, Open Grassy Slopes of N. Himalaya: 3,000-4,000 m.

Part Used : Root.

Utilization : Useful as stomachic, and for scabies, urticaria, kidney trouble, fever and urinary complaints.

References : Agarwal (2003), Bhattacharjee (2001), Jain (1991), Negi *et al.* (1993), Sharma (2003).

Iris nepalensis D. Don

Syn: *I. decora* Wall.; *I. sulcata* Wall.
Family: Iridaceae.

Hindi Name:

Hindi : Chiluchi.

Distribution : W. & E. Himalaya.

Part Used : Root.

Utilization : Reported so far as diuretic, aperient, deobsturent, also useful in billious obstructions, sores, pimples and rheumatic pains.

References : Agarwal (2003), Ambasta (1986), Anonymous (1959), Chopra *et al.* (1956), Kaul (1997), Watt (1889).

Ixora coccinea L.

Syn: *I. bandhuca* Roxb.; *I. grandiflora* Br.; *I. incarnata* DC.; *I. propinqua* Br.; *I. oborata* Heyne; *Pavetta bandhuca* Miq.; *P. coccinea* Blume; *P. incarnata* Blume.

Family: Rubiaceae.

English, Hindi & Sanskrit Names:

Eng. : Jangle Flame Ixora, The Flame Tree of the Woods;
Hindi : Rajana, Rangan, Rookmini;
Sans. : Bandhuka, Raktaka.

Distribution : Gardens Throughout India.

Part Used : Root.

Utilization : Recommended as an antiseptic, astringent, stomachic, sedative; good for sores, chronic ulcers, headache, hiccup, fever, gonorrhoea, loss of appetite, diarrhoea, dysentery, bile action, and stomach pain.

Active Constituents : Possesses methyl ester of palmitic, steraric, oleic and linoleic acids (root oil).

References : Agarwal (2003), Ambasta (1986), Anonymous (1959), Kirtikar & Basu (1935), Rastogi & Mehrotra (2002), Watt (1889).

Ixora lobbii Loud.

Family: Rubiaceae.

Distribution : Indian Gardens.

Part Used : Root.

Utilization : Decoction given before and after delivery, and also good against diarrhoea and headache (poultice).

References : Ambasta (1986), Anonymous (1959), Asolkar *et al.* (1992).

Janakia arayalpatra Joseph. & Chand.

Family: Asclepiadaceae.

Part Used : Root.

Utilization : Boiled juice mixed with coconut milk prescribed twice daily for a period of 15-30 days for peptic ulcer, cancer-like afflictions, health promoter, blood purifier and for cooling effect to the body.

Reference : Radhakrishnan *et al.* (1996).

Jasminum multiflorum (Burm. f.) Andr.

Syn: *Jasminium Congestum* Wall.; *J. hirsutum* Willd.; *J. pubescens* (Retz.) Willd.; *Nyctanthes multiflora* Burm.f.

Family: Oleaceae.

English, Hindi & Sanskrit Names:

Eng. : Common Jasmine, Downy Jasmine;
Hindi : Chameli, Kundaphool, Kundo, Kundphul;
Sans. : Kunda, Kundah, Sadapushpa, Vasanta.

Distribution : Common in the Sub- Himalayan Tract, Moist Forests of the W. Ghats & Gardens Throughout Tropical India.

Part Used : Root.

Utilization : Used as an emmenagogue and antidote to snake bite.

References : Agarwal (2003), Ambasta (1986), Chaterjee & Pakrashi (1997), Chopra *et al.* (1956), Kumar & Lal (2003), Parrotta (2001), Watt (1889).

Jatropha curcas L. (Pl. 6C)

Syn: *J. moluccana* Herb. Russ.

Family: Euphorbiaceae.

English, Hindi & Sanskrit Names:

Eng. : Physic Nut, Poison Nut, Purging Nut;
Hindi : Bagberenda, Baghrandi, Jangliarandi, Safedarand, Pharierand;
Sans. : Dravanti, Kannaerand, Nepala, Paravatayeranda, Sthulaeranda, Vyagrairanda.

Distribution : Cultivated & Naturalized in India & Sri Lanka.

Parts Used : Root. Root bark.

Utilization : Root bark stomachic and astringent. Its extr. in combination with bland oil, applied externally in rheumatism and leprosy and paste used for treatment of sores, swelling, snake bite, and eczema. Decoction good against toothache, sore throat and intestinal ulcers. Also, said to be good abortifacient and anthelmintic.

Active Constituents : Jatrophol and jatrophin (coumarino-lignan-I) are the principle constituents.

References : Agarwal (2003), Anonymous (1959), Jain (1991), Jha & Verma (1996), Khare (2004), Rajore & Batra (2003), Rao *et al.* (1996), Rastogi & Mehrotra (2002), Reddi *et al.* (2005), Samvatsar & Diwanji (1996), Sood & Thakur (2004), Varghese (1996).

Jatropha glandulifera Roxb.

Syn: *J. glauca* Vahl.
Family: Euphorbiaceae.

Hindi & Sanskrit Names:

Hindi : Janglierandi, Undarbibi;
Sans. : Nikumbha.

Distribution : Deccan & Karnataka.

Part Used : Root.

Utilization : Good for piles, abdominal enlargement (root brayed with water and given to suffering children) and glandular swellings.

Active Constituents : Known to possess coumarino-lignan (I), jatropholone A and fraxetin.

References : Agarwal (2003), Ambasta (1986), Anonymous (1959), Chopra *et al.* (1956), Dwivedi (1993), Kirtikar & Basu (1984), Rastogi & Mehrotra (2002).

Jatropha gossypifolia L.

Family: Euphorbiaceae.

English, Hindi & Sanskrit Names:

Eng. : Bellyache Bush, Tua-Tua;
Hindi : Bhrenda, Verenda;
Sans. : Dravanti, Raktaeramada.

Distribution : Naturalized as a Common Weed in Open Degraded Habitats Throughout Most of the Plains of India.

Part Used : Root.

Utilization : Pills made from its grounded root with 60 ml milk and powder of *Acacia catechu* (1 tablet given daily on empty stomach for 3 days) given in leucorrhoea. Also found to be good against leprosy, snake bite, menstrual pain and veterinary bone-fracture.

Active Constituents : Therapeutic properties are due to jatrophone, jatropholones A, B, 2 α-hydroxyjatrophone, 2β-hydroxyjatrophone and 2β-hydroxy-5, 6-isojatrophone.

References : Agarwal (2003), Ambasta (1986), Anonymous (1959), Asolkar *et al.*(1992), Caius (2003), Kothari & Rao (2000), Parrotta (2001), Reddi *et al.* (2005), Sarangi & Sahu (2004), Varghese (1996).

Juncus effusus Linn.

Syn : *J. communis* E.Mey.

Family : Juncaceae.

English Name:

Eng. : Soft or Matting Rush.

Distribution : Sikkim Himalaya (2,000-3,300 m), Khasia (1,600-1,800 m) & Aka Hills.

Part Used : Root.

Utilization : Employed as diuretic especially in strangury.

References : Ambasta (1986), Anonymous (1959).

Jurenella macrocephala (Royle) Aswal *et* Goel.

Family: Asteraceae.

Part Used : Root.

Utilization : Decoction given in colic. Also considered stimulant and given in fever after the child birth.

Reference : Aswal & Mehrotra (1994).

Jurinea dolomiaea Boiss.

Syn: *Dolomaea macrocephala* DC.; *Jurinea macrocephala* Benth.; *Serrutula macrocephala* Wall.

Family: Asteraceae.

Hindi Name:

Hindi : Dhoop.

Distribution : Himalaya: (Kashmir to Kumaon) Upto 3,500 m.

Part Used : Root.

Utilization : Considered stimulant, colic and perpetual fever (decoction), skin eruptions (bruised root applied), gout and rheumatism (aromatic oil from roots), sores, boils, wounds, bodyache and sprains (root extr.), and against small pox (powdered dry root brunt).

Active Constituents : Possesses aromatic resin.

References : Agarwal (2003), Asolkar *et al.* (1992), Chauhan (1999), Chopra *et al.* (1956), Kaul (1997), Kirtikar & Basu (1984), Nautiyal *et al.* (2003), Negi *et al.* (1993), Singh (1996, 1999).

Jussieua suffruticosa L.

Syn: *J. angustifolia* Lamk.; *J. burmanni* DC.; *J. decumbens* Wall.; J. *exaltata* Roxb.; *J. fruiticosa* DC.; *J. longipes* Griff.; *J. octophila* DC.; *J. villosa* Lamk.; *Epilobium fruticosum* Lour.

Family: Onagraceae.

Hindi & Sanskrit Names:

Hindi : Banlunga;
Sans. : Bhulavanga.

Distribution : Throughout the Moist Regions of India..

Part Used : Root.

Utilization : Extr. taken with 50 g 'gur', once a day for 3 days, to cure infantile diarrhoea and fever.

References : Agarwal (2003), Ambasta (1986), Anonymous (1959), Chopra *et al.* (1956), Islam & Jha (2003), Tirkey (2004), Watt (1889).

Justicia adhatoda L.

Syn: *Adhatoda vasica* Nees.; *A. zeylanica*Medic.
Family: Acanthaceae.

English, Hindi & Sanskrit Names:

Eng. : Malabar Nut Tree, Vasaka;
Hindi : Adalsa, Adulasa, Bansa, Bashing, Basuti;
Sans. : Arus, Atarusha, Bashika, Bhishangmata, Kanthiravi, Nasa, Rasadani, Sinhika, Vasa, Vrisha.

Distribution : Roadsides & Wastelands Upto 1,000 m. Throughout Drier Parts of India.

Part Used : Root.

Utilization : Known to be diuretic, antiseptic, anthelmintic, antispasdmodic, sedative, expectorant; useful for eye sores, gonorrhoea, gastric problems (decoction of 20 g root in 150 ml water till reduced to half; 4 times daily), bronchitis, cough, pulmonary affections, sprains, strangury, asthma, phthisis, intermittent fever (50 ml decoction twice daily), eczema, leucorrhoea with blood discharges, bilious vomiting and painful child birth. 10 g dried root powder with 7 peppers given once to expedite delivery. Its decoction prepared by boiling 50g of its root in (125 ml) cow's milk prescribed for three days on empty stomach for treatment of diabetes.

Active Constituents : Active properties attributed to vasicunol (6-hydroxy vasicine), vasicine (peganine), 2′-hydroxy-4-glucosyloxychalcone, vasicine, galactose, o-ethyl-α-galactoside, tritriacontane, β-sitosterol-β-D-glucoside, 9-acetamide-3, 4-dihydropyrido [3, 4-b] indol, deoxyvasicinone, vasicol and vasicinolone.

Biological Activity : Exhibited positive antiviral and hypoglycaemic activities.

References : Chatterjee & Pakrashi (1997), Dash & Mishra (2000), Dhiman (2003), Dutta & Nath (2000), Dwivedi (1993), Girach (1992), Khanna & Kumar (2000), Kirtikar & Basu (1984), Kurian (1995), Mehrotra *et al.* (1995), Nayak *et al.* (2003), Reddi *et al.* (2005), Sood & Thakur (2004), Thakur *et al.* (1989), Varghese (1996), Verma & Chauhan (2006).

Justicia gendarussa Burm.f.

Syn: *Gendarussa vulgaris* Nees.

Family: Acanthaceae.

Hindi & Sanskrit Names:

Hindi : Nilinargandi;

Sans. : Nila-Nirgundi.

Distribution : Common Throughout India in Wasteplaces, Hedges & Village Shrubberies.

Part Used : Root.

Utilization : Utilised so far as an antiseptic, antipyretic, depressant and antidote for snake bite (25 ml root extr. mixed with 50 ml water and given orally every 3 hrs) and in chronic indigestion, dysentery, fever and rheumatism.

Biological Activity : A decoction of alc. extr. of the roots produced slight paralysis in rats in doses of 1-2 g/kg body wt; and in doses of 10-20 g/kg it is antipyretic and depressant producing violent diarrhoea and eventually death.

References : Agarwal (2003), Ambasta (1986), Anonymous (1959), Chatterjee & Pakrashi (1997), Maiti & Mishra (2000).

Kaempferia rotunda Linn.

Family: Zingiberaceae.

Hindi & Sanskrit Names:

Hindi : Bhuichampa;

Sans. : Bhumichampa.

Distribution : Aromatic Herb Distributed Throughout India (Assam, Meghalaya, Mizoram, Tripura).

Part Used : Root.

Utilization : Known to be recorded for stomachache, reducing swellings and to promote suppuration.

References : Agarwal (2003), Ambasta (1986), Aswal (2001), Chopra *et al.* (1956), Dwivedi (1993), Sood & Prakash (2007), Watt (1889).

Knoxia corymbosa Willd.

Syn: *K. sumatrensis* (Retz.) DC.

Family: Rubiaceae.

Distribution : Throughout Tropical India & From Garhwal in the W. Himalaya to Sikkim Ascending to 1,500 m & in Khasia Hills Upto 1,200 m.

Part Used : Root.

Utilization : Given to stop vomiting due to food poisoning.

References : Panda (1996), Rastogi & Mehrotra (1999b).

Kydia calycina Roxb. Pi. Coram

Syn: *K. fraterna* Roxb.; *K.Pulverulenta* Ham.; *K. roxburghiana* Wight.

Family: Malvaceae.

Hindi Names:

Hindi : Choupultea, Pathra, Polao, Pula.

Distribution : Mixed Deciduous; Common in the Sub-Himalayan Tract.

Part Used : Root.

Utilization : Its paste mixed with oil applied as an embrocation for rheumatism, lumbago, oedema and as febrifuge. Root paste applied on oedema.

Active Constituents : Contains hibiscone, hibiscoqunone B and a new quinine.

References : Agarwal (2003), Ambasta (1986), Anonymous (1959), Asolkar *et al.* (1992), Dwivedi (1993), Kumar (2002), Varghese (1996), Watt (1889).

Kyllinga triceps Rottb.

Family: Cyperaceae.

Hindi & Sanskrit Names:

Hindi : Nirbisi;
Sans. : Apavisha.

Distribution : N.W. India, Sind, Rajputana, Gujarat, Deccan and Konkan.

Part Used : Root.

Utilization : Known to promote the action of liver and relieving thirst in fevers and diabetes.

References : Caius (2003), Sood *et al.* (2005a), Chopra *et al.* (1956).

Lablab purpurea (L.) Sweet

Syn: *Dolichos lablab* L.
Family: Fabaceae.

Part Used : Root.

Utilization : Decoction of old roots with salt recommended for the removal of cough.

References : Rastogi & Mehrotra (1998), Singh (2000).

Lagerstroemia parviflora Roxb.

Syn: *Fatioa nepaulensis* Wall.; *Lagerstroe fatioa* Blume
Family: Lythraceae.

Hindi Names:

Hindi : Bakli, Dhaura, Lendia, Sidi.

Distribution : Found Throughout India including Himalaya.

Part Used : Root bark.

Utilization : Root bark decoction taken daily two times to reduce fever.

References : Agarwal (2003), Ambasta (1986), Goud *et al.* (2000), Reddi *et al.* (2005), Watt (1889).

Lagerstroemia speciosa (L.) Pers.

Syn: *L. flosreginae* Retz.; *L. reginae* Roxb.; *Munchausia speciosa* L.

Family: Lythraceae.

English, Hindi & Sanskrit Names:

Eng. : Queen Crape Myrtle;
Hindi : Jarul;
Sans. : Arjuna, Tinisha.

Distribution : Assam to the W. Coast of the Peninsula; Cultivated commonly Along Roadsides & on Railway Platforms.

Part Used : Root.

Utilization : So far considered astringent, febrifuge and stimulant; and its decoction good for jaundice.

References : Agarwal (2003), Chhatterjee & Pakrashi (1997), Chopra *et al.* (1956), Kurian (1995), Lalramnghinglova (1996).

Lantana camara Linn. (Pl. 6D)

Syn: *Camara aculeata* (L.) Kuntze.; *L. aculeata* Linn.; *L. aculeata* ssp. *normalis* Voss.; *L. camara* auct. non L. *L. spinosa* L. *ex* Le Conite.

Family: Verbenaceae.

English, Hindi & Sanskrit Names:

Eng. : Lantana, Wild Sage;
Hindi : Chaneri, Kuri, Lantana, Ralmuniya;
Sans. : Caturangi, Vanacchedi.

Distribution : Distributed Throughout the Country in Areas Receiving 75 to over 500 cm Annual Rainfall on a Variety of Soils.

Part Used : Root.

Utilization : Decoction recommended for use against infection of the respiratory tract, mumps, a good wash for mouth ulcers and in gargles. Root paste applied on boils and given orally to goats to reduce body swellings caused due to consumption of poisonous plants. Its extr. checks toothache and fever. Also, powdered crushed root (one fourth teaspoonful) taken for seven days or the treatment continues if necessary for another 7 days for diabetes.

Active Constituents : Attributed to stachyose, verbacose, ajugose, verbascotetraose, lantaiursolic acid, lantana A, lantana B, theviridoside, theveside, geriposide, shanzhiside methyl ester, lamiridoside, 8-epiloganin, lantanose A and lantanose B.

References : Chatterjee & Pakrashi (1997), Jain (1996), Kurian (1995), Parrotta (2001), Rastogi & Mehrotra (1998), Reddi *et al.* (2005), Singh & Kumar (2000b), Sood & Thakur (2004), Subramaniam (2000).

Lantana indica Roxb.

Family: Verbenaceae.

Distribution : Throughout India & Baluchistan.

Part Used : Root.

Utilization : 200 ml extr. given orally for easy delivery and expelling the placenta and stomachache.

Active Constituents : Found to possess a new oleanane derivative-3β, 24-dihydroxyolean-12-en-28-oic acid, oleanolic acid, ursolic acds and 24-formyl-3-oxo-olean-12-en-28-oic acid.

References : Balasubramanian & Prasad (1996), Chopra *et al.* (1956), Hosagoudar & Henry (1996a), Rastogi & Mehrotra (1998).

Lasia aculeata Lour.

Syn: *L. heterophylla* Schoot; *L. spinosa* Thw.

Family: Araceae.

Hindi Name:

Hindi : Kanta.

Distribution : Swampy Grounds in Tropical Himalaya of E. India.

Part Used : Root.

Utilization : Decoction used as a remedy for affections of the throat and piles.

References : Agarwal (2003), Ambasta (1986), Caius (2003), Chopra *et al.* (1956), Kumar (2002).

Launaea acaulis (Roxb.) Kerr.

Family: Asteraceae.

Part Used : Root.

Utilization : Eaten raw in urinary complaints. Grounded root mixed with goat's milk, said to activate the secretion of milk in women. 5g powdered root given thrice daily with honey to control diabetes.

References : Aminuddin & Girach (1993), Yoganarasimhan (1981).

Launaea procumbens (Roxb.) Ramayya & Rajpal

Family: Asteraceae.

Part Used : Root.

Utilization : Crushed roots kept on the tooth cavities for 2-3 days against toothache.

Reference : Sharma & Singh (2000).

Lavatera cashmeriana Cambess.

Family : Malvaceae.

Distribution : Forest Clearings, Meadows, and Shrubberies of W. Himalaya (Kashmir to Uttaranchal).

Part Used : Root.

Utilization : Extr. diuretic and used in urinary irritation and kidney pain. Decoction efficacious for swelling in abdomen due to stoppage of urine.

References : Kaul (1997), Sharma (2003), Singh (1996).

Lawsonia inermis L.

Syn : *L. alba* Lam.; *L. spinosa* L.
Family : Lythraceae.

English, Hindi & Sanskrit Names:

Eng. : Camphire, Cypress Shrub, Egyptian Privet, Henna, Jamaica, Tree Mignonette;

Hindi : Mehndi, Hena;

Sans. : Madayantika, Mendhi, Nakharanjak, Ragagarbha, Ragangi, Ranjaka, Sahashara, Timira, Yavaneshta.

Part Used : Root.

Utilization : Paste mixed with sugar candy for jaundice. Pills prepared from pounded root with 'neem' leaf, coconut root and given to babies to check diarrhoea. Also, used for leprosy, preventing conception and drying up ulcers of the mouth and gums.

Active Constituents : Mannitol (0.7 %) and lawsaritol are the principle constituents.

References : Islam & Jha (2003), Kirtikar & Basu (1984), Mohanty & Padhy (1996), Parrotta (2001), Patil & Patil (2002), Rastogi & Mehrotra (1998, 2002), Reddi *et al.* (2005), Trivedi (2002).

Leea aequata L.

Syn: *L. hirta* Roxb. *ex* Hornem.
Family: Leeaceae.

Hindi & Sanskrit Names:

Hindi : Kakajangha, Masi;

Sans. : Dasi, Kaakajanghaa, Kaka, Kaki, Nadikaantaa, Sulomaasha.

Distribution : N.E. India, Orissa, and Andaman Islands.

Parts Used : Root. Root bark.

Utilization : Root paste applied externally on wounds. Known in India as bitter, acrid, hot, pungent, anthelmintic, vulnerary, antipyretic, alexiteric; useful in bronchitis, wounds (root paste), anaesthesia of skin, dyspepsia, bilious fevers, leprosy, itching and tuberculous ulcers. Root bark mixed with *Dioscorea pentaphylla* considered a strong antidote for poison.

References : Jamir & Rao (1990), Khare (2004), Kirtikar & Basu (1935).

Leea alata Edgew.

Family: Leeaceae.

Distribution : W. & N. E. Himalaya.

Part Used : Root.

Utilization : Decoction good for cold and cough. Patients suffering from jaundice given a bath of its decoction.

References : Jain *et al.* (1991), Kumar (2002).

Leea asiatica (L.) Rid.

Syn: *L. aspera* Edge.; *L.crispa* L.; *L. edgeworthii* Santapau; *L. herbacea* Ham.; *L. pumilla* Kurz.; *Phytolacca asiatica* L.

Family: Leeaceae.

Distribution : N.E. India, W. Tropical Himalaya & W. & S. Regions of Peninsular India.

Part Used : Root.

Utilization : Used to expel guinea worms. Massage of its juice mixed with mustard oil for a week heals sprain, followed by application of 'Sal-leaf' (*Shorea robusta* Gaertn.) bandage at night.

References : Maliya (2005), Parrotta (2001).

Leea compactiflora Kurz.

Syn: *L. aspera* Wall.; *L. robusta* auct. non Roxb.
Family: Leeaceae.

Distribution : Sikkim Himalaya & Khasia Mountains (330-1,600 m) & W. Peninsula.

Part Used : Root.

Utilization : Dried tuberous roots good against asthma and tuberculosis. Two teaspoonfuls powdered root taken twice a day with honey for one month to cure asthma permanently.

References : Hooker (1872), Sinha & Dixit (2003).

Leea indica Merr.

Syn: *Aquilicia ottilis* Gaertn.; *A. sambucina* Linn.*Gilbertia nalugu* DC.; *Gustonia nalugu* Spreng.; *L. acuminata* Wall.; *L. aequata* Wall. (not of DC.): *L. gigantea* Griff. *L. Laeta* Wall.; *L. ottilis* DC.; *L. staphylea* Roxb.; *L. sambucina* Willd.; *Ottilis zeylanica* Çaertn; *Staphylea indica* Burm.
Family: Leeaceae.

Hindi Name:

Hindi : Kukur Jihwa.

Distribution : Throughout India from Garhwal to Kerala.

Parts Used : Root. Rootstock.

Utilization : Decoction of roots antiemetic, colic, cooling, expectorant and sudorifice; also used for diarrhoea, dysentery and relieving thirst. Root stalk paste applied on boils.

References : Agarwal (2003), Ambasta (1986), Anonymous (1959), Chopra *et al.* (1956), Kothari & Moorthy (1996), Varghese (1996), Yoganarasimhan *et al.* (1981), Watt (1889).

Leea macrophylla Roxb. *ex* Hornem.

Syn: *L. angustifolia* Lawson; *L. cinerea* Lawson; *L. coriacea* Lawson; *L. diffusa* Lawson; *L. itegrifolia* Roxb.; *L. latifolia* Wall. *ex* Kurz.; *L. parallela* Wall. *ex* Lawson; *L. robusta* Roxb.; *L. talbotii* King *ex* Talbot; *L. venkobarowii* Gamble.
Family: Leeaceae.

Hindi & Sanskrit Names:

Hindi	:	Dholasamudra, Hatkam, Samudraka;
Sans.	:	Dholasamudrika, Gajakarna, Hastikarna.

Distribution : Throughout Hotter Parts of India.

Parts Used : Root. Root bark.

Utilization : Extr. of root bark given orally for stiff joints and rheumatic pains (2 spoonfuls once a day, till cure). The powder of root with clarified butter prescribed in the morning as an age-sustainer. In folk medicine, mucilaginous root applied as a paste or poultice to wounds, sores, for guinea worm, ring worm, allaying pain, sprains, bone fracture and chest pain (paste applied externally on chest twice a day for three days). Also, considered haemostatic, anodyne and given to cattle in diarrhoea. Tribals of Udaipur district recommend the crushed roots as a remedy for abdominal tumour and to cause abortion. Root juice diluted with water given for diarrhoea (2 to 3 spoonfuls thrice a day) by Gonds and Kolams. Pounded roots with rice water applied locally for alleviating enlargement of thyroid gland.

References : Ambasta (1986), Anonymous (1952), Chatterjee & Pakrashi (1997), Chopra *et al.* (1956), Dhiman (2003), Jain & Singh (1997), Khare (2004), Parrotta (2001), Rao & Henry (1996), Ravisankar & Henry (1992), Singh & Pandey (1998), Varghese (1996), Watt (1889).

Lepidagathis cristata Willd.

Family: Acanthaceae.

Hindi & Sanskrit Names:

Hindi	:	Bhuiterada;
Sans.	:	Bhuiteradu.

Distribution : Delhi, Rajasthan, Gujarat, Saurashtra & S. India.

Part Used : Root.

Utilization : Root extr. used orally as demulcent in madness and secondary syphilis.

References : Agarwal (2003), Reddi *et al.* (2005), Trivedi (2002).

Lepidagathis cuspidata Nees.

Family: Acanthaceae.

Distribution : Tropical India; extending from W. Himalaya to Coorg & Malabar.

Part Used : Root.

Utilization : Crushed root good against dental caries.

References : Hooker (1885), Sharma & Singh (2000).

Lepidagathis trinervis Nees.

Family: Acanthaceae.

Distribution : Rajputana Desert, M.P., Chota Nagpur, N.W. Himalaya, W. Peninsula & S. India.

Part Used : Root.

Utilization : Powdered root taken orally 2-3 times in dysentery.

References : Chopra *et al.* (1956), Girach *et al.* (1997).

Leptadenia reticulata (Retz.) Wt. & Arn.

Family: Asclepiadaceae.

English, Hindi & Sanskrit Names:

Eng. : Meda;
Hindi : Dori;
Sans. : Dodi-Shaaka, Hemavati, Jivaniya, Jivanti, Jivapushpa, Maangalya, Madhusravaa, Payaswini, Shaakashreshtha.

Distribution : Sub-Himalayan Tract of Punjab, U.P. & Throughout Peninsular India, Upto 900 m.

Part Used : Root.

Utilization : Root given orally as a galactagogue to cattle. Also, found to be good in skin affections.

Biological Activity : Alc. extr. exhibited antibacterial and abortifacient activities.

References : Agarwal (2003), Kharc (2004), Reddi *et al.* (2005), Trivedi (2002).

Leucas aspera (Willd.) Link.

Syn: *L. dimidiata* Benth.; *L. plukenetii* Benth.; *Phlomis aspera*Willd.; *P. esculenta* Roxb.; *P. plukenetii* Roth.; *P. obliqua* Ham.

Family: Lamiaceae.

English, Hindi & Sanskrit Names:

Eng. : White Dead Nettle;

Hindi : Chhota-Italkusa, Chota Halkusa;

Sans. : Dronapushpi.

Distribution : More or Less Throughout India.

Part Used : Root.

Utilization : Continuous use for 40 days makes on fully resistant to snake poison. Used in hemicrania and for cure of dog bite. Thick roots used as tooth brush.

Active Constituents : A new triterpenoid lactone-leucolactone isolated from roots.

References : Agarwal (2003), Jain *et al.* (1991), Khare (2004), Ragupathy & Mahadevan (1991), Rastogi & Mehrotra (1998), Reddi *et al.* (2005), Tewari *et al.* (2004), Watt (1889).

Leucas ciliata Benth.

Family: Lamiaceae.

Distribution : Temperate Himalaya (2,600-3,000 m), Khasia Mountains (1,300-1,700 m), Deccan Peninsula.

Part Used : Root.

Utilization : Paste swallowed in labour pain to check bleeding at the time of delivery.

References : Chaudhury & Neogi (2000), Hooker (1885).

Litsea chinensis Lam.

Syn: *L. glutinosa* (Lour.) C.B.Robins; *L. sebifera* Pers.
Family: Lauraceae.

English, Hindi & Sanskrit Names:

Eng. : Common Tallow Lurel;
Hindi : Garbijaur, Maidalakri, Medh, Menda, Singrauf;
Sans. : Adhavara, Bhavya, Dhira, Drava, Jivani, Madhura, Medini, Rasa, Sadhavi, Shalyaparni, Shreshtha, Snigdha, Vasa, Vibhavari.

Distribution : Throughout India, Ascending Upto 1,350 m in the Outer Himalaya.

Utilization : Known so far as cooling, aphrodisiac, blood purifier, galactagogue, astringent, expectorant and emmenagogue; also found to be useful in biliousness, burning sensations, bronchitis, consumption, fever, leprosy, inflammations, pains in the joints, thirst, throat troubles, diseases of the spleen and paralysis.

References : Agarwal (2003), Ambasta (1986), Anonymous (1952), Kirtikar & Basu (1984), Reddi *et al.* (2005).

Ludwigia adscendens (L.) Hara.

Syn: *Jussiaea repens* L.
Family: Onagraceae.

Distribution : Throughout the Plains of India.

Part Used : Root.

Utilization : About 5 teaspoonfuls of root juice mixed in a glass of warm milk, given on empty stomach for five consective days in leucorrhoea.

References : Borthakur (1993), Jain *et al.* (1991).

Lufffa acutangula (L.) Roxb.

Syn: *Cucumis acutangulus* L.; *L. acutangula* var. *amaura* (Roxb.) Cl.; *L. amara* Roxb.; *L. kleinii* W.& A.; *L. plukenetiana* DC.; *L. umbellata* (Klein *ex* Willd.)Roem.; *Momordica tubiflora* Wall.
Family: Cucurbitaceae.

English, Hindi & Sanskrit Names:

Eng. : Ribbed Gourd;
Hindi : Jhinga Tori, Kali Tori;
Sans. : Dharakoshataki, Jhongaka, Koshataki.

Distribution : Extensively Cultivated in Different Tropical Parts of India.

Part Used : Root.

Utilization : Used in dropsy and as laxative.

Active Constituents : Contains cucurbitacins B and D.

References : Ambasta (1986), Chatterjee & Pakrashi (1997), Watt (1889).

Lycnis coronaria Lamk.

Family: Caryophyllaceae.

English Name:

Eng. : Mullein Pink Rose Campion.

Distribution : Commonly Met Within Kashmir.

Part Used : Root.

Utilization : Reported to be useful for liver and lung problems and infection of mesentery lymph glands. Decoction given to cure mouth ulcers.

References : Ambasta (1986), Anonymous (1952), Kapur *et al.* (1996), Sood & Thakur (2004).

Lygodium flexuosum (L.) Sw.

Family: Schizaeaceae.

Hindi Name:

Hindi : Bhutraj.

Distribution : A Climbing Fern All Over India.

Part Used : Root.

Utilization : Given to cure loss of appetite, cholera, indigestion, anaemia, diarrhoea, stomach pain, cuts, wounds, scabies, ring worms, jaundice (one teaspoonful root juice given twice a daily till recovery), eczema, spermatorrhoea, rheumatism, sprains, debility, carbuncles, irregular menstruation, and as blood purifier. Mixture of root (15 g) and 7 black peppers made into paste and given (twice a day for four days) for dysmenorrhoea.

Active Constituents : Contains quercetin and 3, 5, 7, 3′ 4′-pentahydr oxyflavone-3-O-rutinoside.

References : Agarwal (2003), Ambasta (1986), Chopra *et al.* (1956), Jain *et al.* (1991), Lalramnghinglova (1996), Molla & Roy (1996), Rajendran & Sikarwar (2003), Rastogi & Mehrotra (2002), Reddi *et al.* (2005), Sahoo & Mudgal (1995), Saxena *et al.* (1997), Varghese (1996).

Lygodium pinnatifidum Sw.

Family: Schizaeaceae.

Part Used : Root.

Utilization : Paste applied on the site of snake bite as well as 20 g of its powder given orally every half an hour till recovery.

Reference : Jha & Verma (1996).

Madhuca longifolia (Koen.) Mac.Br. var. *latifolia* (Roxb.) Chevalier

Syn: *Bassia latifolia* Roxb.

Family: Sapotaceae.

English & Hindi Names:

Eng. : Mowra Butter Tree, South Indian Mahua;

Hindi : Mahua.

Part Used : Root.

Utilization : Paste orally taken for swelling and oedema. Also, its mixture with that of *Diospyros melanoxylon* good for local application in scorpion sting.

Reference : Varghese (1996).

Maerua apetala (Roth.) Jacobs.

Syn: *Capparis apetala* Roth.;
Niebuhria apetala (Roth.) Dunn. in Gamble.
Family: Capparidaceae.

Distribution : Hilly Areas of Nallamalais & Yerramalais.

Parts Used : Root. Root bark

Utilization : Paste prepared from its root (50 g) with leaves (20 g) of *Azadirachta indica* and garlic (5 ml) taken (one spoon thrice a day) for 15 days for leudoderma. Root bark paste also used for the same.

References : Rao & Henry (1996), Rao & Pullaiah (2001).

Maerua arenaria Hook.f. & Th.

Family: Capparidaceae.

Hindi Name:

Hindi : Vaka.

Distribution : Punjab, Sind, Gujarat, the Deccan, C. & S. India & Sri Lanka.

Part Used : Root.

Utilization : Used as an alterative, tonic and stimulant.

References : Agarwal (2003), Ambasta (1986), Anonymous (1965), Caius (2003), Chopra *et al.* (1956).

Maesa indica (Roxb.) DC.

Syn: *M. Dubia* (Wall.) DC.
Family: Myrsinaceae.

Hindi Name:

Hindi : Rammjani.

Distribution : Throughout Greater Parts of India Upto 2,000 m.

Part Used : Root.

Utilization : Paste with lemon juice applied on boils; on mixing with cow's milk, it is prescribed orally (thrice a day for a week) to cure syphilis. Root extr. administered orally for two days to cure dysentery.

References : Agarwal (2003), Chopra *et al.* (1956), Hosagoudar & Henry (1996b), Rastogi & Mehrotra (1999b), Singh & Kumar (2000b), Yoganarasimhan *et al.* (1981).

Malachra capitata L.

Family: Malvaceae.

Hindi Name:

Hindi : Banbhindi

Distribution : Naturalised in Hot & Damp Parts of India.

Part Used : Root.

Utilization : One glass of extr. given once a day to check bleeding with urine. Also, used in embrocations for rheumatism, lumbago and fever.

References : Agarwal (2003), Ambasta (1986), Anonymous (1962), Sharma & Singh (2000).

Mallotus leucocarpus (Kurz.) Airy-Shawii.

Family: Euphorbiaceae.

Part Used : Root.

Utilization : Decoction used for stomach problems.

Reference : Lalramnghinglova & Jha (1999).

Mallotus philippensis (Lam.) Muell.Arg.

Syn: *Croton cascarilloids* Rauesch. *ex* Steud.; *C. coccineus* Vahl.; *C. montanus* Willd.; *C. philippense* Lam.; *Mallotus reticulatus* Dunn.; *Rottlera tinctoria* Roxb.

Family: Euphorbiaceae.

English, Hindi & Sanskrit Names:

Eng. : Kamala Tree, Monkey Face Tree, Spoonwood;

Hindi : Kambhal, Kamela, Kamila, Raini, Rohni, Rora, Ruli, Rulu, Wassanthaganda;

Sans. : Kampilla, Kapilla, Karkash, Rechan, Rechanaka.

Distribution : Throughout Tropical India, Upto 1,500 m in the Outer Himalaya.

Part Used : Root.

Utilization : Useful as anthelmintic, tonic, restorative to pregnant woman, and applied externally to cure skin diseases.

References : Agarwal (2003), Caius (2003), Khare (2004), Parrotta (2001), Reddi *et al.* (2005), Sood & Thakur (2004), Subramaniam (2000).

Malva verticillata Linn.

Syn: *M. alchemillaefolia* Wall.; *M. neilgherrensis* Wight.
Family: Malvaceae.

Hindi Name:

Hindi : Lapha.

Distribution : Temperate Himalaya (2,000-4,000 m). Sikkim to Kumaon & Lahoul. Corn Fields of Nilgiris.

Part Used : Root.

Utilization : Used as an emetic in whooping cough.

References : Agarwal (2003), Ambasta (1986), Anonymous (1962), Viswanathan (2000), Watt (1889).

Manihot esculenta Crantz.

Syn: *Jatropha manihot* L.; *Manihot aipi* Pohl.; *M. dulcis* Pax.; *M. palmata* Muell.-Arg.; *M. utilissima* Pohl.
Family: Euphorbiaceae.

English & Hindi Names:

Eng. : Cassava, Manioc, Tapioca;
Hindi : Alukand, Simla Alu.

Distribution : Cultivated in Warmer & Moist Parts of India for its Rhizome.

Part Used : Root.

Utilization : Given in dryness of throat, giddiness, stomach pain, diarrhoea, chest pain, gastric disorders, sleeplessness and vomiting.

References : Agarwal (2003), Baburaj (2000), Rajan *et al.* (2000), Sood & Prakash (2007).

Manilkara hexandra (Roxb.) Dub.

Syn: *Mimusops hexandra* Roxb.
Family: Sapotaceae.

English, Hindi & Sanskrit Names:

- **Eng.** : Milk Tree;
- **Hindi** : Khair, Khirni;
- **Sans.** : Rajadani.

Distribution : C. India & Deccan; Often Cultivated in Drier Regions.

Part Used : Root.

Utilization : Paste of root in water applied externally on forehead for headache and improving eye sight.

Active Constituents : α-spinasterol, α-& β-amyrin acetates and cinnamates, taraxerol, taraxeryl acetate, ursolic acid, β-sitosterol glucoside, quercitol, quercetin, dihydroquercetin, cinnamic acid and hentriacontane are the active principles.

References : Agarwal (2003), Ambasta (1986), Rastogi & Mehrotra (1999b), Reddi *et al.* (2005), Shah & Singh (1990).

Marsdenia roylei Wight

Family: Asclepiadaceae.

Part Used : Root.

Utilization : Aq. extr. along with butter and black pepper given as an antidote to snake bite. For boils and blisters, its paste along with black pepper and cow's milk is efficacious.

Reference : Singh & Kumar (2000b).

Marsdenia tenacissima (Roxb.) Moon.

Family: Asclepiadaceae.

Hindi Names:

Hindi : Chiti, Jiti, Tongus.

Distribution : Himalaya & Assam.

Parts Used : Root. Root bark.

Utilization : Powdered root bark (1 teaspoon 2 times a day) given for 7 days for the treatment of syphilis. Root extr. (1-2 teaspoonfuls given 3 times a day for 3 days) employed in the treatment of high fever. Also, considered colic, strong purgative and useful in colds, asthma, cough especially in whooping cough and urinary trouble (root decoction).

Active Constituents : Reported to possess 17β-marsdenin, 17α-marsdenin, D-cymarose, β-D-glucosyl-L-thevetose and cinnamic acid.

Biological Activity : Alc. extr. of root abolished contractions induced by histamine and acetylcholine in guinea pigs ileum. Also, it induced CNS depression activity.

References : Agarwal (2003), Ambasta (1986), Anonymous (1962), Jain *et al.* (1991), Painuli & Maheshwari (1996), Rastogi & Mehrotra (1999a), Singh & Kumar (2000a), Singh & Prakash (1994).

Martynia annua L.

Syn: *Carpoceras angulata* A. Rich.; *Disteria angulosa* Raf.; *Martynia angulosa* Lam.; *M. diandra* Gloxin; *Vatkea diandra* O.Hoffm.

Family: Pedaliaceae.

English, Hindi & Sanskrit Names:

Eng. : Devil's Claw, Snake's Head, Tiger's Claw;
Hindi : Bichu, Hathajori;
Sans. : Kakanasa, Nrtyakundala.

Distribution : Naturalised in Many Tropical Areas, Roadsides & Rubbish Heaps.

Part Used : Root.

Utilization : Aq. extr. given orally to cure bronchial disorders, toothache, ansd to relieve painful urination associated with hardening of the urinary bladder sphincter due to severe attack of cold among the 'Baigas' in northeastern M.P.

References : Parrotta (2001), Rao & Henry (1996), Reddi *et al.* (2005), Sharma & Singh (2000), Verma *et al.* (1995).

Maytenus senegalensis (Lam.) Excell.

Syn: *Catha montana* (Roth.) G. Don; *Celastrus emarginatus* Willd.; *C. montanus* Roth. in Roem. & Schult.; *C. senegalensis* Lam.; *Gymnosporia montana* (Roth.) Benth.; *G. emarginata* (Willd.) Thw.; *Maytenus emarginata* (Willd.) Ding Hou.

Family: Celastraceae.

English, Hindi & Sanskrit Names:

Eng. : Thorny Staff Tree;
Hindi : Baikal, Kingani, Tondarsaijhad, Vingar;
Sans. : Vikankala, Vikantaki.

Distribution : Scrub Forests Throughout India.

Part Used : Root.

Utilization : Powder given with honey to cure duodenal ulcer, piles, gastric complaints, scorpion sting, rheumatism, swelling and migraine (paste).

Active Constituents : Contains iquesterin, pristimerin, tingenone, β-amyrin, β-sitosterol, lupenone and dulcitol.

References : Bhatt *et al.* (2002), Bhogaonkar & Devarkar (2002), Kothari & Londhe (2000), Parrotta (2001), Rastogi & Mehrotra (2001), Reddi *et al.* (2005).

Meconopsis aculeata Royle

Family: Papaveraceae.

Part Used : Root.

Utilization : Extr. (5-10 ml thrice a day) taken for the treatment of jaundice for a week.

Reference : Singh & Kumar (2000b).

Megacarpaea polyandra Benth.

Family: Brassicaceae.

Distribution : Occasionally Found on Open Sunny & Loose Slopes of N. Himalaya: 3,500-4,000 m.

Part Used : Root.

Utilization : Considered febrifuge and tonic; also used for dyspepsia.

References : Bhattacharjee (2001), Negi *et al.* (1993).

Melia azedarach Linn.

Syn: *M. bukayun* Royle; *M. sempervirens* Sw.

Family: Meliaceae.

English, Hindi & Sanskrit Names:

Eng. : Barbados Liliac, Bead Tree, Indian Liliac, Persian Liliac, Pride of China, Pride of India;

Hindi : Bakain, Bakarja, Bakayan, Betain, Deikna, Drewk, Mahanimb;

Sans. : Akshadru, Brihannimba, Dreka, Giripatra, Himadruma, Kaitarya, Karmuka, Mahadroksha, Nimbaka, Parvata, Sakaleyaka, Vishamushtika.

Distribution : Sub-Himalayan Tracts: 600-1,000 m.

Part Used : Root.

Utilization : Known in India as acrid, bitter, cooling, astringent to the bowels, anthelmintic; also for biliousness, tumours, pain in the heart; vomiting, leucoderma, belching, blood impurities, ulcers, headache, uterine pains after delivery, fever, burning sensations, urinary discharges, lung complaints and rat-bite poisoning.

Active Constituents : Roots afford apigennin-5-O-β-D-galactopyranoside.

References : Agarwal (2003), Dastur (1970), Kirtikar & Basu (1935), Kurian (1995), Rastogi & Mehrotra (2002), Watt (1889).

Melochia corchorifolia L.

Syn: *M. affinis* Wall.; *M. concatenata* Wall.; *M. pauciflora* Wall.; *M. supina* Linn.; *M. truncata* Willd.; *Riedleia concatenata* DC.; *R. corchorifolia* DC.; *R. supina* DC.; *R. truncata* W. & A.; *Sida cuneifolia* Roxb.; *Visenia corchorifolia* Spreng.

Family: Sterculiaceae.

Hindi Name:

Hindi : Bilpat.

Distribution : Hotter Parts of India on Roadsides.

Part Used : Root.

Utilization : About 5g powder given with a glass of water in snake bite and dysentery (root extr.).

References : Agarwal (2003), Shukla & Verma (1996), Watt (1889).

Melothria heterophylla Boiss.

Syn: *Solena amplexicaulis* (Lamk.) Gandhi; *S. heterophylla* (Lour.) Cogn.

Family: Cucurbitaceae.

Hindi Name:

Hindi : Tarali.

Distribution : Abundant in Various Parts of India.

Part Used : Roots.

Utilization : Known to be as invigorating, purgative, stimulant; useful in dysuria, spermatorrhoea, diabetes and gonorrhoea. Its juice along with cumin and sugar given in cold milk as remedy for spermatorrhoea, and paste for curing mouth ulcers, gastric problems and discomfort in stomach (when mixed with turmeric powder: 2:1).

Active Constituents : Contains columbin.

References : Agarwal (2003), Anonymous (1962), Chatterjee & Pakrashi (1997), Chopra *et al.* (1956), Rawat & Chowdhury (1998), Sood & Thakur (2004).

Melothria maderaspatana (Linn.) Cogn.

Syn: *Mukia scabrella* Arn.
Family: Cucurbitaceae.

Hindi Names:

Hindi : Agumaki, Bilari.

Distribution : Throughout India Ascending Upto 1,800 m.

Part Used : Root.

Utilization : Chewed for relief from toothache; its decoction given in flatulence.

Active Constituents : Columbin, m.p.195^0, isolated from roots.

References : Ambasta (1986), Anonymous (1962), Chopra *et al.* (1956), Rastogi & Mehrotra (1999b).

Melothria perpusilla (Blume.) Cogn.

Syn: *Zehneria hookeriana* Arn.
Family: Cucurbitaceae.

Hindi Name:

Hindi : Agumaki.

Part Used : Root.

Utilization : Used in diarrhoea and fever.

References : Ambasta (1986), Chopra *et al.* (1956).

Memecylon umbellatum Burm.f.

Syn: *M. edule* Roxb.; *M. tinctorium* Koen. *ex* Wight & Arn.
Family: Melastomaceae.

English, Hindi & Sanskrit Names:

Eng. : Ironwood Tree;
Hindi : Anjani;
Sans. : Anjani.

Distribution : Coastal Regions of W. Peninsular India, Forests of S. India, Sri Lanka, Andaman Islands, Orissa, Assam Bangladesh.

Parts Used : Root. Root bark.

Utilization : Bark extr. good for leucorrhoea whereas decoction of the root in small doses cures menorrhagia and gonorrhoea.

References : Agarwal (2003), Chatterjee & Pakrashi (1956), Chopra *et al.* (1956), Parrotta (2001), Rao & Henry (1996).

Merremia tridentata (L.) Hallier f. ssp. *hastata* (Desr.)

Syn: *Ipomoea angustifolia* C.B. Clarke non Jacq.; *Merremia hastata* Hallier f.

Family: Convolvulaceae.

Hindi & Sanskrit Names:

Hindi : Prasarini;
Sans. : Prasarini.

Distribution : Upper Gangetic Plains, Bihar, Bengal, Orissa (near Chilka Lake) S. India, Deccan & Gujarat.

Part Used : Root.

Utilization : Recorded so far to cure rheumatism (decoction), toothache, body pain in human beings and cows (paste), kidney diseases (extr.), hemiplegia, piles, swellings and urinary disorders.

References : Agarwal (2003), Ambasta (1986), Anonymous (1962), Samvatsar & Diwanji (2000), Sur & Halder (2002).

Meyna laxiflora Robyns.

Syn: *Vangueria spinosa* Hook.f. in part.

Family: Rubiaceae.

Hindi Names:

Hindi : Mainakanta, Mainphal, Muduna, Muyna.

Distribution : U.P., Bengal, Assam & Deccan.

Part Used : Root.

Utilization : Colic pain.

References : Agarwal (2003), Ambasta (1986), Kothari *et al.* (2003).

Mezoneurum cucullatum (Roxb.) W. & A.

Family: Fabaceae.

Distribution : Moist Forests From Kumaon Eastwards to Assam, Khasia & Lushai hills, Deccan, W. Ghats & Andaman Islands.

Part Used : Root.

Utilization : Root pills given (one pill prescribed twice daily) in stomach pain and gastric disorder.

References : Anonymous (1962), Molla & Roy (1996).

Michelia champaca L.

Family: Magnoliaceae.

English, Hindi & Sanskrit Names:

Eng. : Champaca;
Hindi : Champa, Champaca;
Sans. : Campakah.

Distribution : E. Himalaya, N.E. India & W. Ghats.

Part Used : Root. Root bark.

Utilization : Both considered purgative and emmenagogue; also good to cure abscess, inflammation, constipation, amenorrhoea and dysmenorrhoea. Infusion of root recommended as tonic, stomachic, carminative, diuretic, cooling, given in dyspepsia, fever, renal diseases and gonorrhoea.

References : Agarwal (2003), Anonymous (1962), Chopra *et al.* (1956), Dastur (1970), Dhiman (2003), Rastogi & Mehrotra (2002), Reddi *et al.* (2005).

Milletia pachycarpa Benth.

Family: Fabaceae.

Hindi Name:

Hindi : Bishloti.

Distribution : E. Himalaya.

Parts Used : Root. Root bark.

Utilization : Root bark chewed against toothache. Infusion of root considered antispasmodic and used as lotion for wounds, swellings and sprains.

Active Constituents : Attributed to cis-12a-hydroxyrot-2′-enonic acid, rotenone, cis-12a- hydroxyl-rotenone and rot-2′-enonic acid.

References : Agarwal (2003), Lalramnghinglova & Jha (1999), Rastogi & Mehrotra (2001).

Mimosa hamata Willd.

Family: Fabaceae.

Distribution : Punjab, C. & S. India.

Part Used : Root.

Utilization : Regarded as antidiabetic.

References : Bhatt *et al.* (2003), Rastogi & Mehrotra (2002).

Mimosa himalayana Gamble

Syn: *M. rubicaulis* Lam. ssp. *himalayana* (Gamble) Ohashi

Family: Fabaceae.

Hindi Name:

Hindi : Shiakanta.

Distribution : A Straggling Shrub Found Throughout India.

Part Used : Root.

Utilization : Paste of powdered root prepared in mustard oil applied on affected joints to cure pain. Powder also employed to check vomiting.

References : Agarwal (2003), Girach & Aminuddin (1995), Rastogi & Mehrotra (2002).

Mimosa invisa Mart.

Family: Fabaceae.

Part Used : Root.

Utilization : Decoction employed for dissolution of stones in gall bladder and kidney.

Reference : Lalramnghinglova & Jha (1997, 1999).

Mimosa pudica L. (Pl. 7A)

Family: Fabaceae.

English, Hindi & Sanskrit Names:

Eng. : Sensitive Plant, Touch-Me-Not;

Hindi : Chui-Mui, Lajwanti;

Sans. : Khadirkaa, Lajjalu, Lajjahah, Lajvanti, Namaskaari, Samkochini.

Distribution : Gardens & Fields Throughout India as a Hedge.

Parts Used : Root. Root bark.

Utilization : Bark used to cure diabetes. Decoction of roots efficacious in liver problems (once daily for 7 days), toothache (gargle), gravel and other urinary complaints. Its paste externally applied on wounds; on mixing with sulphur and coconut oil, it is applied to cure alti (reddish small boil like structure). Water extr. of root (10g) given orally (2 teaspoonfuls for 4 days) as diuretic, and to ladies suffering from leucorrhoea. Also, powdered roots given to cure malarial fever, diarrhoea, dysuria, urinary complaints, whooping cough, rickets, and as antiseptic, alterative, blood purifier and oral contraceptive.

Biological Activity : Root toxic in large quantities.

References : Agarwal (2003), Ambasta (1986), Anonymous (1962), Balasubramanian & Prasad (1996), Bhatt *et al.* (2003), Borthakur *et al.* (2004), Chatterjee & Pakrashi (1997), Chopra *et al.* (1956), Gogoi & Das (2003), Gogoi *et al.* (2003b), Islam & Jha (2003), Jain

(1991), Katewa *et al.* (2001), Khare (2004), Kirtikar & Basu (1935), Kurian (1995), Mohanty & Padhy (1996), Panthi & Chaudhery (2003), Parrotta (2001), Reddi *et al.* (2005), Sen & Behera (2003), Sharma (2000, 2003), Singh *et al.* (2003), Watt (1889).

Mirabilis jalapa L.

Family: Nyctaginaceae.

English, Hindi & Sanskrit Names:

Eng. : Four O' Clock Flower, Marvel of Peru;

Hindi : Gulabash, Gulabbas, Krishna-Keli;

Sans. : Krshnakeli, Sandhyakali, Sandhya-Raga, Tridosha, Trisondhi.

Distribution : Spread Throughout India, Often Planted.

Part Used : Root.

Utilization : Tuberous root considered purgative and aphrodisiac. 8g roots made into paste and applied externally in piles and contusions (once daily for three days). Extr. of bulbous root taken for diabetes whereas its paste given orally to cattle for cooling effect against sunstroke and for resistance against infectious diseases.

Active Constituents : A linear D-glucan characterised from root bulb.

References : Agarwal (2003), Chatterjee & Pakrashi (1997), Chopra *et al.* (1956), Dastur (1970), Islam (1996), Jain (1991), Jain *et al.* (1991), Kurian (1995), Lalramnghinglova (1996), Parrotta (2001), Rastogi & Mehrotra (2002), Saxena *et al.* (1997), Singh & Kumar (2000b), Sood & Thakur (2004), Upadhye & Kulkarni (2004), Watt (1889).

Mitragyna parviflora (Roxb.) Korth.

Syn: *Stephegyne parviflora* Korth.

Family: Rubiaceae.

Hindi Names:

Hindi : Kadassa, Kaim, Kalmi.

Distribution : Throughout Greater Parts of India.

Part Used : Root.

Utilization : Extr. given orally in snake bite; also for fevers and colic pains.

Active Constituents : Isolated constituents are isorhynchophylline and rhynchophylline.

References : Agarwal (2003), Ambasta (1986), Anonymous (1962), Rastogi & Mehrotra (1999a, b), Venkataratnam & Raju (2004).

Mollugo cerviana Serv.

Syn: *M. umbellata* Seringe; *Pharnaceum cerviana* Linn.

Family: Aizoaceae.

Hindi Name:

Hindi : Ghimasak.

Distribution : Dry Sandy Regions of Gangetic Plains, Rajasthan, Gujarat, Deccan, Karnataka & Orissa.

Part Used : Root.

Utilization : Paste in hot oil (mustard) used for gout and rheumatism.

References : Agarwal (2003), Ambasta (1986), Anonymous (1962), Watt (1889).

Momordica charantia L.

Syn: *Cucumis africanus* Bot. Reg.; *Momordica humulis* DC.; *M. muricata* DC.; *M. senegalensis* Lamk.

Family: Cucurbitaceae.

English, Hindi & Sanskrit Names:

Eng. : Bitter Gourd, Carilla Fruit;
Hindi : Bara Masiya, Karela, Poti-Kakar, Uchche;
Sans. : Karabella, Sushavi.

Distribution : Cultivated in Different Parts of India.

Part Used : Root.

Utilization : Reported to be astringent and used in haemorrhoids.

Active Constituents : These are 5, 24-dienol, glutinol and simiarenol.

References : Ambasta (1986), Chatterjee & Pakrashi (1997), Chopra *et al.* (1956), Dastur (1970).

Momordica dioica Roxb. *ex* Willd.

Syn: *Bryonia grandis* Wall.; *Momordica balsamina* Wall. not of others; *M. missionis* Wall.; *M. renigera* Wall.; *M. subangulata* Blume; *M. wallichii* Roem.; *Trichosanthes russeliana* Wall.

Family: Cucurbitaceae.

English, Hindi & Sanskrit Names:

Eng. : Small Bitter Gourd;
Hindi : Bankarela, Golkandra, Kaksa, Khaksa;
Sans. : Kakota, Karkotaki, Vahisi.

Distribution : Tropical Parts of India.

Part Used : Root.

Utilization : Reported to be used for diphtheria, poisonous animal bitten spot as an antidote, abortion, swollen breasts, bleeding piles, fevers, bowel troubles, urinary complaints, scorpion sting, cough, sedative and malarial fever.

References : Agarwal (2003), Ambasta (1986), Anonymous (1962), Chatterjee & Pakrashi (1997), Chopra *et al.* (1956), Jadeja (1999), Jain (1991), Jain *et al.* (1991), Kothari & Moorthy (1996), Rajendran & Sikarwar (2003), Reddi *et al.* (2005), Singh (2003), Singh & Maheshwari (1993), Srivastava *et al.* (2003), Watt (1889).

Monochoria vaginalis Presl. *ex* Kunth.

Family: Pontederiaceae.

Hindi Name:

Hindi : Nanka.

Distribution : Swampy Marshes of India.

Part Used : Root.

Utilization : Considered good for stomach and liver complaints, asthma and toothache.

References : Agarwal (2003), Ambasta (1986), Anonymous (1962), Chopra *et al.* (1956).

Monotropa uniflora Linn.

Family: Pyrolaceae.

Distribution : Temperate Himalaya (2,000-2,700 m), Khasia Mountains (1,600-2,000 m).

Part Used : Root.

Utilization : Considered sedative, antispasmodic and a nervine tonic.

References : Ambasta (1986), Anonymous (1962), Hooker (1888).

Morina longifolia Wall. *ex* DC.

Family: Dipsacaceae.

Hindi Name:

Hindi : Dipsa.

Distribution : Himalaya, Occurring at Chyangeel of Uttarkashi, Above 3, 000 m in the Grassy Slopes.

Part Used : Root.

Utilization : Known in India to remove intestinal worms in animals (extr.), suppuration of boils (poultice), killing worms in wounds (root paste) and pused ear (infusion of the root).

References : Agarwal (2003), Bhattacharjee (2001), Chandrasekar & Srivastava (2003), Singh & Kumar (2000b).

Morinda angustifolia Roxb.

Syn: *M. squarrosa* Hom.

Family: Rubiaceae.

Hindi Name:

Hindi : Barrhaldi.

Distribution : E. Himalaya, Assam, Bengal, Orissa & Andhra.

Part Used : Root.

Utilization : 100 ml extr. of 100 g pounded roots boiled in water taken orally daily for 15 days to reduce jaundice.

References : Agarwal (2003), Pandey *et al.* (1996), Watt (1889).

Morinda tinctoria Roxb. Hort. Beng.

Syn: *M. angustifolia* Roth.; *M. coreia* Buch.-Hans.; *M. stenophylla* Spr.; *M. tomentosa* Heyne *ex* Roth.

Family: Rubiaceae.

Hindi & Sanskrit Names:

Hindi : Achohhuka;
Sans. : Achchhuka.

Distribution : Throughout Greater Parts of India.

Part Used : Root.

Utilization : Considered styptic, astringent to the bowels, tonic (infusion as a bath), and dries boils. Decoction of its roots mixed with that of *Saraca indica* (1:1 part) honey is prescribed orally in leucorrhoea.

Active Constituents : Morindone diglucoside (morindonin) is the active principle.

References : Agarwal (2003), Ambasta (1986), Chopra *et al.* (1956), Hooker (1886), Kirtikar & Basu (1984), Rastogi & Mehrotra (1999a, b), Sarangi & Sahu (2004).

Moringa longifolia Wall. *ex* DC.

Family: Moringaceae.

Part Used : Root.

Utilization : Extr. used as a tonic.

Reference : Katewa *et al.* (2001).

Moringa pterygosperma Gaertn.

Syn: *Guilandina moringa* Linn.; *Hyperanthera decandra* Wall.; *H. moringa* Vahl.; *Moringa oleifera* Lam.; *M. polygona* DC.; *M. zeylanica* Pers.

Family: Moringaceae.

English, Hindi & Sanskrit Names:

Eng. : Drumstick Tree, Horse Radish Tree, Oil of Ben Tree;

Hindi : Mungna, Sahjan, Sainjana, Sanjana, Sohajna;

Sans. : Akshiva, Haritashaaka, Mochaka, Murangi, Raktaka, Shobhaanjana, Sigra, Sigruh.

Distribution : Sub-Himalayan Tracts from Chenab to U.P.; Cultivated Throughout India.

Parts Used : Root. Root bark.

Utilization : Root given to cure intermittent fever, epilepsy, hysteria, earache (juice), chronic rheumatism, dropsy, enlargement of spleen , soreness of mouth and throat, pain in gums due to dental caries, dyspepsia, fainting, giddiness, nervous debility, spasmodic affections of the bowels, flatulence, intermittent fever, intestinal worms, asthma, delivery problems, liver complaints and to induce vomiting; also used as counter irritant, blistering agent (fresh root mixed with mustard seeds and ginger for external use), abortifacient, rubifacient and stimulant in paralysis (fresh root). Its paste plastered with bamboo strips for setting bone fracture for 15 days and when mixed with salt considered efficacious to inflammatory swellings, rheumatic joints, snake bite and paralytic parts. About 50g of the paste administered orally, once only after delivery to expel placenta, and its decoction beneficial for body or chest pain, cold and fever. A decoction of root bark recommended for ascites, enlarged spleen or liver, internal or deep seated inflammation and calculus affections.

Active Constituents : These are traces of essential oils, phytosterols, resins, moringin which is identical to benzylamine and moringinine (root bark), spirochin, pterygospermin, isothiocynate, moringine and moringinene (root).

Biological Activity : Spirochin paralyses vagus nerve and has antibiotic action mainly in Gram +ve infections and used as an antiseptic and prophylactic on infected wound. Athomin showed bacterial action against *Vibrio cholerae* in both *in vitro* and *in vivo* conditions. Pterygospermin from the plant exhibited antibiotic action against both gram positive, gram negative and acid fast bacteria.

References : Agarwal (2003), Chatterjee & Pakrashi (1997), Chopra *et al.* (1956), Dastur (1970), Hosagoudar & Henry (1996a), Jain & Singh (1997), Jha & Kumar (2003), Khare (2004), Kurian (1995), Mughal *et al.* (2000), Parrotta (2001), Punjani (2003), Rastogi & Mehrotra (2001), Reddi *et al.* (2005), Sharma (2003), Singh (2002), Singh *et al.* (2003), Srivastava *et al* . (2003), Varghese (1996), Watt (1889).

Morus indica L.

Syn: *M. acidosa* Griff.; *M. alba* var. *cuspidata* Bureau; *M. alba* var. *indica* Bureau; *M. cuspidata* Wall.

Family: Moraceae.

English & Hindi Names:

Eng. : Mulberry;

Hindi : Shahtut, Tut.

Distribution : Temperate & Sub-Tropical Himalaya from Kashmir to Sikkim, Ascending to 2,100 m.

Parts Used : Root. Root bark.

Utilization : Pulverised roots given with black pepper and water in diptheria. Tea made from its root used to treat diarrhoea; also considered anthelmintic and astringent. Root bark regarded as diuretic, expectorant and tonic in oriental medicine.

Active Constituents : Therapeutic properties attributed to mulberrin, mulberrochromene, cyclomulberrin, cyclomulber rochromene (roots), morusin, cyclomorusin, morusinol, kuwanons A, B, C, oxydihydro-morusin, 2-arylbenzofuran-mulberrofuran A, and mulberrofuran A (root bark).

References : Rastogi & Mehrotra (1999a, 2001), Sharma (2003), Singh & Maheshwari (1993), Watt (1889).

Mucuna imbricata DC.

Syn: *Carpopogon imbricatum* Roxb.; *Mucuna monosperma* Wall.
Family: Fabaceae.

Distribution : E. Himalaya, Khasia, Asşam, Chittagong, Hills of W. Peninsula & Sri Lanka Upto1,300 m.

Part Used : Root.

Utilization : One cup decoction given 2 to 3 times a day for 15 days to cure madness and hysteria.

References : Hooker (1879), Tirkey (2004).

Mucuna pruriens (L.) DC.

Syn: *Carpopogon niveum* Roxb.; *C. pruriens* Roxb.; *Dolichos pruriens* L.; *Mucuna nivea* (Roxb.) DC.; *Stizolobium pruriens* Medicus
Family: Fabaceae.

English, Hindi & Sanskrit Names:

Eng. : Common Cowitch, Cow Hage;
Hindi : Baikhujani, Erka, Goncha, Kaunch, Kiwach;
Sans. : Atmagupta, Kapikaachhu, Kapikahciliu.

Distribution : Forests in Plains of India.

Part Used : Root.

Utilization : A paste prepared from the powdered root used externally to treat dropsy, falling hair, elephantiasis, itching, prickly heat and wounds. 5g powdered root given once daily with water to relieve constipation and to increase sexual vigour. Its extr. with honey given in cholera, liver, urine and kidney troubles. Inhaling vapours of powdered roots boiling in an earthen pot cures bodyache due to cold and cough. Also, considered stimulant, diuretic, tonic, purgative, emmenagogue and used in diseases of the nervous system such as facial paralysis and hemiplegia.

References : Agarwal (2003), Ambasta (1986), Aminuddin & Girach (1991), Anonymous (1962), Chatterjee & Pakrashi (1997), Chopra *et al.* (1956), Dastur (1970), Deshmukh *et al.* (2000), Girach *et al.* (1997), Islam & Jha (2003), Kumar & Lal (2002), Kurian (1995), Parrotta (2001), Sharma & Sood (1997), Singh (1999), Trivedi (2002), Varghese (1996), Watt (1889).

Mukia maderaspatana (L.) Cogn.

Syn: *Bryonia altheoids* DC.; *B. Maderaspatana* DC.; *B. scabrella* Linn.f.; *B. wightiana* Wall.; *Cucumis maderaspstana* Linn.; *Mukia scabrella* Arn.; *Trichosanthes dioca* Wall.

Family: Cucurbitaceae.

Distribution : Common Throughout India.

Part Used : Root.

Utilization : So far known to cure headache (root paste applied on forehead), toothache, colic and flatulence (decoction).

Active Constituents : Columbine, m.p. 195^0, characterised in the roots.

References : Balasubramanian & Prasad (1996), Hooker (1879), Punjani (2006), Rastogi & Mehrotra (1999b), Varghese (1996), Yoganarasimhan *et al.* (1981).

Mundulea sericea (Willd.) A. Cheval.

Syn: *M. suberosa* DC. Benth.

Family: Fabaceae.

Hindi Names:

Hindi : Supti, Verribilludu.

Distribution : Rocky Hills of W. & S. India.

Part Used : Root bark.

Utilization : 200 ml decoction of paste prepared by pounding its bark with stem bark of *Sapindus emarginatus* and *Azadirachta indica* and garlic in equal quantities given orally for about 3 days to cure post-delivery pains, and as an antiseptic and general tonic.

Active Constituents : Contains, unetone and mundulane, m.p. 179°.

References : Agarwal (2003), Ambasta (1986), Kumar & Pullaiah (1998), Rastogi & Mehrotra (1999a).

Murraya koenigii (L.) Spreng.

Syn: *Bergera koenigii*

Family: Rutaceae.

English, Hindi & Sanskrit Names:

Eng. : Curry Leaf Tree, Limblee Tree;

Hindi : Barsanga, Gandhala, Kathnim, Kurry Patta, Mitha Neem.

Sans. : Karuveppilai.

Distribution : Foot Hills of Himalaya, Assam, Sikkim, Kerala, Tamil Nadu, A.P. & Maharashtra.

Parts Used : Root. Root bark.

Utilization : Juice of the root taken as mild purgative and to relieve pain associated with kidney. Bark considered antiemetic, carminative, stomachic, tonic and on grounding with jaggery (*Saccharum officinarum*) said to be good for dysentery (2 spoonfuls twice a day till cure).

Active Constituents : Pharmacologically active principles are girinibine, mukoline, mukolidine, mahanimbine, isomahanimbine, murrayacine and koenoline (root bark).

Biological Activity : Koenoline exhibited cytotoxic activity against KB-cell culture.

References : Ambasta (1986), Anonymous (1962), Bhattacharjee (2001), Chatterjee & Pakrashi (1997), Kurian (1995), Parmar & Kaushal (1982), Rao & Henry (1996), Rastogi & Mehrotra (1999a, b, 2002), Watt (1889).

Murraya paniculata (L.) Jack.

Syn: *Chalcas paniculata* L.; *Murraya exotica* sensu Trimen; *M. Odorata* Blanco; *M. sumatrana* Roxb.

Family: Rutaceae.

English, Hindi & Sanskrit Names:

Eng. : Honey Bush, Orange Jasmine;
Hindi : Bibsar, Juti, Kamini, Marchula;
Sans. : Ekangi, Muramamsi.

Distribution : Throughout India & Andaman Islands to 1,500 m.

Parts Used : Root. Root bark.

Utilization : Paste of powdered roots and lemon juice applied to cure aching tooth. Root bark considered anodyne and effective in cough, hysteria, rheumatism, and also taken internally as well as rubbed externally to rellieve bodyache.

Active Constituents : Properties attributed to de-N-methyl-acronycine, de-N-methylnoracronycine, noracronycine, skimmianine, yuehchukene, a fatty acid ester of murrangatin (paniculatin) and murraxonin.

Biological Activity : Root extr. showed anti-implantation activity in rats.

References : Chatterjee & Pakrashi (1997), Hosagoudar & Henry (1996b).

Musa sapientum L.

Syn: *M. paradisiaca* L.
Family: Musaceae.

English, Hindi & Sanskrit Names:

Eng. : Adams's Fig, Banana, Fig of India, Plantain;
Hindi : Amrit, Kachkula, Kela, Maozkula;
Sans. : Alabu, Ambusara, Amsumatphala, Balakapriya, Bhanuphala, Chamannvati, Dirghhapatra, Guchhaphala, Hastirishani, Mocha, Nisara, Rambha, Sakritphala.

Distribution : Some Parts of Chindwara District of M.P.

Part Used : Root.

Utilization : Its amulet tied around the neck during early morning on a Sunday to cure goitre; considered anthelmintic, tonic, appetizer, drink for persons under the influence of alcohol, enema (infusion given in diarrhoea), antiscorbutic (juice given in retention of urine and with ghee, sugar in gonorrhoea), alterative for goitre and other glandular diseases, biliousness, pain in the ear, menstrual disorders, diseases of the blood, diabetes insipidus, venereal diseases, anaemia, dyspepsia, leprosy, haemorrhages from the genital and air passages (the juice of tender roots). Decoction (20 ml) taken in the morning for one month as tonic in blood and venereal diseases. Juice beneficial as a tonic for congestion of the liver, to prevent or cure scurvy (when mixed with a little pure honey) and in case of inflamed mucous membranes of the urethra and vagina.

References : Agarwal (2003), Chopra *et al.* (1956), Dastur (1970), Dhiman (2003), Hembrom (1996), Kirtikar & Basu (1984), Kurian (1995), Sharma (2003).

Mussaenda erythrophylla Schum & Thom.

Family: Rubiaceae.

Part Used : Root.

Utilization : Useful for cough and chewed as an appetiser.

Reference : Ambasta (1986).

Mussaenda frondosa Linn.

Syn: *M. corymbosa* Roxb.; *M. flavescens* Ham.; *M. formosa* Linn.; *M. frendosa* var. *glabrata* Hook.f.; *M. glabrata* (Hook.f.) Hutch. *ex* Gamble; *M. villosa* Wall.; *M. zeylanica* Burm.

Family: Rubiaceae.

English, Hindi & Sanskrit Names:

Eng. : White Lady;
Hindi : Bebina, Bedina;
Sans. : Srivati.

Distribution : Tropical Parts of Himalaya from Nepal Eastwards. Also in Assam and Khasia Mountains.

Parts Used : Root. Root bark

Utilization : Bark chewed for gastric ulcer, heart burn, sensation in heart and hiccups while the roots used as a diuretic, demulcent, alterative, eye troubles, white leprosy, cough and fever.

References : Agarwal (2003), Ambasta (1986), Anonymous (1962), Chatterjee & Pakrashi (1997), Chopra *et al.* (1997), Jain (1991), Jamir (1997), Maya *et al.* (2003), Saklani & Jain (1994), Watt (1889).

Nardostachys grandiflora DC.

Syn: *N. jatamansi* DC.; *Valeriana jatamansi* auct. non Jones.

Family: Valerianaceae.

English, Hindi & Sanskrit Names:

Eng. : Indian Nard, Musk-Root;

Hindi : Bal-Chir, Bhootjata, Jatamansi;

Sans. : Bhuutajataa, Jatamansi, Jatilaa, Maansi, Naladaa, Sulomashaa, Tapaswini.

Distribution : Alpine Zones of C. & E. Himalaya Upto 6,000 m.

Part Used : Root.

Utilization : So far known as tonic, stimulant, diuretic, emmenagogue, deobstruent, carminative, pectoral, stomachic, laxative; and for increasing lustre of the eyes, spasmodic pain, hysteria, promoting growth and blackness of the hair, gleet, cough, asthma, pain in the chest, intestinal inflammation, kidney and lumbar troubles and dry wounds. Its extr. considered a very good medicine for cholera, burn and when mixed with 'ghee' for joint pain and rheumatism. Dried root boiled in coconut oil used as a hair tonic; decoction as a tonic and for epilepsy, gastric problem, and also infusion given to cure palpitation of heart, menopause and various other nervous diseases.

Active Constituents : Active constituents are jatamansone, jatamansic acid, lupeol, jatamansinone, jatamansinol, jatamansine, angelin, dihydrojatamansin, orosol, oroselol, angelic acid, seselin, α and β pinene, Δ^3 carenene, b-eudesmol, elemol, b-sitosterol, valeranone, valeranal, nardol, nardostachone, nardosinon, nardostachnol, 9-dehydroaristolene, actinidin, spirojatamol, carotene, aristolen-1- α-ol, 1 (10)- aristolen-2-one, calarene, calarenol, 1 (10)-dehydroaristolene, elemol, erythro-1-(3, 4-dimethoxyphenyl)-2-(2-methoxy-4 (E)-prophenylphenoxy)-propan-1-ol, erythro-1-(4-hydroxy-3-methoxyphenyl)-2-(2-methoxy-4 (E)-propenylphenoxy), propan-1-ol, n-hexaosane, n-hexacosanal, n-hexaosanylarachidate, isovalerate, 2α-maleine, β-maliene, a C_{15}-alcohol and its isovaleric ester, (+)-1-hydroxypinoresinol, jatamols A and B, jatamansic acid, nardol, patchouli, alcohol, α- & β-patchoulenes, (+),-pinoresinol, norseychelanone, seychellane, seychellene, spirojatamol, virolin, β-sitosterol, 9-aristolen-1-α-ol, and 1 (10)-aristolen-2-one.

Biological Activity : Various extr. (light petroleum, benzene, ethanol) of root showed sedative action in rats and hypotensive activity in cats.

References : Agarwal (2003), Ambasta (1986), Chatterjee & Pakrashi (1997), Chauhan (1999), Chhetri (2005), Chopra *et al.* (1956), Jain (1968, 1991), Khare (2004), Nautiyal *et al.* (2003), Negi *et al.* (1993), Rastogi & Mehrotra (1999b, 2001), Reddi *et al.* (2005), Said (1997), Sarin (1996), Sharma (2000, 2003), Sinha & Sinha (2001), Thakur *et al.* (1989).

Naregamia alata W. & A.

Family: Meliaceae.

English, Hindi & Sanskrit Names:

Eng. : Goanese Ipecacuanha;

Hindi : Tinpani;

Sans. : Amlavalli, Brihatpatra, Chhinagrantika, Drumaruha, Kandabahula, Kandalu, Triparnika.

Distribution : W. Ghats & Karnataka.

Part Used : Root.

Utilization : Considered sweet, cooling, alexiteric, vulnery, emetic, expectorant, cholagogue; also useful to cure asthma, biliousness, ulcers, acute dysentery, chronic bronchitis, rheumatism, enlarged spleen, malaria and chronic fevers.

Active Constituents : Contains heneicosane, m.p. 40^0, β-sitosterol, stearic and palmitic acids.

References : Agarwal (2003), Ambasta (1986), Chopra *et al.* (1956), Kirtikar & Basu (1935), Rastogi & Mehrotra (1999b), Thomas & Britto (2003).

Naringi crenulata (Roxb.) Nicol.

Syn: *Hesperethusa acidissima* Roem.; *H. ambigua* Roem.; *H. crenulata* Roxb.; *Limonia acidissima* auct. (non L.); *L. crenulata* Roxb.
Family: Rutaceae.

Distribution : Dry Hills of N.W. Himalaya Shimla & Kumaon, Ascending to 1,700 m, Assam, W. Peninsula.

Part Used : Root.

Utilization : Paste externally applied to check body pain.

Active Constituents : Sitosterol, marmesin, suberosin, suberenol and 7-methoxy-6-(2, 3-epoxy-3-methyl-butyl), coumarin, xanthotoxin, osthenol, osthol, aurapten, stigmasterol, isopimpinellin and integriquinolone are the active ingredients.

References : Balasubramanian & Prasad (1996), Hooker (1872), Jain (1991), Rastogi & Mehrotra (1999a, 2002).

Nelsonia campestris R.Br.

Syn: *N. canescens* (Lam.) Spreng.
Family: Acanthaceae.

Hindi Name:

Hindi : Patta Kamraj.

Distribution : Throughout India.

Part Used : Root.

Utilization : Extr. taken as an aphrodisiac tonic whereas its paste said to be good for treating boils, blisters on the tongue, throat, gastric complaints and wounds.

References : Ambasta (1986), Jain (1991), Jain *et al.* (1991), Shah & Singh (1990), Varghese (1996).

Nelumbo nucifera Gaertn.

Syn: *Cyamus nelumbus* Smith; *Nelumbium nelumbo* Druce.; *N. speciosum* Willd.; *Nelumbo indica* Pou.; *Nymphaea asiaticum* Rich.; *N. nelumbo* Linn.

Family: Nymphaeaceae.

English, Hindi & Sanskrit Names:

Eng. : Chinese Water Lily, Egyptian Bean, Indian Lotus, Lotus, Pythagorean Bean, Sacred Lotus;

Hindi : Ambuj, Kamal, Kanwal Kakri, Lalkamal, Padam, Padma;

Sans. : Abja, Arvinda, Jalajanma, Kunja, Pankaja, Pushkara, Putaka, Rajiva, Sarojanna, Sharada, Sharapadma, Shatapatra, Shuklapadma, Tumarasa.

Distribution : Throughout the Ponds in Warmer Parts of India, Upto 1, 800 m.

Part Used : Roots.

Utilization : Powdered root prescribed for piles, dysentery, dyspepsia, and its paste applied locally in ringworm and other cutaneous affections.

References : Caius (2003), Dastur (1970), Islam (1996), Khare (2004), Kirtikar & Basu (1935), Sharma (2003), Sood & Prakash (2007), Watt (1889).

Nerium oleander L.

Syn: *N. indicum* Nill.; *N. odorum* Soland.; *Thevitia peruviana* (Pers.) Merill.

Family: Apocynaceae.

English, Hindi & Sanskrit Names:

Eng. : Indian Oleander, Sweet Scented Oleander;

Hindi : Kaner, Karaber, Karber, Kuruvira;

Sans. : Asvamarakah, Gauripushpa, Hayamaara, Karavira, Raktaprasava, Raktapushpa, Siddhapushpa.

Distribution : Distributed in the Himalaya (Nepal to Kashmir Upto 1,950 m) Extending to Baluchistan & Afghanistan.

Parts Used : Root. Root bark.

Utilization : Root juice applied to haemorrhoids, ulcers, leprosy, snake bite, eczema, skin diseases, scorpion sting, inducing abortion, alleviating venereal diseases and ophthalmia. An oil extr. from root bark used in skin diseases of scaly nature. Powdered root bark (2g) given twice daily with honey for 30 days in paralysis.

Active Constituents : Found to possess plumericin, alpha-amyrin, beta-sitosterol, kaemferol, cardioactive glycosides named odorosides A-H (root bark), neriumosides A-1, A-2, B-1, B-2 & C-1 (cadenolide pigments), odorisides A, B, D-H, K-M, pregnenolone, karabin, neriodore, neriodorin, digitoxigenin-β-gentiotriosyl (1→4) β-D-digitaloside, uzarigenin- β-gentiobiosyl (1→4)-β-D-digitaloside, 5α-oleandrigenin,its β-D-digitaloside, β-D-glucosyl (1→4)-β-D-dignoside, β-D-glucosyl (1→4)-β-D-digitaloside, two pregnanolone glycosides-5α-pregnanolone-O-β-D-glucosyl [(2→1)-β-D-glucosyl] (6→1)-β-D-glucoside (III) and pregnenolone-β-D-apiosyl (1→6)-β-D-glucoside (IV).

References : Ambasta (1986), Chatterjee & Pakrashi (1997), Dastur (1970), Dhiman (2003), Katewa *et al.* (2001), Parrotta (2001), Patil & Patil (2002), Khare (2004), Rastogi & Mehrotra (1998), Sarin (1996), Sharma (2004), Thakur *et al.* (1989), Watt (1889).

Nymphaea pubescens Willd.

Syn: *Castalia pubescens* (Willd.) Blume Bijdr.; *Nymphaea lotus* Hook.f. & Thoms. non Linn.; *N. nouchali* Burm.f.; *N. rubra* Roxb. *ex* Salisb.

Family: Nymphaeaceae.

English, Hindi & Sanskrit Names:

Eng. : Indian Red Water Lily;

Hindi : Alli, Bhenght, Kanval, Koi, Koka;

Sans. : Alagandha, Alipriya, Alohita, Aravinda, Bhadra, Hallaka, Indivara, Kalharamu, Kamala, Kumuda, Raktambhoja, Shnapadma.

Distribution : All Over Warmer Parts of India.

Part Used : Roots.

Utilization : Powder prescribed for piles as a demulcent and for dysentery, diarrhoea and dyspepsia.

References : Agarwal (2003), Ambasta (1986), Kirtikar & Basu (1935), Srivastava *et al.* (2003).

Nymphaea stellata Willd.

Syn: *Castalia stellata* (Willd.) Blume Bijdr.

Family: Nymphaeaceae.

English, Hindi & Sanskrit Names:

Eng. : Indian Blue Water Lily;
Hindi : Lilophal, Nilkamal, Nilpadma;
Sans. : Asitapala, Indiwar, Kandota, Kuvalaya, Mridutpala, Nilaliya, Nilapadma, Nilotpala, Nilpankaja, Saugandhika, Utpala, Utpalaka.

Distribution : Warmer Parts of India.

Part Used : Rootstock.

Utilization : Efficacious in dyspepsia, diarrhoea and piles.

References : Kaul (1997), Kirtikar & Basu (1935), Sharma (2003).

Ochna obtusa DC. var. *gamblei* (King. *ex* Brandis.)

Syn: *O. beddomei* Gamble; *O. gamblei* King. *ex* Brandis; *O. squarrosa* auct non L.

Family: Ochnaceae.

English, Hindi & Sanskrit Names:

Eng. : Golden Champak;
Hindi : Rakatrohan;
Sans. : Kanakachampa.

Distribution : N. & Peninsular India from M. P. & Bihar Southwards to Karnataka & N. Tamil Nadu.

Part Used : Root.

Utilization : Extr. given twice a day for two days for dysmenorrhoea, and infusion drunk in snake bite. A decoction reported to be used to ease menstrual discomfort, asthma and consumption.

Active Constituents : Affords 5, 7, and 8-trimethoxy-3, 4 – methylene dioxyisoflavone.

References : Parrotta (2001), Rastogi & Mehrotra (1998), Reddi *et al.* (2005), Sahoo & Mudgal (1995).

Ochna pumilla Ham. *ex* D. Don

Family: Ochnaceae.

Distribution : Foot of Himalaya (Kumaon to Sikkim), Bihar, Chota Nagpur.

Part Used : Root.

Utilization : One glassful decoction of its root mixed with a pinch of black pepper powder taken for treatment of urinary troubles (twice a day for 12 days). Fresh root paste (one teaspoonful) taken with a cup of milk in early morning for one month to cure weak eye sight. Also, roots known to be useful for epilepsy, piles (extr. given for 15-20 days), antidote to snake bite, menstrual complaints (decoction), consumption, asthma, dysentery and diarrhoea (extr. given 3 times for 3 days).

References : Ambasta (1986), Anonymous (1966), Chopra *et al.* (1956), Singh & Prakash (1994), Tirkey (2004).

Ochna squarrosa L.

Syn: *O. lucida* Lamk.; *O. nitida* Thunb.; *O. obtusata* DC.

Family: Ochnaceae.

Sanskrit Name:

Sans. : Kanakchampa.

Distribution : Bengal, Assam, S.India, Sri Lanka & Myanmar.

Part Used : Root.

Utilization : Root paste pills (two pills taken twice a day for one month) good against tuberculosis. Also, massaged on chest twice a day and/or half teaspoonful paste prescribed orally with a cup of water (once a day for one week) to treat chest pain. Decoction used for asthma, consumption and menstrual complaints.

References : Ambasta (1986), Tirkey (2004), Watt (1889).

Ocimum sanctum L.

Syn: *O. inodurum* Burm.; *O. tenuiflorum* L.
Family: Lamiaceae.

English, Hindi & Sanskrit Names:

Eng. : Holy Basil, Monk's Basil, Rough Basil, Sacred Basil;
Hindi : Branda, Kali Tulsi, Krishna Tulsi, Tulsi;
Sans. : Arjaka, Bhutagni, Bhutapatri, Brinda, Devadundubhi, Divya, Haripriya, Krishnamula, Krishnatulsi, Suravalli, Vaishnavi, Vishnupatni, Vrinda.

Distribution : Throughout India: Planted in the Temples & Upto 1, 800 m in the Himalaya.

Part Used : Root.

Utilization : About 25 ml root paste with about 50 ml water given orally thrice a day in snake bite. Its decoction given as a diaphoretic in malarial fevers and poultice of the fresh roots applied to the affected parts for reducing pain caused by stings of bees, wasps and the problems caused by worms and leeches. Root dipped in water overnight and one glassful taken in early morning for 7 days in menorrhagia and as aphrodisiac tonic.

Active Constituents : Reported to contain alkaloids, glycosides, saponins and tannins.

References : Agarwal (2003), Ambasta (1986), Anonymous (1966), Bhattacharjee (2001), Caius (2003), Chatterjee & Pakrashi (1997), Chopra *et al.* (1956), Dastur (1970), Jain (1991), Khare (2004), Kurian (1995), Maiti & Mishra (2000), Parrotta (2001), Reddi *et al.* (2005), Saini (1996a), Sood *et al.* (2005b),Thakur *et al.* (1989), Watt (1889).

Olax scandens Roxb.

Family: Olacaceae.

Hindi Name:

Hindi : Dheniani.

Distribution : Banks in the Sub-Himalayan Tracts from Kumaon to E. India, Deccan & W. Ghats.

Part Used : Root.

Utilization : Extr. taken orally to give relief from muscle catch and heat stroke.

References : Agarwal (2003), Anonymous (1966), Seetharam *et al.* (1999).

Onosoma echioids Linn.

Syn: *O. hispidum* Wall. *ex* D. Don

Family: Boraginaceae.

Hindi Names:

Hindi : Newar Maharangi, Ratanjyoti, Rattanjot.

Distribution : W. Himalaya from Kashmir to Kumaon.

Part Used : Root.

Utilization : Bruised root applied to various forms of cutaneous eruptions, wounds, cuts, ulcers and swellings.

Active Constituents : Possesses ursolic acid, two naphthoquinones-onosone A and onosone B.

References : Agarwal (20030, Anonymous (1966), Chopra *et al.* (1956), Jain (1991), Rastogi & Mehrotra (1998), Sharma & Rana (2000).

Operculina turpethum (L.) Silva Manso

Syn: *Ipomoea turpethum* R.Br.

Family: Convolvulaceae.

English, Hindi & Sanskrit Names:

Eng. : False Jalap, Indian Jalap, Rhubarb, Turbeth Root, Turbith;

Hindi : Nisotar, Nisoth, Pitohri, Tarbal;
Sans. : Ardhachandra, Aruma, Kalameshi, Kalingika, Malavika, Nandi, Pari Pakini, Rechani, Saha, Sara, Shyama, Tribhandi, Triputa, Vidala.

Distribution : Grown in the Indian Gardens as an Ornamental Plant.

Part Used : Root.

Utilization : Considered bitter, anthelmintic, purgative, antipyretic, alexiteric; also useful in ascites, leucoderma, itch, ulcers, constipation, abdominal troubles, inflammation, anaemia, fevers, biliousness, piles, erysipelas, tumours, jaundice, consumption, paralysis, bronchitis, pains in the joints and chest, ophthalmia and bites from insects..

Active Constituents : Turpethin.

References : Agarwal (2003), Anonymous (1966), Bhattacharjee (2001), Chopra *et al.* (1956), Dhiman (2003), Dwivedi (1993), Jain (1968), Kirtikar & Basu (1984), Mitaliya & Bhatt (2003).

Ophiorrhiza mungos L.

Family: Rubiaceae.

English, Hindi & Sanskrit Names:

Eng. : Mungoose Plant;
Hindi : Gandhanakuli, Sarahati;
Sans. : Sarpakshi.

Distribution : Khasia Hills. Assam. W. Peninsula, Ascending From 650-700 m. Andaman & Nicobar Islands.

Part Used : Root.

Utilization : Its paste applied on snake bitten spot as an antidote. Roots intensely bitter and used as stomach tonic and for cancer treatment.

Active Constituents : Contains 5α-ergost-7-en-3β-ol, 5α-ergost-8 (14) en-3β-ol and β-sitosterol.

References : Agarwal (2003), Ambasta (1986), Anonymous (1966), Chatterjee & Pakrashi (1997), Chopra *et al.* (1956), Rajendran & Henry (1994), Watt (1889).

Oreocnide integrifolia (Gaud.) Miq.

Syn: *Villebrunea integrifolia* Gaudich.
Family: Urticaceae.

English Name:

Eng. : Risa.

Distribution : Khasia Hills, Sikkim, Manipur.

Part Used : Root.

Utilization : Mashed with ginger and consumed to cure rashes and skin diseases.

References : Ambasta (1986), Khakongor & Joseph (1997), Kumar (2002).

Oroxylum indicum (L.) Vent.

Syn: *Bignonia indiaca* L.; *Calosanthes indica* (L.) Bl.; *Spathodea indica* Pers.
Family: Bignoniaceae.

English, Hindi & Sanskrit Names:

Eng. : Indian Trumpet Tree;

Hindi : Arlu, Assarsauna, Khakath, Pharrai, Pnarkhat, Sauna, Shyona, Sonapatha;

Sans. : Aratu, Bhalluka, Bhutapushpa, Mandukparna, Ruksha, Shyonak, Tuntuka.

Distribution : Greater Parts of India Upto 1,200 m, (Ravines & Moist Places in the Forests).

Parts Used : Root. Root bark.

Utilization : Decoction of root bark used in dropsy, eruptive fevers, diarrhoea, dysentery, ear complaints, rheumatism, fractures, leucoderma, leprosy, and as tonic, antifertility agent, astringent, bitter, aphrodisiac, diaphoretic, stomachic and anodyne whereas powdered roots considered good for checking pneumonia.

Active Constituents : Bark gave chrysin, baicalein, 6-methyl ether of baicalein= oroxylium A and oroxylin.

References : Ambasta (1986), Chatterjee & Pakrashi (1997), Dastur (1970), Dey (1973), Dhiman (2003), Jain *et al.* (2003), Khare (2004), Kirtikar & Basu (1984), Kothari & Londhe (2000), Kothari & Moorthy (1996), Kurian (1995), Mukherjee & Namhata (1990), Parrotta (2001), Rao & Henry (1996), Rastogi & Mehrotra (1998), Reddi *et al.* (2005), Sarin (1996), Saxena *et al.* (1997), Singh (1999), Singh *et al.* (2003), Varghese (1996), Watt (1889).

Orthosiphon rubicundus (D. Don) Benth.

Family: Lamiaceae.

Hindi Name:

Hindi : Tikipota.

Distribution : N.W. Himalaya (Kashmir to Nepal). Deccan, W. Ghats Upto 1,800 m.

Part Used : Root.

Utilization : Mixture prepared from equal quantity of root of this plant and those of 'Swet madar', (white flowered *Calotropis gigentia*), 'Devanasan' (*Croton roxburghii*), 'Bhutnasan' (*Zingiber montanum*) and 'Raktamahajal' (*Tectaria cicutaria*) given with cow's milk to ladies (10g twice a day, for 7 days after menstruation) to induce conception. To allay body heat and check blood discharge from urine, paste of mixture of its roots along with the roots of 'Muthuri', (*Smilax ovalifolia*), 'Akanbindi' (*Cissampelos pareira)* and a few black peppers, given twice a day for 3-4 days

References : Brahmam & Saxena (1990), Sood & Prakash (2007).

Osbeckia nepalensis Hook.

Family: Melastomaceae.

Hindi Name:

Hindi : Phatkala.

Distribution : Sub-Tropical Himalaya from Nepal to E. India.

Part Used : Root.

Utilization : Decoction taken orally in kidney troubles. Root also used for bodyache and swellings.

References : Agarwal (2003), Chhetri (2005), Saklani & Jain (1994).

Osbeckia stellata Wall.

Syn: *Melastoma crinita* Roxb.; *Osbeckia crinita* Naud.

Family: Melastomaceae.

Hindi Name:

Hindi : Tsulesi.

Distribution : Found on the Himalaya from Shimla to Bhutan, Bihar, Bengal & Khasia Hills.

Part Used : Root.

Utilization : Pounded roots used for relieving rheumatic pains. Root extr. stomachic, accelerate digestion and in cough.

References : Agarwal (2003), Jain (1991), Sood & Thakur (2004).

Osyris wightiana Wall. *ex* Wight.

Syn: *O. arborea* Wall.

Family: Santalaceae.

Hindi Name:

Hindi : Dalmi.

Distribution : Sub-Himalayan Tracts from Kangra to Assam. C. & S. India.

Part Used : Root bark.

Utilization : Hot paste of old root bark applied for 7-10 days in fracture or dislocation of veins.

References : Agarwal (2003), Ambasta (1986), Panthi & Chaudhary (2003).

Ougenia oojeinensis (Roxb.) Hochr.

Syn: *O. dalbergioids* Benth.

Family: Fabaceae.

English, Hindi & Sanskrit Names:

Eng. : Sandan;

Hindi : Panan, Panjan, Sandan, Tinsa;

Sans. : Tinisah, Tinisha.

Distribution : Throughout Outer Himalaya from Jammu to Bhutan, C. India & Deccan.

Part Used : Root.

Utilization : Aq. extr. given for 2 months for treating menstrual disorder and tumour.

References : Agarwal (2003), Ambasta (1986), Anonymous (1966), Reddi *et al.* (2005), Singh & Prakash (1994).

Oxyspora paniculata DC.

Family: Melastomaceae.

Distribution : Common at Lower Elevation Forming Gregarious Patches.

Part Used : Root.

Utilization : Decoction taken orally in kidney troubles.

References : Chhetri (2005), Rawat & Chowdhury (1998).

Paederia scandens (Lour.) Merr.

Syn: *P. foetida* L.

Family: Rubiaceae.

Hindi & Sanskrit Names:

Hindi : Gandhali, Somraji;

Sans. : Prasarani.

Distribution : Throughout N.E. Region (Bushes & Outskirts of Forest).

Part Used : Root.

Utilization : Juice found to be beneficial in piles, spasm, gout, rheumatism, inflammation of spleen, pain in chest and liver; and as aperient, anthelmintic, emetic, convalescent, carminative, emollient and colic.

References : Agarwal (2003), Ambasta (1986), Anonymous (1966), Chatterjee & Pakrashi (1997), Jain (1991), Kumar (2002), Watt (1889).

Paeonia indica Wall.

Syn: *P. officinalis* H.f. & T.

Family: Ranunculaceae.

English, Hindi & Sanskrit Names:

Eng. : Himalayan Peony, Paeony Rose;
Hindi : Mamekh, Udsalap;
Sans. : Svetamula.

Distribution : N. W. Himalaya, (Kashmir to Kumaon): 2,000-3,000 m.

Part Used : Root.

Utilization : Tuberous roots used for intestinal diseases, swellings, uterine diseases, colic pain, biliousness, nervous affections, dropsy, epilepsy, convolution and hysteria; also considered as blood purifier.

Active Constituents* : Known to contain starch (9.5 %), sucrose (5.4 %), malic acid (0.47 %), oxalic acid (0.36 %), tartaric acid (0.34 %), benzoic acid, n-alkanes, beta-amyrin, butyrospermol, cycloartenol, lupeol, linoleic acid, 24-methylenecycloartenol, cholesterol, campesterol, sitosterol, octanoic decanoic, lauric, myristic, myristoleic, palmitic, palmitoleic, stearic, oleic, linoleic acids and ethyl gallate.

References : Agarwal (2003), Ambasta (1986), Anonymous (1966), Chatterjee & Pakrashi (1997), Chauhan

(1999), Khare (2004), Kirtikar & Basu (1935), Sharma (2003).

Panax pseudoginseng Wall.

Family: Araliaceae.

Hindi Name:

Hindi	: Tapmari.

Distribution : Endemic in Sikkim Himalaya between 1,800-3,600 m.

Part Used : Root.

Utilization : Dried root powder taken as a stimulant, adaptogenic, anti-ageing medicine, thymoleptic, sedative, demulcent, stomachic, aphrodisiac, and used to cure neurasthenia, neuralgia, insomnia, hypotonia, amnesia, headache, convulsion, dysentery, palpitation, asthma, dyspepsia and cancer.

References : Chhetri (2005), Karnick (1994), Sharma (2003).

Pandanus fascicularis Lam.

Syn: *P. amaryllifolius* Roxb.; *P. laevis* Miq.; *P. latifolius* Hassk.; *P. odoratissimus* Linn.f.; *P. tectorius* Soland *ex* Parkin; *P. variegatus* Miq.

Family: Pandanaceae.

English, Hindi & Sanskrit Names:

Eng.	: Screw-Pine, Thatch Screw-Pine;
Hindi	: Vilayati Keora;
Sans.	: Ketaka, Ketaki, Suuchikapushpaa.

Distribution : Sea Coast of Indian Peninsula & Andaman Islands.

Part Used : Root.

Utilization : Brayed root in milk good against threatened abortion. Its juice administered in gynaecological disorders. An extr. of root applied topically for arresting premature greying of hair. Root oil considered antispasmodic and stimulant.

References : Agarwal (2003), Ambasta (1986), Khare (2004).

Pandanus lerum Jones.

Family: Pandanaceae.

English & Hindi Names:

Eng. : Nicobar Bread Fruit;
Hindi : Nicobari Keoraphal.

Distribution : Andaman & Nicobar Islands.

Part Used : Root.

Utilization : Paste in coconut oil applied on urinogenital organs against various venereal diseases.

References : Agarwal (2003), Dagar & Dagar (2000).

Paracalyx scariosus (Roxb.) Ali.

Syn: *Cylista scariosa* Roxb.

Family: Fabaceae.

Distribution : Peninsular India.

Part Used : Root.

Utilization : Externally applied (with other medicines) to reduce tumours; and its decoction considered a good remedy for leucorrhoea and dysentery.

References : Jain *et al.* (1991), Kothari *et al.* (2003), Rao & Henry (1996), Sood *et al.* (2005c).

Paramignya monophylla Wight.

Family: Rutaceae.

Hindi Name:

Hindi : Kadukangi.

Distribution : E. Himalaya, Orissa, A. P. & W. Ghats.

Part Used : Root.

Utilization : Employed as alterative, diuretic and given to cattle in bloody fluxes from abdomen.

References : Agarwal (2003), Ambasta (1986), Chopra *et al.* (1956), Watt (1889).

Paris polyphylla Smith.

Family: Liliaceae.

Distribution : Temperate Himalaya from Shimla to Bhutan.

Part Used : Root.

Utilization : Two doses (2-3g) of its powder given as anthelmintic to cure diarrhoea.

Active Constituents : Monoglucoside of diosgenin, m.p. 274^0 isolated from roots.

References : Agarwal (2003), Panthi & Chaudhary (2003), Rastogi & Mehrotra (1999b).

Parthenium hysterophorus Linn.

Family: Asteraceae.

English & Hindi Names:

Eng. : Congress Grass;
Hindi : Partheni.

Distribution : Throughout India.

Part Used : Root.

Utilization : Decoction used as a tonic and in dysentery. Poultice in 'ghee' recommended for slipped navel.

References : Agarwal (2003), Ambasta (1986), Chatterjee & Pakrashi (1997), Sood & Thakur (2004).

Passiflora nepalensis Wall.

Syn: *P. geminiflora* Don

Family: Passifloraceae.

Distribution : Sikkim (1,700-2,000 m), Khasia mountains (1,300-2,000 m).

Part Used : Root.

Utilization : Decoction taken orally in fever.

References : Chhetri (2005), Hooker (1879).

Pavetta indica L. var. *tomentosa* Roxb.

Syn: *Ixora tomentosa* DC.; *Pavetta burnonis* Wight; P. *canescens* Wall.; *P. crassicaulis* Bremek; *P. mollis* Wall.; *P. tomentosa* Roxb. *ex* Sm.; *P. velutina* Wall.

Family: Rubiaceae.

English, Hindi & Sanskrit Names:

Eng. : White Pavetta;
Hindi : Canchra, Caricara, Papri;
Sans. : Papata, Pappana.

Distribution : Spread Throughout India.

Parts Used : Root. Root bark.

Utilization : Supposed to possess purgative, aperient, diuretic and tonic properties and also prescribed in visceral obstruction, headache, urinary diseases, boils and itches (paste), constipation, dropsical affections, jaundice (½-1 teaspoon twice daily for 12-15 days), and dysentery (root extr.). Its extr. administered for toothache and tightening of teeth and poultice applied in boils and itches. Extr. of root bark crushed with that of *Alangium salvifolium* and garlic (*Allium sativum*) given for epilepsy (3 spoonfuls twice a day for 21 days).

Active Constituents : D-mannitol isolated from root bark.

References : Agarwal (2003), Ambasta (1986), Anonymous (1966), Chatterjee & Pakrashi (1997), Chopra *et al.* (1956), Hosagoudar & Henry (1996b), Kshirsagar & Singh (2000b), Kumar (2000), Rao & Henry (1996), Sood & Thakur (2004), Venkataratnam & Raju (2004), Yoganarasimhan *et al.* (1981), Watt (1889).

Pavonia odorata Willd.

Syn: *Hibiscus odoratus* Roxb.; *Pavonia romborua* Wall.; *P. rosea* Wall.; *P. sidoides.*

Family: Malvaceae.

English, Hindi & Sanskrit Names:

Eng. : Fragrant Sticky Mallow;

Hindi : Bala, Sugandha Bala;
Sans. : Ambu, Baalaka, Barhishtha, Hrivera, Jala, Kacha, Nira, Murdhaja, Toya, Udichi, Valakah.

Distribution : Open Woods & Wasteplaces in the Deccan Peninsula, Parts of W. Bengal, Bihar, Orissa, U. P. & Rajasthan.

Part Used : Root.

Utilization : Known in India as refrigerant, antipyretic, stomachic and astringent. In combination with fruits of *Aegle marmelos*, it is found to be good for dysentery, intestinal inflammations, and haemorrhages. Bruised roots mixed with 'ghee' applied in the form of a poultice to treat erysipelas.

Active Constituents : Found to contain isovaleric acid, isovaleraldehyde, aromadendrene, pavonene, alpha-terpinene, azulene and pavonenol.

References : Agarwal (2003), Ambasta (1986), Anonymous (1966), Chatterjee & Pakrashi (1997), Chopra *et al.* (1956), Khare (2004), Parrotta (2001), Watt (1889).

Pentanema cernuum (Dalz.) Ling.

Family: Asteraceae.

Part Used : Root.

Utilization : Powder given for abortion.

Reference : Bhogaonkar & Devarkar (2002).

Pentanema indicum (L.) Ling.

Syn: *Inula indica* L.; *Vicoa auriculata* Cass.; *V. indica* (L.) DC.

Family: Asteraceae.

Distribution : Throughout Indian Plains, Ascending Upto 1,800 m.

Part Used : Root.

Utilization : Powdered root or its decoction together with *Madhuca longifolia* taken to induce sterility in women. Its paste applied by the tribals of Udaipur on forehead against nervousness and on the wounds against snake bite. Extr. given for 2-3 days to cure tetanus and urinary disorders. Also found to be good in kidney, toothache, stomach and gum complaints.

References : Jain (1991, 2003), Rastogi & Mehrotra (2001), Singh & Pandey (1998), Varghese (1996).

Pentapetes phoenicea Linn.

Family: Sterculiaceae.

Hindi & Sanskrit Names:

Hindi : Dopahariya;

Sans. : Arkavallabha.

Distribution : Rice Fields, Marshy & Wet Places Throughout the Hotter Parts of India.

Part Used : Root.

Utilization : Said to be astringent, antibilious, antiphlegmonous and alleviative of wind and fever.

Reference : Anonymous (1966).

Pentatropis spiralis Decne.

Syn: *P. cynanchoids* R.Br.

Family: Typhaceae.

Hindi Names:

Hindi : Amberbel, Kauathodi.

Distribution : Punjab, Delhi, Upper Gangetic Plain, Rajasthan & Gujarat (Saurashtra).

Part Used : Root.

Utilization : Dry roots given in decoction as astringent, refrigerant, alterative and in gonorrhoea.

References : Anonymous (1966), Chopra *et al.* (1956).

Pepromia pellucida A. Dietr.

Family: Piperaceae.

Hindi Name:

Hindi : Krimi Nas Sag.

Distribution : Many Parts of India.

Part Used : Root.

Utilization : Roots taken orally to check miscarriage during pregnancy.

References : Agarwal (2003), Sarangi & Sahu (2004).

Pergularia daemia (Forssk.) Chiov.

Syn: *Asclepias daemia* Forssk.; *Cynanchum extensum* Jacq.; *Daemia extensa* R.Br.

Family: Asclepiadaceae.

English & Hindi Names:

Eng. : Whitlow Plant;
Hindi : Dudhi, Jutuk, Sagovani, Utranajutuka.

Distribution : Throughout India in Hedges & Shrubberies.

Part Used : Root. Root bark.

Utilization : Bark mixed with cow's milk prescribed as purgative and in rheumatic cases. 2g root powder taken with water to induce abortion within 2-3 months of pregnancy. Also, used to cure stomach pain, rheumatic swellings and diabetes. Decoction of its root with niger seed taken orally for abortion.

Active Constituents : Therapeutic properties due to α-amyrin, lupeol, calactin, calotropin, β-sitosterol and its glucoside.

References : Agarwal (2003), Anonymous (1966), Chatterjee & Pakrashi (1997), Chopra *et al.* (1956), Dastur (1970), Goud & Pullaiah (1996), Rao & Henry (1996), Rastogi & Mehrotra (1999a, b), Reddi *et al.* (2005), Sarangi & Sahu (2004), Tewari *et al.* (1996).

Peristylus constrictus (Lindle) Lindle

Family: Orchidaceae.

Part Used : Root.

Utilization : Powdered tuberous roots given with milk and sugar in sexual debility.

Reference : Bhogaonkar & Devarkar (2002).

Phoenix humilis Royle

Family: Arecaceae.

English & Hindi Names:

Eng. : Dwarf Date Palm;
Hindi : Pahari Khajur.

Distribution : Throughout Subtropical Regions of India.

Part Used : Root.

Utilization : Decoction given for cooling.

References : Agarwal (2003), Panthi & Chaudhary (2003).

Phoenix sylvestris (L.) Roxb.

Syn: *Elate sylvestris* Linn.
Family: Arecaceae.

English, Hindi & Sanskrit Names:

Eng. : Date-Sugar Palm, Indian Wine Palm, Sugar Palm, Wild Date Palm;
Hindi : Khaji, Khajur, Khajuri, Salma, Sendhi, Thakil, Thalma;
Sans. : Bhumikharjurika, Duraruha, Haluka, Kapila, Kharju, Mriduchhada, Nishreni, Skandhappala.

Distribution : Throughout India 1,500 m.

Part Used : Root.

Utilization : A small root piece wrapped in a cloth is tied to the hair of the expectant mother to start labour pain and to facilitate delivery and removed immediately after it. Also, chewed for quick relief in toothache and nervous debility. To relieve nausea, root paste of young plant is reportedly given to relieve nausea in S. W. Bengal.

References : Ambasta (1986), Bhatt *et al.* (2002), Caius (2003), Chopra *et al.* (1956), Dhiman (2003), Gosh & Sensarma (1997), Jain (1991), Pandey & Verma (2002), Parmar & Kaushal (1982), Parrotta (2001), Punjani (2002), Singh & Pandey (1998), Sood & Thakur (2004).

Phyllanthus amarus Schum & Thom.

Syn: *P. fraternus* Webster; *P. niruri* auct. non L.; *P. urinaria* linn.

Family: Euphorbiaceae.

Hindi & Sanskrit Names:

Hindi : Bhonyabali, Bhuinanvalah, Jangli Amla, Jaramal, Jaramla;

Sans. : Ajata, Amlika, Amritaa, Bahuphala, Bhumy Amalaki, Dridhapadi, Jada, Sahupushpi, Tamalaki.

Distribution : Most Widespread Weed Throughout India.

Part Used : Root.

Utilization : Roots and leaves made into a poultice with rice water for application on oedema and ulcers. Fresh roots used extensively in jaundice and with milk as a galactagogue. Its juice good against fever, stomachache, urinary diseases, and for treating camels suffering from digestive troubles in Rajasthan. In S. Orissa, the root paste is given to children under 2 years of age to treat diarrhoea.

Active Constituents : Known to yield 3, 5, 7-trihydroxyflavonol-4′-O-α-L-(-)-ramnopyranoside, 5, 3′, 4′-trihydroxyflavonone-7-O-α–L(-)-rhamnopyranoside, lupa-20(29)-ene-3β-ol, nirurin (5, 6, 7, 4′-tetrahydroxy-8-(3-methyl but-2-aryl) flavone-5-rutinoside, estra diol, kaempferol-4′-rhamnopyranoside and eriodictyol-7-rhamnopyranoside and lup-20 (29)-en-3β-ol.

References : Agarwal (2003), Ambasta (1986), Anonymous (1969), Chatterjee & Pakrashi (1997), Hosagoudar & Henry (1996a), Jha & Kumar (2003), Khan *et al.* (2003), Khare (2004), Kirtikar & Basu (1984), Kumar & Pullaiah (2000), Parrotta (2001), Rastogi & Mehrotra (1999a, b), Reddi *et al.* (2005), Thakur *et al.* (1989), Upadhyay & Chauhan (2000).

Phyllanthus reticulatus Poir.

Syn: *Cicca microcarpa* Benth.; *C. reticulata* Kurz.; *Kirganelia reticulata* (poir.) Baill.; *Phyllanthus kiraganelia* Ham. & Roxb.; *P. microcarpus* Muell.; *P. multiflorus* Willd.; *Rhamnus zeylanicus* Burm.

Family: Euphorbiaceae.

English, Hindi & Sanskrit Names:

Eng. : Blackhoney Shrub;
Hindi : Kala-Maimuda, Panjhuli, Panjoli;
Sans. : Krishna-Kamboji.

Distribution : Shrubberies & Hedges in the Plains of Tropical India.

Parts Used : Root. Root bark.

Utilization : Root cures cough and asthma while bark considered alterative, astringent, attenuant and diuretic.

Active Constituents : Active properties attributed to octacosanol, taraxeryl acetate, friedelin, epifriedelinol, taraxerone, betulin, b-sitosterol, glochidinol, and 21α-hydroxyfriedelan-3-one.

References : Agarwal (2003), Chatterjee & Pakrashi (1997), Rastogi & Mehrotra (2001), Reddi *et al.* (2005), Watt (1889).

Phyllanthus urinaria

Family: Euphorbiaceae.

Part Used : Root.

Utilization : Used for fracture and toothache for which it is tied around the affected part (fracture) and the neck (toothache).

Reference : Varghese (1996).

Phyllanthus virgatus Forst.f.

Syn: *P. simplex* Retz.
Family: Euphorbiaceae.

Part Used : Root.

Utilization : Root seasoned in mustard oil good for migraine and other headaches; also used for abscess (mammary).

Reference : Varghese (1996).

Physalis minima Linn.

Family: Solanaceae.

English, Hindi & Sanskrit Names:

Eng.	:	Country Gooseberry;
Hindi	:	Bondapariya, Chirpoti, Tulatipati;
Sans.	:	Lakshmipriya, Tankari.

Distribution : Throughout India; a Common Weed in Cultivated Fields & Other Open, Disturbed Sites Upto 2,000 m in the Himalaya.

Part Used : Root.

Utilization : Good for burning sensation in stomach, gastric disorders; its paste also given in dog bite.

References : Jain *et al.* (1991), Parrotta (2001), Reddi *et al.* (2005), Saklani & Jain (1994), Sur & Halder (2002).

Picrorhiza kurrooa Royle *ex* Benth.

Family: Scrophulariaceae.

English, Hindi & Sanskrit Names:

Eng.	:	Picrorhiza;
Hindi	:	Katki, Kuru, Kutki;
Sans.	:	Anjani, Arishta, Ashoka, Dvijanghni, Katuka, Katuki, Katurohini, Shatparvaa.

Distribution : N.E. Region: 2,700-4,500 m.

Part Used : Root.

Utilization : Infusion taken in case of stomachache and dysentery. Reported to have beneficial effect in abdominal pain, anaemia, asthma, cold, diarrhoea, dyspepsia, dysentery, malaria, fever, urinary discharges, hiccups, influenza, blood troubles, burning sensations, dropsy, leucoderma, colic, intestinal catarrh, jaundice, disorders of stomach and liver; also as purgative, antiperiodic, cholagogue, stomachic, laxative (small doses) and cathartic (large doses).

Active Constituents : These are 1, 6-cinnamoyl catapol, 25-acetoxy-2-β-glucosyloxy-3, 16, 20-trihydroxy-9-methyl-19-norlanosta-5, 23-dien-22-one (I), kutkin, picroside I, picriside II, D-mannitol, kutkiol, kutkisterol, 4-hydroxy-3-methoxyacetophenone (apocyanin), vanillic acid, cinnamoyl-α- and 6-cinnamoyl-β-D-glucopyranose, phenol glucosides, androsin, picein iridoiod glycosides, 6-feruloylcatapol, kutkin (a stable mixed crystal of two (-9 iridoid glycosides), 6′- O-trans-cinnamoyl catapol (picroside I), 10-O-vaniloyl catapol (kutikoside), 6-vanilloul catapol (picroside II), 6′-(4-hydroxy-3-methoxycinnamoyl) catapol (picroside III), minecoside, picrorhizin, veronicoside, cucurbitacin glycosides, arvenin-III, 25 acetoxy-2β-glucosyloxy-3, 16-dihydroxy-9-norlanosta-5, 23-dien-22-one, 25-acetoxy-2β-glucosyloxy-16, 20-dihydroxy-9-methyl-19-norlanosta-5, 23-diene-3, 11, 22-trione, 25-acetoxy, 2β-glucosyloxy-3, 16, 20-trihydroxy-9-methyl-19-noelanosta-5-en-22-one, 25-acetoxy-2β-glucosyloxy-3, 16, 20-trihydroxy-9-methyl-19-norlanosta-5, 23 (Z)-dien-11-22-one, 25-acetoxy-2β-glucosyloxy-3, 16 dihydroxy-4, 4, 9, 14-tetramethyl-19-norpregn-5-en—22-one, 2β-glucosyloxy-16, 20-dihydroxy-9-methyl-19-norlanosta-5, 24-dien-3, 11, 12-trione, 2β-glucosyloxy-16, 20, 22-trihydroxy-9 methyl-19-norlanosta-5,24-dien-3, 11-dione, 2β-glucosyloxy-3, 16, 2o, 25-tetrahydroxy-9-methyl-19-norlanosta-5en-22-one, 2-(6-O-cinnamoyl-β-glucosyloxy)-3, 16, 20, 25-tetrahydroxy-9-methyl-19 norlanosta-4-en-22-one, 2β-glucosyloxy-3, 16, 20, 25-tetrahydroxy-9-methyl-19-norlanosta-5, 23-dien-22-one and 2β-glucosyloxy-3, 16, 20, 25-pantahydroxy-9-methyl-19-norlanosta-5-en-2-one.

Biological Activity : Aq. extr. showed moderate antibacterial activity against *S. aureus* and *S. typhi* and marked inhibition against *E. coli*. Kutkin showed significant hepatoprotective activity in hepatic damage induced by galactosamine in rats and mastomys.

References : Ambasta (1986), Anonymous (1969), Arya & Prakash (2000), Aswal & Mehrotra (1994), Chatterjee & Pakrashi (1997), Chhetri (2005), Chopra *et al.* (1956), Aswal & Mehrotra (1994), Jain (1991), Jain & Goel (2005), Kala & Rawat (2001), Kapur & Singh (1996), Kaul (1997), Khare (2004), Kirtikar & Basu (1984), Nautiyal *et al.* (2003), Negi & Pant (1990), Rastogi &

Mehrotra (1999a, b, 2001, 2002), Samant *et al.* (2001), Sarin (1996), Sharma (2000, 2003), Sharma *et al.* (2003), Thakur *et al.* (1989), Watt (1889).

Pimpinella heyneana (DC.) Kurz.

Family: Apiaceae.

Distribution : Konkan, N. Kanara, Deccan, Circas and Chittagong.

Part Used : Root.

Utilization : Powder taken internally to cure rheumatism and fever.

References : Ambasta (1986), Chopra *et al.* (1956), Samvatsar & Diwanji (2000), Yoganarasimhan *et al.* (1981).

Piper diffusum Vahl.

Family: Piperaceae.

Part Used : Root.

Utilization : Given for indigestion in Assam.

Reference : Jain & Saklani (1992).

Piper longum L.

Syn: *Chavica roxburghii* Miq.; *C. sarmentosa* Miq.; *P. latifolium* Hunter; *P. sarmentosum* Wall.

Family: Piperaceae.

English, Hindi & Sanskrit Names:

Eng. : Indian Long Pepper, Joborandi, Long Pepper;

Hindi : Pipali, Piplamool, Pipli;

Sans. : Bhaarprakaasha, Chapalaa, Kanaa, Maagadhaa, Maagadhika, Pippali, Vaidehi.

Distribution : Wild in U.P., Assam, W. Bengal & S. India.

Part Used : Root.

Utilization : Infusion prescribed for expulsion of the placenta in Travancore and its extr. given orally for asthma and perpeural fever. Also, roots said to be beneficial for bronchial troubles, jaundice, paralysis, hair fall, asthma, throat infection, stomach pain, epilepsy, cough, fever, and as abortifacient, carminative and antidote to snake bite.

Active Constituents : Root gave the alkaloids piperine, piplartine (piperlongumine), piperloguminine, sesamine, and methyl-3, 4, 5-trimethoxycinnamate.

Biological Activity : Exhibited antifertility activity.

References : Agarwal (2003), Anonymous (1969), Chatterjee & Pakrashi (1997), Khare (2004), Rao & Henry (1996), Rastogi & Mehrotra (1999a, b), Reddi *et al.* (2005), Thakur *et al.* (1989), Watt (1889).

Piper nigrum L.

Family: Piperaceae.

English, Hindi & Sanskrit Names:

Eng. : Black Pepper;
Hindi : Golki, Golmorich, Kala-Morich, Kolukung, Murichung, Vellajung;
Sans. : Krishnahapusha, Marricha, Suvrrita, Uushana, Vellaja.

Distribution : Cultivated in the Hot & Moist Parts of India & Sri Lanka.

Part Used : Root.

Utilization : So far known in India to cure dysentery, gumache and vomiting. Decoction used as laxative, diuretic and for softening inflamed parts.

Active Constituents : Piperlongumine (piplartine) and piperlongumine characterised the roots.

Biological Activity : Piperlonguminine showed insecticidal activity against *Musca domestica*.

References : Ambasta (1986), Kurian (1995), Khare (2004), Rao & Henry (1996), Rastogi & Mehrotra (1999b), Reddi *et al.* (2005), Watt (1889).

Pithecellobium heterophyllum (Roxb.) Haridasan & Rao

Family: Fabaceae.

Part Used : Root.

Utilization : Poultice employed for applying to boils, swellings and pustules of small-pox.

Reference : Prakash & Mehrotra (1991).

Plantago major L.

Syn: *P. asiatica* Linn.; *P. erosa* Wall.; *P. longiscapa* Jacq.

Family: Plantaginaceae.

English & Hindi Names:

Eng. : Common Plantain, Englishman's Foot, Ribgrass, Wild Ispaghula;

Hindi : Luhhuriya, Sirapottagida.

Distribution : Moist & Wastelands in the Entire N.E. Region & Temperate Hills of S. India.

Part Used : Root.

Utilization : So far considered astringent and febrifuge; decoction used for cough and pneumonia, and poultice good as plaster on bone fracture.

Active Constituent : Aucubin.

References : Agarwal (2003), Ambasta (1986), Chatterjee & Pakrashi (1997), Chopra *et al.* (1956), Jain & Saklani (1992), Jain & Sharma (2000), Kumar (2002), Panda (1996), Sood & Thakur (2004).

Plantago ovata Forssk. (Pl. 7B)

Syn: *P. ispaghula* Roxb.

Family: Plantaginaceae.

English, Hindi & Sanskrit Names:

Eng. : Blond Psyllium, Ispaghula, Spogel Seeds;

Hindi : Isapgol, Isubgol;

Sans. : Isabgola, Ishadgola.

Distribution : Cultivated in N. Gujarat, Rajasthan & Haryana.

Part Used : Root.

Utilization : Beneficial in the treatment of diabetes.

References : Ambasta (1986), Beg *et al.* (2005), Khare (2004),

Pluchea indica Less.

Syn: *Baccharis indica* Linn.; *Conyza corymbosa* Roxb.;*C. indica* Miq.; *Pluchea foliolosa* DC.
Family: Asteraceae.

Hindi Name:

Hindi : Munjhu.

Distribution : Hedges, Wasteplaces, Salt Marshes in Tropical Parts of India.

Part Used : Root.

Utilization : Known to be considered as an astringent, antipyretic and diaphoretic in fever.

Active Constituents : These are 2-(prop-1-inyl)-5-(5, 6-dihydroxyhexa-1, 3-diinyl)-thiophene & 2-(prop-1-inyl)-5-(6-actoxy-5-hydroxyhexa-1, 3-dinyl)-thiophene, 17(21)-en-3β-yl acetate and boehmeryl acetate.

References : Agarwal (2003), Chatterjee & Pakrashi (1997), Chopra *et al.* (1956), Rastogi & Mehrotra (1998), Watt (1889).

Pluchea lanceolata Clarke

Syn: *Berthelotia lanceolata* DC.; *Conyza lanceolata* Wall.
Family: Asteraceae.

English, Hindi & Sanskrit Names:

Eng. : Indian Groundsel;
Hindi : Rasana, Rashana;
Sans. : Raasnikaa, Rasaa, Rasna, Rasnaa.

Distribution : Sandy or Saline Soils of Punjab, Upper Gangetic Plains, Rajasthan & Gujarat.

Part Used : Root.

Utilization : Decoction given in rheumatism and paralysis.

Active Constituents : Roots afford sorghumol and boehmeryl acetate.

Biological Activity : Petroleum ether and chloroform extr. of roots exhibited marked anti-inflammatory activity.

References : Ambasta (1986), Jadeja (1999), Khare (2004), Rastogi & Mehrotra (1998), Watt (1889).

Plumbago indica L.

Syn: *P. coccinea* Boiss.; *P. rosea* L.; *Thela coccinea* Lour.

Family: Plumbaginaceae.

English, Hindi & Sanskrit Names:

Eng. : Fire Plant, Officinal Leadwort, Rose-Coloured Leadwort;

Hindi : Chitra, Lalchita, Lalchitarak, Lalchitra, Raktacitra;

Sans. : Chitraha, Dahaka, Davanah, Dipika, Hrasvagni, Marjara, Pavaka, Raktasikha, Vahnimula, Vyala.

Distribution : An Ornamental Plant Cultivated Throughout India; Wild in Sikkim & Khasia Hills.

Part Used : Root.

Utilization : 2-3 pellets of dried root powder given early in the morning on empty stomach with boiled cow milk on the 5th day onwards of menstruation for 8 days to control conception and permanent sterility (medicine continued for 6 consecutive months). Also, known to cure rheumatism (root mixed with bland oil, applied externally), ulcers, syphilis, piles, paralytic affections, boils, skin diseases, leprosy, bronchitis, dyspepsia, cough, leucoderma, expelling worms in horses, scabies, amenorrhoea, anaemia, jaundice (2-3 pieces of root about 4 cm in length worn as an amulet around neck for 15- 20 days), elephantiasis, intermittent fever, anorexia, dyspepsia, flatulence and abortifacient; and as acrid, alterative, appetizer, gastric stimulant, sialagogue, vesicant and acronarcotic.

Active Constituents : Aracidy alcohol, 2-methyl-5-hydroxy-1, 4-naphthoquinone (plumbagin), β-sitosterol, α-naphthoquinone and α-naphthylamine are the chief constituents.

References : Agarwal (2003), Ambasta (1986), Anonymous (1969), Chatterjee & Pakrashi (1997), Chopra *et al.* (1956), Dastur (1970), Girach *et al.* (1997), Kirtikar & Basu (1984), Kurian (19950, Nayak *et al.* (2003), Pandey *et al.* (1996), Prasad *et al.* (2002), Rahman (2000), Rajendran *et al.* (2001), Reddi *et al.* (2005), Watt (1889).

Plumbago zeylanica L. (Pl. 7C)

Syn: *P. auriculata* Blume; *Thela alba* Lour.

Family: Plumbaginaceae.

English, Hindi & Sanskrit Names:

Eng. : Ceylon Leadwort, White Leadwort;
Hindi : Chita, Chiti, Chitra, Chitrak;
Sans. : Agni, Agnisikha, Bhali, Chitraka.

Distribution : Found Wild on Wastelands of Sub-Arid Areas of India & Often Cultivated in Gardens.

Parts Used : Root. Root bark.

Utilization : Two teaspoons of root infusion given twice daily for 10 days to treat spleen enlargement. About 5g of root powder or root paste (10g) in cow's milk given for 7 days twice daily to check conception. One teaspoonful of its paste in goat milk good against stomach disorder. In jaundice, 20-30-ml of root decoction with bark of *Oroxylum indicum* and roots of *Carissa congesta* given twice a day for 5-6 days. Pills made from grounded fresh roots, common salt and jaggery given orally to cure ring worm. Root paste with coconut oil applied on measles, leucodermic spot and itchy skin, also given to cure leprosy, dyspepsia, scabies, ringworm, pyorrhoea, rheumatism, piles, bladder stones, promote the appetite, liver enlargement, diarrhoea, dysentery, asthma, fever, cancer, filariasis, tumours, snake bite (decoction), and ulcers in the stomach and intestine. For curing hydrocele, root is tied to the opposite arm of the affected organ. Bark juice applied externally 2 times a day with thorough rubbing for severe headache and tincture employed as an antiperiodic.

Active Constituents : Plumbagin, 3-chloroplumbagin, drose-vrone, 3, 3'-biplumbagin (chitranone), zeylanone, isozeylanone acoumarin, elliptinone, slliptinone, chitranone with zeylinone and droserinone are the active ingredients of roots.

Biological Activity : Hot aq. extr. has been prescribed as an abortifacient.

References : Agarwal (2003), Ambasta (1986), Anonymous (1969), Balasubramanian & Prasad (1996), Bhandary *et al.* (1996), Bhattacharjee (2001), Brahmam & Saxena (1990), Chatterjee & Pakrashi (1997), Chauhan (1999), Chopra *et al.* (1956), Gogoi & Das (2003), Gogoi *et al.* (2003a), Hembrom & Goel (2005), Jain & Patole (2001), Jha & Verma (1996), Khare (2004), Khare & Khare (2000), Kirtikar & Basu (1984), Kumari & Narasimhan (2003), Nayak *et al.* (2003), Pullaiah & Kumar (1996), Rajendran *et al.* (2001), Rao & Henry (1996), Reddi *et al.* (2005), Reddy *et al.* (1996), Sarin (1996), Sharma (2003, 2004), Sharma & Singh (2000), Sikarwar (1996), Singh (1999), Singh & Kumar (2000b), Singh & Pandey (1998), Singh *et al.* (2003), Sinha & Sinha (2001), Sood & Thakur (2004), Tewari *et al.* (2001), Tirkey (2004), Varghese (1996), Verma & Chauhan (2006), Venkataratnam & Raju (2004), Watt (1889).

Plumeria alba L.

Family: Apocynaceae.

English & Hindi Names:

Eng. : Pagoda Tree, White Champa;
Hindi : Safed Champa.

Distribution : Occasionally Grown in Gardens.

Part Used : Root bark.

Utilization : Recorded so far for gonorrhoea, leprosy, blennorrhagia, herpes, stomachache and as abortifacient, anthelmintic, purgative, alterative, detergent; also its extr. used internally and externally for syphilitic ulcers.

Active Constituents : Plumieride characterised from roots.

Biological Activity : Plumieride exhibited antibacterial activity against gram positive and gram negative organisms.

References : Ambasta (1986), Anonymous (1969), Chopra *et al.* (1956), Rastogi and Mehrotra (1999a, b), Reddi *et al.* (2005).

Plumeria rubra L.

Syn: *P. acuminata* Ait.; *P. acutifolia* Poir.; *P. rubra* L. var. *acutifolia* Bailey.

Family: Apocynaceae.

English, Hindi & Sanskrit Names:

Eng. : Frangipani, Jasmine Tree, Pagoda Tree;
Hindi : Goburchampa, Golainchi, Gulanchin;
Sans. : Alariya, Devaganagalu.

Distribution : Grows Wild in Many Parts of India.

Parts Used : Root. Root bark.

Utilization : Crushed root cooked in castor oil applied to cure bruises. Also, reported to be powerful cathartic; useful in gleet, urinary discharges, tumours and rheumatic pains. Bark considered antherpetic, purgative, venereal sores and gonorrhoea.

Active Constituents : Reported to possess β-O-Caffeoyl plumieride, 13-deoxyplumieride, β-dihydroplumericinic acid glucosylester (plumenoside), 8-isoplumieride, 1α-plumieride, 1α-protoplumericine A, fulvoplumierin, plumericin, â-dihydroplumericinic acid, plumieride and 13-O-coumaroylplumieride.

References : Chatterjee & Pakrashi (1997), Chopra *et al.* (1956), Dhiman (2003), Kirtikar (2003), Rastogi & Mehrotra (1999a, b, 2002), Sen & Behera (2003), Sood *et al.* (2005c).

Podophyllum hexandrum Royle (Pl. 7D)

Syn: *P. emodi* Wall. *ex* Hook.f. & Thoms.;
P. emodi var. *hexandrum* (Royle) Chatt. *et* Muker.
Family: Berberidaceae.

English & Hindi Names:

Eng. : Duck's Foot, Himalayan May Apple, Indian Podophyllum, May Apple, Podophyllum;

Hindi : Bakrachimaka, Bhavanbakra, Bankakri, Papra, Papri.

Distribution : Found at 3,000-4,200 m in the Inner Himalaya from Kashmir to Sikkim.

Part Used : Root.

Utilization : Paste of the root applied on boils and given with luke warm water as an appetizer. Also, used for urticaria, dyspepsia, destructive action on cancerous tissues, stomachache, vomiting, cough, cuts, wounds, fever, gastric ulcer, hepatic diseases, tumour, skin diseases, and diarrhoea.

Active Constituents : Pharmacologically active principles are podophyllin, podophyllotoxin, 4'-demethylpodophyllotoxin, deoxypodophyllotoxin, β-peltatin, α-peltatin, podophyllotoxone, 4'-demethyliso-picropodophyllone, quercetin, kaempferol, iso-rhamnetin, quercetin-3-galactoside, dehydropodohyllotoxin, picropodophyllin, deoxypodophyllotoxin, podophyllotoxone, podophyllotoxin and 4'-demethylpodophyllotoxin.

Biological Activity : Many podophyllotoxins have been shown to possess anti-mitotic activity.

References : Ambasta (1986), Bhattacharjee (2001), Jha & Kumar (2003), Lal *et al.* (1996), Negi & Pant (1990), Rastogi & Mehrotra (2001), Samant *et al.* (2001), Sharma &

Sood (1997), Singh (2000), Singh & Kumar (2000b), Sinha & Sinha (2001), Sood *et al.* (2001), Sood *et al.* (2005c), Thakur *et al.* (1989).

Podophyllum peltatum L.

Syn: *P. emodi* Wall. *ex* Hook.f. & Thoms.
Family: Berberidaceae.

English Name:

Eng. : Common May Apple.

Part Used : Root.

Utilization : Tea prepared from roots effective in controlling constipation; also used in jaundice, syphilis, fever, liver ailments and cancer.

Reference : Bhattacharjee (2001).

Pogostemon auricularius (L.) Hassk.

Syn: *Dysophylla auricularia* (L.) Blume.; *Mentha auricularia* Linn.; *M. foetida* Burm.
Family: Lamiaceae.

Distribution : From Sikkim & Assam Through E. Bengal to Penang & Singapore. Deccan Peninsula From Goa Southwards.

Part Used : Root.

Utilization : Decoction good in flatulence.

References : Barua *et al.* (2000), Hooker (1885), Rastogi & Mehrotra (1998).

Pogostemon benghalensis (Burm.f.) Kuntz.

Syn: *P. plectranthoides* Desf.
Family: Lamiaceae.

Distribution : W. Himalaya, Lower Bengal, Bihar, Gujarat, Konkan, N. Kanara, N. Circars & Deccan.

Part Used : Root.

Utilization : Paste applied on body of children to alleviate fever. Ash of roots useful in whooping cough, and its poultice effective against joint pains.

References : Ambasta (1986), Aminuddin & Girach (1996), Chopra *et al.* (1956), Sharma (2004).

Pogostemon parviflorus Benth.

Syn: *P. frutescens* Grah.; *P. intermedius* Wall.; *P. pubescens* Benth.; *P. purpuricaulis* Dals.

Family: Lamiaceae.

Hindi Name:

Hindi : Phangla.

Distribution : Kumaon Hills to Assam & W. Ghats.

Part Used : Root.

Utilization : Known to be stimulant, antihaemorrhagic (uterine haemorrhage), antidote for snake bite and scorpion sting.

References : Agarwal (2003) Ambasta (1986), Anonymous (1969), Chopra *et al.* (1956), Kothari *et al.* (2003).

Polygala chinensis Linn.

Syn: *P. brachystachya* DC. non Blume; *P. glaucoids* Wight; *P. grandiflora* Hb. Wight; *P. prostrata* Willd.; *P. rothiana* W. & A.; *P. telephioids* Willd.; *P. tranquebarica* Mart.

Family: Polygalaceae.

English & Hindi Names:

Eng. : Common Indian Milkwort;
Hindi : Meradu, Miragu.

Distribution : Throughout India Upto 1,600 m.

Part Used : Root.

Utilization : Given medicinally in case of cough, fever, dyspepsia, dizziness, menorrhagia, gingivitis, pharyngitis, bronchitis and nervous disorders. 10g root powder given twice a day for a month to cure asthma, also considered astringent, anti-diarrhoeal, antiseptic, and antihaemorrhagic.

Active Constituents : Reported to contain rutin (quercetin-3-rutinoside), arctigenin-4'-gentiobioside, arctigenin-4'-glucoside (arctiin), kaempferol-3-rhamnoside (afzeline) and myrricetin-3-rhamnoside.

References : Agarwal (2003), Ambasta (1986), Anonymous (1969), Caius (2003), Chatterjee & Pakrashi (1997), Chopra *et al.* (1956), Karnick (1994), Kirtikar & Basu (1935), Rastogi & Mehrotra (1999a, b), Shukla & Verma (1996), Varghese (1996), Yoganarasimhan (1981), Watt (1889).

Polygala elongata Klein *ex* Willd.

Family: Polygalaceae.

Distribution : W. Peninsula, From the Concan Southwards.

Part Used : Root.

Utilization : Paste rubbed on body to bring down temperature in fever accompanied by shivering.

References : Hooker (1872), Kshirsagar & Singh (2000a).

Polygonatum multiflorum All.

Syn: *Convallaria govaniana* Wall.; *C. multiflora* Linn.

Family: Liliaceae.

English & Hindi Names:

Eng. : Solomen's Seal;
Hindi : Seel Mul.

Distribution : Himalaya from Kashmir to Kumaon Hills, Manipur.

Part Used : Root.

Utilization : Used against suzak (passing spermatozoa with urine), female weakness and flour allbus. Also, reported to be astringent, demulcent and tonic.

References : Agarwal (2003), Chopra *et al.* (1956), Kaul (1997), Watt (1889).

Polygonum alpinum All.

Syn: *P. acidum* Pall.; *P. divariccatum* Vill.; *P. polymorphum* Ledeb.; *P. sibirium* Linn.
Family: Polygonaceae.

Hindi Name:

Hindi : Pahari Machoti.

Distribution : N.W. Himalaya (Kashmir to Kumaon): 2,100-3,600 m.

Parts Used : Root. Rootstock.

Utilization : Rice prepared in water extr. of roots of this plant fed to patients for one month as a potent home remedy against all joint pains especially bodyache. Also reported to be used to control dysentery in calves and fawns. Rootstocks considered cooling, blood purifier, astringent, tonic, diuretic, depurative and sudorific.

Active Constituents : Rootstock contains 16-22 % tannin, 14-18 % soluble non tannins, 5 % sugars, leucoanthocyanidins, gallic acid, hyperoside, rutin, avicularin and quercetin.

References : Agarwal (2003), Hooker (1890), Kaul (1997), Rastogi & Mehrotra (1999a, b), Sharma (2003), Sood *et al.* (2001).

Polygonum amphibium Linn.

Family: Polygonaceae.

Distribution : Kashmir to Kumaon :1,800-2,100 m.

Part Used : Rootstock.

Utilization : Considered tonic, diuretic, depurative, sudorific, cooling, blood purifier and substituted for sarsaparilla.

References : Ambasta (1986), Anonymous (1969).

Polygonum amplexicaule Don

Family: Polygonaceae.

Hindi Names:

Hindi : Kukur Makri.

Distribution : Kashmir to Sikkim Upto 5,000 m.

Part Used : Root.

Utilization : Decoction of 1-2 g of fresh roots boiled with tea good for cold, cough and to restore body strength.

References : Agarwal (2003), Arya & Prakash (2000), Kumar & Rao (2001).

Polygonum chinensis L.

Family: Polygonaceae.

Hindi Name:

Hindi : Paral.

Distribution : All Over India.

Parts Used : Root. Rootstock.

Utilization : Root prescribed for dysentery and stomachache while rootstock regarded as antiscorbutic and vulnery tonic.

References : Agarwal (2003), Jamir & Rao (1990).

Polygonum glabrum Willd. Gomez.

Syn: *P. persicaria* Wall.; *P. poiretii* Meissn.

Family: Polygonaceae.

Hindi Name:

Hindi : Bihagni.

Distribution : Distributed Throughout India.

Part Used : Root.

Utilization : Pounded roots given against piles, bacteria, urinary stones, jaundice, general debility, consumption and to the victim of snake bite.

References : Agarwal (2003), Ambasta (1986), Anonymous (1969), Bhogaonkar & Devarkar (2002), Watt (1889).

Polygonum nepalense Meissn.

Syn: *Persicaria nepalensis* (Meissn.) H. Gross.
Family: Polygonaceae.

Part Used : Root.

Utilization : Paste applied on sprain for one week.

Reference : Singh & Kumar (2000b).

Polygonum polystachyum Wall. *ex* Meissn.

Syn: *P. molle* wight.
Family: Polygonaceae.

Distribution : Temperate Himalaya (Mishmi to Kashmir): 2,300-4,000 m

Part Used : Root.

Utilization : Crushed root mixed with mustard oil rubbed on skin to relieve pain due to burns and scalds.

References : Negi *et al.* (1999), Watt (1889).

Polygonum posumbi Buch.Ham.

Family: Polygonaceae.

Distribution : Entire N.E. Region.

Part Used : Root.

Utilization : Considered astringent and used in diarrhoea, dysentery and sores.

Reference : Kumar (2002).

Polygonum verticillatum Allioni

Family: Polygonaceae.

Part Used : Corm.

Utilization : Fleshy corms used for curing spermatorrhoea and sexual debility in males.

Reference : Sarin (1996).

Polygonum viviparum Linn.

Syn: *P. affine* Wall.; *P. angustifolium* Don; *P. bistorta* Gaercke; *P. bracteatum* Spreng.; *P. bulbiferum* Royle

Family: Polygonaceae.

English Names:

Eng. : Alpine Knotweed, Viviperous Bistort;

Distribution : Kashmir to Sikkim Upto 5,000 m.

Part Used : Root.

Utilization : So far used for leucorrhoea, excellent gargle in relaxed sore throat and spongy gums, lotion for ulcers, intermittent fever (mixed with Gentian), passive haemorrhage, diarrhoea, dysentery and haemoptysis. Also, useful as tonic, astringent and said to be applied to abscesses.

Active Constituents : Contains large quantity of tannic and gallic acids.

References : Agarwal (2003), Ambasta (1986), Chopra *et al.* (1956), Kirtikar & Basu (1984), Watt (1889).

Polypodium wallichii R.Br.

Family: Polypodiaceae.

Part Used : Rootstock.

Utilization : Extr. taken in diabetes.

Reference : Chhetri (2005).

Pongamia pinnata (L.) Pierre.

Syn: *Dalbergia arborea* Willd.; *Derris indica* (Lamk.)Bennet; *Galedupa Arborea* Roxb.; *G. indica* Lam.; *Robinia mitis* Linn.; *Pongamia glabra* Vent.

Family: Fabaceae.

English, Hindi & Sanskrit Names:

Eng. : Indian Beech, Karang, Pongam Oil Tree;

Hindi : Karaja, Karanj, Karanjaka, Kiramal;

Sans. : Grita-Kaaranja, Karaja, Karanjah, Naktaahva, Naktamaala, Prakiryaa, Udakiryaa.

Distribution : Native of W. Ghats; Found Along Banks of Streams & River, Beach & Tidal Forests.

Part Used : Root.

Utilization : Paste reported to cure ulcers, dandruff, scrofulous enlargements, and paralysis while its powder given in arthritis, rheumatism, earache, cleaning teeth, wash for wounds, fistula and blood purification. Decoction beneficial for stomach pain (1 glass twice daily for 7 days) and gonorrhoea (mixed equally with coconut milk and lime water), and its juice useful for strengthening gum, fistulous sores, cleaning foul ulcers and sexual diseases.

References : Agarwal (2003), Anonymous (1969), Bhattacharjee (2001), Chatterjee & Pakrashi (1997), Khare (2004), Kirtikar & Basu (1935), Kothari & Londhe (2000), Kurian (1995), Reddi *et al.* (2005), Subramaniam (2000), Varghese (1996), Watt (1889).

Porana paniculata Roxb.

Family: Convolvulaceae.

English & Hindi Names:

Eng. : Bridal Creeper;
Hindi : Bhowri.

Distribution : Throughout India.

Part Used : Root.

Utilization : Poultice applied on affected parts for 4-5 days to treat wounds. One teaspoonful of its paste taken with a cup of milk once daily for 5 days to terminate 3-4 months old pregnancy.

References : Agarwal (2003), Ambasta (1986), Tirkey (2004).

Potentilla fragarioides Linn.

Family: Rosaceae.

English Names:

Eng. : Agrimony, Cocklebur, Herberola, Stickwort.

Distribution : Temperate Himalaya (Kashmir to Bhutan), Nilgiris, 2, 300 m.

Part Used : Root.

Utilization : Useful in India as a mild astringent, tonic, diuretic and its decoction useful against coughs, simple diarrhoea and relaxed bowels.

References : Chopra *et al.* (1956), Kirtikar & Basu (1984).

Potentilla fruticosa L.

Syn: *P. arbuscula* Don; *P. rigida* Wall.

Family: Rosaceae.

Hindi Name:

Hindi : Merino.

Distribution : Kashmir to Sikkim Upto 4,800 m.

Part Used : Root.

Utilization : Tea prepared by boiling its roots in water given to patients suffering from asthma, cold and fever.

References : Agarwal (2003), Kumar & Rao (2001), Singh (1996), Watt (1889).

Potentilla fulgens Wall.

Family: Rosaceae.

Distribution : Temperate Himalaya (Kumaon to Sikkim) & Khasia Hills: 1,200-3,600 m.

Parts Used : Root. Rootstock.

Utilization : Decoction taken orally in piles. Fresh roots recommended for the treatment of stomachache and toothache. Also, given to strengthen gums and teeth (paste filled in hollow cavity of teeth to check the infection), uterine disorders, wounds and tiger bite. Rootstocks given against diarrhoea.

References : Ambasta (1986), Arya and Prakash (2000), Anonymous (1969), Bhattacharjee (2001), Chauhan (1999), Chhetri (2005), Kala & Rawat (2001), Negi *et al.* (1999), Rana *et al.* (1996), Sharma (2003).

Potentilla leuconata D. Don

Family: Rosaceae.

Distribution : Alpine Himalaya. Garhwal (3,300 m), Sikkim (3,300-5,300 m).

Part Used : Root.

Utilization : Good for cleaning and strengthening teeth.

References : Hooker (1879), Negi *et al.* (1999).

Potentilla nepalensis Hook.

Syn: *P. bifurcata* Wall.; *P. coccinea* Hoffm.; *P. colorata* Lehm.; *P. formosa* Don

Family: Rosaceae.

Distribution : W. Temperate Himalaya (1,600-3,000 m): Marri to Kumaon.

Parts Used : Root. Rootstock.

Utilization : Tea prepared from its root taken for cold, chest pain, fevers and its powder known to cure stomach disorder, cuts, burns and wounds. Rootstock considered depurative and its ash mixed with oil applied to burns.

References : Ambasta (1986), Anonymous (1969), Chauhan (1999), Chopra *et al.* (1956), Kapur & Singh (1996), Kumar & Rao (2001), Singh (1996, 2000), Watt (1889).

Pouzolzia auriculata Wight.

Syn: *P. propinqua* Blume; *P. rheedi* Wight.

Family: Urticaceae.

Distribution : C. India, E. & S. Deccan.

Part Used : Root.

Utilization : Paste applied on fresh wounds for healing. For muscular knots extr. administered orally, once a day till cure

References : Hooker (1890), Hosagoudar & Henry (1996), Rajendran *et al.* (2001).

Prangos pabularia Lindley

Family: Apiaceae.

Hindi Name:

Hindi : Komal.

Distribution : Kashmir & H.P. Upto 4,000 m.

Part Used : Root.

Utilization : Considered diuretic, emmenagogue, and in external application for treatment of itches, indigestion (infusion) and irregularity in menses.

Active Constituents : Regarded a good source of caumarins, prangenin [(S) heraclenin], osthol, oxypeucedanin (R) prangolarin, peucedanin, aviprin, komalin, isoimperatorin, bergapten, heraclenin, pabulin, prangolarin, merancin, psoralen, xanthotoxol, xanthotoxin, imperatorin, prangenin, pabularinone, imperatorin and isooxypeucedanin.

References : Agarwal (2003), Ambasta (1986), Anonymous (1969), Chopra *et al.* (1956), Kaul (1997), Rastogi & Mehrotra (1999a), Sharma (2003), Watt (1889).

Premna corymbosa Rottl.

Syn: *Cornutia corymbosa* Burm.f.; *Premna corymbosa* (Burm.f.) Miq.; *P. integrifolia* L.; *P. obtusifolia* R.Br.

Family: Verbenaceae.

Hindi & Sanskrit Names:

Hindi : Agetha, Arani, Arni, Ustabunda;

Sans. : Tarkari.

Distribution : Forests of S. India & N. Bengal.

Part Used : Root.

Utilization : Useful in anaemia, diabetes, constipation, stomachache, chyluria, inflammations, swellings, bronchitis, dyspepsia, piles, chronic fever, liver complaints, urticaria (paste with water and butter fat) and as tonic.

Active Constituents : Ganiarine, ganikarine, 6-C-β-D-glucopyranosyl-8-C-β-D-xylopyranosylapigenin, caryophyllen-3-one, 5-hydroxy-4,5-dihydrocaryophyllen-3-one, 11β-hydroxy-7-ketosandaracopimar-8 (14), 14-diene, premnenol, premnaspirodiene, sandaracopima-8(14), 15-diene, 7-α-hydroxy-sandaracopimar-8(14), 15-diene; 11, 12, 16-trihydroxyabieta-5, 8, 11, 13-tetraen-7-one and 11, 14-dihydroxy-12, 16-epoxyabieta-5, 8, 11, 13-tetraen-7-one are the active ingredients.

References : Agarwal (2003), Ambasta (1986), Chatterjee & Pakrashi (1997), Kirtikar & Basu (1984).

Premna herbacea Roxb.

Syn: *P. pygmaea* Wall.; *Pygmaeopremna herbacea* Mold.

Family: Verbanaceae.

Hindi & Sanskrit Names:

Hindi : Bharanghi;

Sans. : Bhumijambu.

Distribution : Forests of Kumaon & Terai of Sub-Tropical Himalaya: 200-1,000 m.

Parts Used : Root. Rootstock.

Utilization : Fresh rootstocks and roots given along with ginger in asthma, rheumatism, dropsy and ulcers. One glassful root decoction taken orally before bed time daily for one month to treat tuberculosis. Root paste applied twice daily on affected parts for 6-7 days and one cupful taken twice daily in joint pains; also found to be alterative, bitter, stomachic, tonic, stimulant and beneficial in asthma, cough, fever, scrofulous diseases and dropsy. Rootstock good against toothache.

References : Agarwal (2003), Ambasta (1986), Anonymous (1969), Chatterjee & Pakrashi (1997), Chopra *et al.* (1956), Khanna (2002b), Tirkey (2004), Watt (1889).

Premna serratifolia Linn.

Syn: *P. corymbosa* auct. non Rottl.; *P. integrifolia* Linn.; *P. obtusifolia* R.Br.; *P. ovalifolia* Wall.; *P. spinosa* Roxb.

Family: Verbenaceae.

English, Hindi & Sanskrit Names:

Eng.	: Headache Tree;
Hindi	: Bhut-Bhirravi;
Sans.	: Agnimatha, Jayee, Shriparni.

Distribution : Andaman Coasts, Plains of Assam & Khasia Hills.

Parts Used : Root. Root bark.

Utilization : Prescribed as a stomachic, laxative, cordial, tonic, febrifuge (obstinate fever) and to cure urticaria and roseola (paste taken with clarified butter for a week).

Biological Activity : Hepatoprotective, antiarrythmic (root), antirheumatic and astringent properties (root bark) confirmed.

References : Khare (2004), Watt (1889).

Prinsepia utilis Royle *ex* Hook.

Family: Rosaceae.

Hindi Names:

Hindi : Bekkra, Bhekal, Cherara, Dhatila, Jhatela, Karanga, Mhat.

Distribution : Khasia Hills, Assam, Nilgiris.

Parts Used : Root. Root bark.

Utilization : Bark reported to check stomach disorders while root paste used for healing wounds, suppurations and healing of boils (ointment prepared by mixing its root paste with black pepper and cow's urine).

References : Agarwal (2003), Ambasta (1986), Bhattacharjee (2001), Negi *et al.* (1999), Singh & Kumar (2000b), Verma & Chauhan (2006).

Psidium guajava Linn.

Family: Myrtaceae.

English, Hindi & Sanskrit Names:

Eng. : Common Guava, Guava;
Hindi : Amrood, Safed Safari;
Sans. : Mansala.

Distribution : Widely Cultivated Throughout India.

Part Used : Root bark.

Utilization : Pounded root bark plastered on bone fracture and in sprains, also considered astringent and useful in chronic diarrhoea.

References : Agarwal (2003), Ambasta (1986), Chopra *et al.* (1956), Jamir (1997), Watt (1889).

Pueraria tuberosa DC.

Syn: *Hedysarum tuberosum* Roxb.
Family: Fabaceae.

English, Hindi & Sanskrit Names:

Eng. : Indian Kundzu;
Hindi : Bankumra, Bharda, Bilaikand, Sural, Tirra;
Sans. : Bharda, Kohnda.

Distribution : Throughout India: 300-1,200 m.

Part Used : Root.

Utilization : Said to be demulcent, lactagogue, sweet, cooling; and given in fevers, joint swellings, cataplasm and to relieve renal complaints.

Active Constituents : β-sitosterol, stigmasterol, diadzein, puerarin, C-glycoside-4′, 6″-diacetylpuerarin and tuberosin are the active principles.

References : Agarwal (2003), Ambasta (1986), Anonymous (1969), Chopra *et al.* (1956), Islam & Jha (2003), Rastogi & Mehrotra (1999a, b), Sharma (2003), Sood & Prakash (2007).

Punica granatum L.

Family: Punicaceae.

English, Hindi & Sanskrit Names:

Eng. : Pomegranate;
Hindi : Anar;
Sans. : Dadimah.

Distribution : Widely Cultivated for its Fruits Throughout India.

Part Used : Root bark.

Utilization : Root bark used for deworming.

Active Constituents : Affords pelletierine, isopelletierine, pseudopelletierine, methylpelletierine and methylpelletierine.

References : Agarwal (2003), Chopra *et al.* (1956), Rastogi & Mehrotra (1998, 1999b), Reddi *et al.* (2005).

Quercus semecarpifolia Smith

Family: Fagaceae.

English & Hindi Names:

Eng. : Brown Oak of Himalaya, Kharshu Oak;
Hindi : Karshu.

Distribution : Throughout Temperate Himalaya & Manipur Upto 3,600 m.

Part Used : Root.

Utilization : Crushed root applied on cuts and wounds to hasten healing.

References : Agarwal (2003), Ambasta (1986), Negi *et al.* (1999).

Radermachera xylocarpa (Roxb.) Schum.

Syn: *Stereospermum xylocarpum* Benth. & Hook.f.

Family: Bignoniaceae.

English & Hindi Names:

Eng. : Padri Tree;
Hindi : Paral.

Distribution : Throughout Deccan, M.P. & E. India.

Parts Used : Root. Root bark.

Utilization : Root paste given orally and as well as applied locally in snake, mad dog and cat bites as an antidote. Bark astringent and antibacterial.

References : Agarwal (2003), Ambasta (1986), Brahmam & Saxena (1990).

Randia ulginosa DC.

Syn: *Xeromphis ulginosa* (Retz.) Maheshwari

Family: Rubiaceae.

English, Hindi & Sanskrit Names:

Eng. : Grey Emetic Nut;

Hindi : Bharani, Katul, Panar, Pendual, Pinda, Pindalu;

Sans. : Devatamalla, Gangati Pindalu, Pinditak.

Distribution : Moist Places in S., C. & E. India including Assam & Sikkim.

Parts Used : Root. Root bark.

Utilization : Wounds and sores washed with boiled extr. of root bark while root considered sweet, cooling, diuretic, aphrodisiac, haematinic, astringent, antidiarrhoeal, antidysenteric and good for thirst, heart, boils in children, biliousness, dysuria and strangury.

References : Agarwal (2003), Ambasta (1986), Anonymous (1969), Chatterjee & Pakrashi (1997), Chopra *et al.* (1956), Jadeja (1999), Kirtikar & Basu (1984), Rastogi & Mehrotra (1999a, b).

Raphanus sativus L.

Family: Brassicaceae.

English, Hindi & Sanskrit Names:

Eng. : Garden Radish, Rabone, Radish;

Hindi : Fugil, Mula, Muli, Muro, Sweti Sarisha, Tora;

Sans. : Bhumikakshara, Dirghakandaka, Dirghamulaka, Hariparna, Kandamula, Ksharamula, Mulaka, Nilakantha, Shankhamula, Shimbiphala, Sita.

Distribution : Cultivated Throughout India Upto 3,000m in the Himalaya & Other Hilly Regions.

Part Used : Root.

Utilization : Extr. (1-2 teaspoonfuls) given on empty stomach to cure acidity, flatulence, hiccups, influenza, dysentery, urinary troubles, colic, erysipelas, oedema from pulmonary diseases and earache (external application of the extr.). Raw roots consumed to get relief within a few hours in case of congestion or dry cough. Juice of roots and leaves taken for curing jaundice, anaemia and to increase secretion of bile and saliva. Also, roots considered useful for urinary complaints, piles, gastrodyna, syphilitic diseases, neuralgia, headache, sleeplessness, chronic diarrhoea and as diuretic and laxative.

Active Constituents : A good source of ascorbic acid (15 mg/100g), trace elements viz. aluminium, barium, lithium, manganese, silicon, titanium, fluorine and iodine (upto 18 mg/100g). White radish root contains vitamin A 5.0 I.U., thiamine 0.06, riboflavin 0.02, nicotinic acid 0.5, vit.c 15.0/100 g, pungent flavour due to volatile oil isothiocyanate (mustard oils) and it is reported to contain butyl-crotonyl isothiocyanate-sulphide as the chief constituent.

References : Agarwal (2003), Ambasta (1986), Anonymous (1969), Caius (2003), Chatterjee & Pakrashi (1997), Chopra *et al*. (1956), Dastur (1970), Khare (2004), Kirtikar & Basu (1935), Kumar & Chauhan (2005), Nadeem *et al*. (2001), Pandey (2003), Rajan *et al*. (2000), Rastogi & Mehrotra (1998), Saini (1996b).

Raphia ruffia Mart.

Syn: *R. pedunculata* Beauv.; *Sagus ruffia* Jacq.

Family: Arecaceae.

English Names:

Eng. : Raffia, Raffia Palm.

Part Used : Root.

Utilization : Given for toothache.

Reference : Caius (2003).

Rauvolfia serpentina (L.) Benth. *ex* Kurz.

Syn: *Ophioxylon album* Gaertn; *O. serpentinum* L.; *O. treifoliatum* Gaertn.; *Rauvolfia trifoliata* (Gaertn.) Baill.

Family: Apocynaceae.

English, Hindi & Sanskrit Names:

Eng. : Rauvolfia Root, Serpentine Root, Snake Root;
Hindi : Chandrabhaga, Chhota Chand;
Sans. : Chandrika, Nakuli, Sarpachhatrikaa, Sarpagandha.

Distribution : Wild Upto 1,200 m in Deciduous Forests of the Sub-Himalayan Tracts Upto Nepal, Gangetic Plains, E. & W. Ghats.

Part Used : Root.

Utilization : About 20 g of its powder given for fever (1 teaspoon thrice a day for 3 days), and for expelling thread worms in children (3g paste) and orally every half an hour as antidote to snake bite and extr. administered for chest pain, fever, rabies (5 spoonfuls twice a day for 21 days), and wounds. Its decoction good for rheumatism (1 teaspoon twice a day for 7 days), increasing uterine contractions, expulsion of foetus in difficult cases, intestinal disorders particularly diarrhoea and dysentery; and as anthelmintic, bitter tonic and febrifuge. Also, reported to be used in cholera, pneumonia, psycosis, excitement, anxiety, insomnia, epilepsy, nervous disorders, snake bite, stomach pain (fresh root juice ingested) and herpes (root juice externally applied, 2-3 times a day).

Active Constituents : These are reserpine, deserpidine, serpajamaline, resajmaline, rescinnamine, ajmalexine, ajmalicidine

(1-carbomethoxy-17α-hydroxy-16-decarbomethoxy-16, 17-dihydroajmalicine), ajmalicine, corynanthine (rauhimbine), indobine, indobinine, (cyclohexylester of indole propionic acid), resinnamidine, rescinnaminol, renoxidine, reserpiline, sarpagine, serpentine, serpentinine, yohanabinine (5β-methyl-pseudoyonimbine), yohimbine, α-yohimbine (rauwolscine), 3-epi-α-yohimbine, ajmalimine, (21-trimethoxybenzoyl ajmaline), ajmaline, ajmalinimine, ajmalinine, isoajmaline, rauwolfinine, (perakenine), sandwicolidine, sandwicoline (21-monohydro-N-methylsandwicine), tetraphyllicine(serpinine), serposterol (17β-methyl-5-androsten-3β-ol), β-sitosterol, serpoterpene, yohambanoid, yohambinine, indobinine, ajmalicidine, 12-hydroxyajmaline, 3-epi-α-yohimbine and18-β-hydroxy-3-epi-α-yohimbine.

References : Agarwal (2003), Ambasta (1986), Anonymous (1969), Balasubramanian & Prasad (1996), Bhandary *et al.* (1996), Bhattacharjee (2001), Chatterjee & Pakrashi (1997), Chopra *et al.* (1956), Dastur (1970), Dwivedi (1993), Jain (1968, 2003), Jha & Verma (1996), Kurian (1995), Nayak *et al.* (2003), Parrotta (2001), Rajendran (1997), Rao & Henry (1996), Rastogi & Mehrotra (1998, 1999a, b), Said (1997), Saini (1996b), Saxena *et al.* (1997), Shah (1995), Sharma (2003), Singh & Kumar (2000a), Singh *et al.* (2003), Thakur *et al.* (1989), Verma & Chauhan (2006).

Rauvolfia tetraphylla L.

Syn: *R. canescens* L.; *R. heterophylla* Roem. & Schult.

Family: Apocynaceae.

Hindi Names:

Hindi : Barachandrika, Barichadrika.

Distribution : Throughout S. India, M.P. & E. India.

Part Used : Root.

Utilization : Few drops of root juice given for insomnia. Paste of root in mustard oil applied topically in skin troubles. Also, the roots often used as a substitute or adulterant of those of *R. serpentina.*

References : Agarwal (2003), Ambasta (1986), Anonymous (1969), Banerjee (2000).

Reinwardtia indica Dumort.

Syn: *Kiltelocharis trigyna* Alef.; *Linum repens* Don; *L. trigynum* Roxb.; *Macrolinum trigynum* Reichb.; *Reinwardtia repens* Planch.; *R. trigyna* Planch.

Family: Linaceae.

Hindi Name:

Hindi : Basanti.

Distribution : Kashmir to Sikkim upto 2,000 m.

Part Used : Root.

Utilization : Decoction alongwith a pinch of black pepper given in the treatment of jaundice and colds.

References : Agarwal (2003), Sood & Thakur (2004).

Remusata vivipara Schott.

Family: Araceae.

Distribution : Sub-Tropical Himalaya (600-900m) from Kumaon to Sikkim, Khasia Hills, Bihar, Maharashtra, Mysore & W. Ghats.

Part Used : Root.

Utilization : Ointment prepared by grinding aromatic root with turmeric used as a remedy for itch and its juice considered alexipharmic.

References : Anonymous (1969), Chopra *et al.* (1956).

Rheum emodi Wall. *ex* Meissn.

Syn: *R. australe* D. Don

Family: Polygonaceae.

English, Hindi & Sanskrit Names:

Eng. : Himalayan Rhubarb, Indian Rhubarb, Rhubarb;
Hindi : Dolu, Hindirevandchini, Revandchini, Revatchini;
Sans. : Amlaparni, Amlavetasa, Pitamuuli, Revatikaa.

Distribution : Wild in the Alpine Himalaya at about 4,000 m.

Part Used : Root.

Utilization : So far found to be useful in rheumatic pains, cleaning teeth, internal injury, fever, asthma, bronchitis, cuts, wounds, ulcers, dysentery, skin diseases, tumours, constipation, headache, swelling throat, tonsilitis, mumps, loss of appetite, and as purgative, stomachic and cathartic. Powdered root with mustard oil applied topically for joint pains. Paste of root applied on affected nipple to cure mastits in cows. Paste also applied on infected hooves for the treatment of hoof diseases.

Active Constituents : Contains in addition to 0.05 % of an essential oil (Eugenol), an unidentified terpenic alcohol and a constituent believed to be methyl-n- heptyl ketone.

References : Agarwal (2003), Ambasta (1986), Beigh *et al.* (2003), Bhattacharjee (2001), Chhetri (2005), Dwivedi (1993), Kala & Rawat (2004), Kaul (1997), Khare (2004), Kumar (2002), Negi & Pant (1990), Samant *et al.* (2004), Said (1997), Sharma (2003), Singh (1996), Sinha & Sinha (2001), Sood *et al.* (2005c), Thakur *et al.* (1989), Tiwari & Pande (2006).

Rheum moorcroftianum Royle

Family: Polygonaceae.

Distribution : Uttaranchal & H. P.: 3,000-5,200 m.

Part Used : Root.

Utilization : Used for tumours, sprains and contusions, also, applied on cuts and wounds to hasten healing.

References : Aswal & Mehrotra (1994), Negi & Pant (1990), Negi *et al.* (1999), Sharma (2003), Sood *et al.* (2005c).

Rhinacanthus nasuta (L.) Kurz.

Syn: *Jasticia nasuta* L.; *Rhinacanthus communis* Nees.; *R. rottlerianus* Nees.
Family: Acanthaceae.

English, Hindi & Sanskrit Names:

Eng. : White Flowered Justicia, Snake Jasmine;
Hindi : Palakjuhi;
Sans. : Yuthikaparni, Yuthukaparni.

Distribution : Gardens & Village Shrubberies Throughout India. Also Found Wild in Deccan Peninsula.

Part Used : Root.

Utilization : Paste applied to cure eczema, ringworm, Dobi's itch, goitre (applied on neck), snake bite, impotency in males, herpes and as antiseptic.

Active Constituents : Found to contain oxy-methylanthroquinone and rhinacanthin.

Biological Activity : Rhinacanthin accredited with antiseptic and antiparasitic properties.

References : Agarwal (2003), Ambasta (1986), Chatterjee & Pakrashi (1997), Chopra *et al.* (1956), Balasubramanian & Prasad (1996), Reddi *et al.* (2005), Reddy *et al.* (1996), Watt (1889), Yoganarasimhan (1981).

Rhynchosia sericea (Span.) DC.

Syn: *R. rothii* Benth. *ex* Ait.
Family: Fabaceae.

Distribution : W. & C. Himalaya & Hills of W. Peninsular India: 1,500-1,800 m.

Part Used : Root.

Utilization : Considered antidiabetic.

References : Bhatt *et al.* (2003), Rastogi & Mehrotra (2002).

Rhynchosia viscosa DC.

Syn: *Dolichos glutinosus* Roxb.; *R. stipulosa* A. Rich.
Family: Fabaceae.

Distribution : Plains from the Himalaya to Sri Lanka.

Part Used : Root.

Utilization : Known to be antidiabetic.

References : Bhatt *et al.* (2003), Hooker (1879).

Rhynchostylis retusa (L.) Bl. (Pl. 8A)

Family: Orchidaceae.

Distribution : Tropical Himalaya from Kumaon to Bhutan; Often Cultivated.

Part Used : Root.

Utilization : Infusion (100 g) prescribed once a day for 3 days in jaundice.

References : Agarwal (2003), Dutta & Nath (2000).

Ricinus communis L.

Syn: *Croton spinosus* Linn.; *Ricinus inermis* Jacq.; *R. lividus* Jacq.; *R. spectabilis* Blume; *R. speciosus* Burm.; *R. viridis* Willd.
Family: Euphorbiaceae.

English, Hindi & Sanskrit Names:

Eng. : Castor Oil Plant, Castor Seed, Palma Christi;
Hindi : Arand, Arend, Erand, Rand;
Sans. : Amanda, Amangala, Chankuka, Chitrabija, Eranda, Ishta, Kanta, Vatari.

Distribution : All Over N.E. Region.

Parts Used : Root. Root bark.

Utilization : 20-30 ml of root juice given twice/thrice as abortifacient for inducing abortion in early months of pregnancy; also considered carminative, purgative, febrifuge, and useful in leprosy, eye diseases, jaundice, nervous disorders, rheumatic affections such as lumbago, pleurodynia, sciatica, inflammations, pains, asthma, bronchitis, amenorrhoea, diseases of rectum and the head. Its decoction given in rheumatism and applied topically in skin diseases. Root mixed with warm milk prescribed for missed peristalsis, constipated bowels, diarrhoea with blood and mucus. A decoction of root and bark beneficial in rheumatic disorders, gastrointestinal affections like flatulence and constipation. Root bark also found to be purgative, alterative, good in skin diseases and lumbago.

Active Constituents : Root bark shows the presence of steroid and considerable amount of gallo-tannins along with many inorganic ions.

Biological Activity : Root antiamoebic, anticancer, hypoglycaemic. The root bark extracted with different solvents, exhibited anti-inflammatory actitity in various animal models. Crude alcohol extr. showed 72.2% inhibition of carrageenin- induced oedema in rats.

References : Bhattacharjee (2001), Caius (2003), Chatterjee & Pakrashi (1997), Khare (2004), Khare & Khare (2000), Kirtikar & Basu (1984), Kumar (2002), Panthi & Chaudhary (2003), Mitaliya *et al.* (2003), Parrotta (2001), Rawat & Chowdhury (1998), Reddi *et al.* (2005), Sarin (1996), Sharma & Singh (2000), Singh (1999), Singh & Pandey (1998), Sood & Thakur (2004), Varghese (1996), Verma & Chauhan (2006).

Rivea cuneata Wight.

Syn: *Argyreia cuneata* Ker.
Family: Convolvulaceae.

Hindi Name:

Hindi : Bondwail.

Distribution : A Climber Grown in Maharashtra & S. India.

Part Used : Root.

Utilization : Its extr. along with the root of *Marsdenia tenacissima*, (2 tablespoons, once a day in the morning for 21 days) given for the treatment of arthritis.

References : Agarwal (2003), Singh & Prakash (1994).

Rivea hypocrateriformis (Derr.) Choisy.

Syn: *Convolvulous candicans* Wall.; *C. hypocrateriformis* Lamk.; *Lettsomia uniflora* Roxb.; *Rivea fragrans* Nimmo; *R. ornata* Ait.

Family: Convolvulaceae.

English & Hindi Names:

Eng. : Midnapur Creeper, The Midnapore Clove-Scented Creeper;

Hindi : Kalmilata, Phang.

Distribution : Almost Throughout India.

Part Used : Root.

Utilization : Decoction of roots brayed in milk given orally as purgative. Also given to woman in child birth to clean the womb.

References : Agarwal (2003), Ambasta (1986), Venkataratnam & Raju (2004), Watt (1889).

Roscoea alpina Royle (Pl. 8B.)

Family: Zingiberaceae.

Distribution : Uttaranchal to S.E. Tibet: 1,500-3,000 m.

Part Used : Root.

Utilization : About 3-5g of its dose with sugar and milk taken as tonic in general debility and impotency; also useful in diabetes, diarrhoea, leucorrhoea and dysentery.

References : Chauhan (1999), Sharma (2003).

Roscoea purpurea Sm. var. *gigantea* Wall.

Syn: *R. procera* Wall.

Family: Zingiberaceae.

Distribution : Sikkim, 3,500-5,000 m.

Part Used : Root.

Utilization : Used in malaria as a tonic in hills of U.P.

References : Jain (1995), Kumar (2002).

Rourea santaloids Wight & Arn.

Syn: *Connarus monocarpus* W. & A. not of Linn.; *C. santaloids* Vahl.; *Rourea acuminata* Hook.f.; *R. caudata* Planch; *R. commutata* Planch; *R. minor* (Gaertn.) Alston; *R. pulchella* Planch; *Santaloids erectum* Schellenb.; *S. floridum* Kuntze.; *S. minus* Schellenb.

Family: Connaraceae.

Hindi & Sanskrit Names:

Hindi : Kalavidhara, Vidhara;

Sans. : Ajanti, Antri, Avegi, Chhagala, Chhagalantrika, Dirghavallari, Jantuka, Junga, Kotarapushpi, Rukshagandha, Vridha, Vridhadaruka, Vrishyagandha.

Distribution : W. Peninsula From the Konkan to Travancore & in Sri Lanka.

Part Used : Root.

Utilization : Regarded pungent, acrid, alterative, tonic, laxative, appetiser, aphrodisiac; and reported to cure bronchitis, dyspepsia, anaemia, tuberculosis, urinary discharges, inflammations, pain in the joints, bitter tonic in rheumatism, scurvy, diabetes, pulmonary complaints, tonic in syphilis, ulcers and other skin affections.

Active Constituents : Therapeutic active ingredients are due to β-sitosterol-β-D-glucoside, hentriacontane, meso-inositol, rapanone, leucopelargonidin and β-sitosterol.

References : Ambasta (1986), Kirtikar & Basu (1935), Rastogi & Mehrotra (1999a, b), Watt (1889).

Rubia cordifolia L.

Syn: *R. alata* Wall.; *R. javanica* DC.; *R. chinensis* Rogel & Maack.; *R.Manjith* Roxb. *ex* Fleming; *R. mitis* Miq.; *R. mungisth* Desv.; *R. pururea* Decne.; *R. scadens* Zoll & Morr.; *R. secunda* Moon.

Family: Rubiaceae.

English, Hindi & Sanskrit Names:

Eng. : Indian Madder;
Hindi : Majit, Manjista, Manjit;
Sans. : Aruna, Gandira, Jingi, Kaalameshika, Manjistha, Raktaangi, Samangaa, Vikasaa.

Distribution : Himalaya, Nilgiris and Other Hilly Areas of India: 1,000-2,500 m.

Part Used : Root.

Utilization : Credited with tonic, astringent, diuretic, emmenagogue, analgesic, lactagogue, antidysentric, stomachic, deobstruent, antibacterial, antperiodic, and antiseptic properties; also used to cure liver complaints, skin diseases, spleen disorders, paralysis, fever, ulcers in mouth, obstruction in urinary passages, inflammatory conditions of the chest, rheumatism, eye sores, lethargy, pains in the joints, leucoderma, diseases of the uterus, vagina, ear, blood, erysiphelas, piles, anaemia and as vermifuge and blood purifier. Decoction of roots prescribed for menstrual disorders and leucorrhoea in women (2 tablespoons twice daily till cure), jaundice and diabetes.

Active Constituents : Root contains alizarin and its glucoside, 1-hydroxy-2-methylanthraquione, nor-damnacanthal, physcion, 1, 4-dihydroxy-6-mehylanthraquinone, 1,4-dihydroxy-2-methylanthraquinone, 1,5-dihydroxy-2-methylanthraquinone, 3-prenyl-5-methoxyanth raquinone, 1-hydroxy-2-methoxyanth raquinon, 1,4-dihydroxy-2-methyl-5-methoxy-anthroquinone, 1,3-dimethoxy-2-carboxyanth raquinone, 1-hydroxy-2-methyl-9-10-anthraquinone, lucidine, primevoside, ruberythric acid, 2-methyl-1,3,6-trihydroxy-9, 10-anthraquinone, 2-methyl-1,3, 6-trihydroxy-9,10-anthraquinone-3-O-(6′-O-acetyl)-α-rhamnosyl (1→2) glucoside & 2-methyl-1,3,6-trihydroxy-9,10-anthraquinone-3-O-α-rhamnosyl (1→2)-β-D-glucoside, antitumour cyclic hexapeptides-RA-V & RA-V^{II}, purpurin, munjistin, xanthopur purein or purpuroxanthin, pseudopurpurin, 1-acetoxy-6-

hydroxy-2-methylanthraquinone-3-O-α-rhamnosyl(1>4)-α-glucoside, 1, 4-dihydroxy-2-carbomethoxyanthraquinone, 1-hydroxy-2-carboxy-3-methoxyanthraquinone, 1-hydroxy-2-methyl-6-or-7-methoxyanthraquinone, β-sitosterol, scopoletol & oleanolic acid acetate, a new fernane derivative-rubiatriol, 2-methyl-1, 3, 6,-trihydroxyanthraquinone, more new anthraquinones-1,3-dihydroxy-2-methoxymethylanthraquinone, 1-methoxy-2-methoxymethyl-3-hydroxyanthraquinone, 4-hydroxy-2-carboxyanthraquinone, 1, 4-dihydroxy-2-hydroxy methylanthraquinone and 1-hydroxy-2-hydroxy methylanthraquinone.

Biological Activity : Methanol extr. prepared from dried roots have shown antineoplastic activity against sarcoma 180 ascites in mice and also exhibited anti-inflammatory activity.

References : Agarwal (2003), Ambasta (1986), Augustine & Sivadasan (2004), Chhetri (2005), Chopra *et al.* (1956), Khare (2004), Kharkongor & Joseph (1997), Kirtikar & Basu (1984), Kumar (2002), Nadeem *et al.* (2004), Rana *et al.* (1996), Rao & Henry (1996), Rastogi & Mehrotra (2001, 2002), Rawat & Chowdhury (1998), Reddi *et al.* (2005), Sarin (1996), Singh (1999), Sinha & Sinha (2001), Sood & Thakur (2004), Srivastava *et al.* (2003), Thakur *et al.* (1989).

Rubus ellipticus Smith

Syn: *R. flavus* Ham.; *R. gowraephul* Roxb.; *R. rotundifolius* Wall.; *R. sessilifolius* Miq.

Family: Rosaceae.

English & Hindi Names:

Eng. : Himalayan Yellow Rasberry;

Hindi : Anchu, Hansalu, Hasalu, Hinsalu, Lalanchhu, Lalhasalu.

Distribution : Throughout Himalayas & Hilly Regions of India, alt. 1,200-2,400 m.

Part Used : Root.

Utilization : Decoction good for cough (2 tablespoons twice a day till relief), colic, stomach troubles and diarrhoea. Extr. applied on cuts and wounds. Paste made out of its root powder administred orally once in a day for 15 days to cure jaundice and applied over cracked heels. Pounded roots given to cattle for promoting lactation. For malarial fever, decoction of its root boiled along with root of *Leucas zeylanica* Br. in water taken twice daily till cure.

Active Constituents : Octacosanol, β-sitosterol, its glucoside, octacosanoic, ursolic and rubitic acids are the active principles.

References : Bhardwaj & Gakhar (2003), Bisht & Jain (2006), Dash *et al.* (2003), Jain *et al.* (1991), Lalramnghinglova (1996), Rao & Henry (1996), Rastogi & Mehrotra (2002), Sood & Thakur (2004).

Rubus niveus Thunb.

Syn: *R. albescens* Roxb.; *R. gracilis* Roxb.; *R. lasiocarpus* Hook.f. in part; *R. mysorensis* Heyne.

Family: Rosaceae.

English & Hindi Names:

Eng. : Mahabaleshwar Raspberry, Mysore Raspberry;

Hindi : Gowriphal, Kalianchhi, Kala Hinsalu.

Distribution : Himalaya from Punjab to Assam.

Part Used : Root.

Utilization : Taken in stomachache and colic pains. Extr. applied on cuts and wounds.

References : Agarwal (2003), Ambasta (1986), Bisht & Jain (2006), Parmar & Kaushal (1982), Rana *et al.* (1996).

Ruellia patula Jacq.

Syn: *Dipteracanthus patulus* (Jacq.) Nees.

Family: Acanthaceae.

Distribution : Throughout India.

Part Used : Root.

Utilization : Powder given (3g per dose for 8-10 days) with cow's milk as a tonic for children in the age group 5-10 years.

Reference : Sood & Thakur (2004).

Ruellia suffruticosa Roxb.

Syn: *Dipteracanthus sibua* Nees.;
D. suffruticosus Voigt.; *Ruellia repens* Wall.
Family: Acanthaceae.

Distribution : Lower Bengal (Throughout Chutia Nagpur) & Upper Gangetic Plains.

Part Used : Root.

Utilization : Used medicinally by the Santhals in gonorrhoea, syphilis and renal affections. If taken in the quantity of 2 oz. by a pregnant woman, it is said to cause abortion. Also, used as a medicine for sore eyes (macerated in water and then strained for using as an eye wash).

References : Ambasta (1986), Kirtikar & Basu (1984), Watt (1889).

Ruellia tuberosa L.

Family: Acanthaceae.

English & Hindi Names:

Eng. : Menow-Weed;
Hindi : Sirkaba.

Distribution : Indian Gardens.

Part Used : Root.

Utilization : Tuberous root paste prescribed to cure body swellings and to cause abortion, and its decoction considered diuretic, aphrodisiac, and taken internally for urinary trouble and gonorrhoea.

References : Agarwal (2003), Jain & Sikarwar (1998), Reddi *et al.* (2005).

Rumex dentatus L.

Syn: *R. klotzschianus* Meissn.; *R. obtusifolius* Hefb. Ham.
Family: Polygonaceae.

Hindi Names:

Hindi : Ambavah, Amrule.

Distribution : Plains of India (Assam & Sylhet to the Indus), ascends Himalaya to 3,300 m.

Part Used : Root.

Utilization : Decoction used externally to treat skin diseases. Its poultice applied to skin itch, joint pain, ulcers and skin diseases.

References : Ambasta (1986), Chopra *et al.* (1956), Kala & Manjrekar (2000), Sharma (2003), Watt (1889).

Rumex hastatus D. Don

Family: Polygonaceae.

Distribution : N. W. Himalaya (Kumaon to Kashmir): 300 -2,600 m.

Part Used : Root.

Utilization : Its paste prepared in mustard oil and rubbed on the skin to relieve pain due to burns and scalds.

Active Constituents : Yields tannins (21-23%).

References : Ambasta (1986), Negi *et al.* (1999), Watt (1889).

Rumex nepalensis Spreng.

Syn: *R. ramulosus* Meissn. in DC.; *R. roxburghianus* Schultes (non Wall.); *R. peregrinus* Boiss.; *R. tuberosus* Roxb.; *R. uncinatus* Hort.

Family: Polygonaceae.

Distribution : Temperate Himalaya from Bhutan to Kashmir: 1,300 and 3,000 m. Ghats of the W. Peninsula from Konkan to the Nilgiris.

Part Used : Root.

Utilization : About 50 ml thick decoction of root administred twice daily for 5 days to cure food poisoning. Root also given in constipation in doses of 10 g to 20 g. Pounded root boiled in mustard oil and applied topically for use in relieving rheumatic pains. Infusion used in hepatitis as a liver tonic

Active Constituents : Reported to contain tannin (12.8%), nepodin, crysophanic acid, chrysophanol, emodin, physicion, lupeol, sitosterol, itsglucoside, musizin-1-O-β-D-glucopyranoside, orientalone, two new compounds-3-methoxy-5, 6-methylenedioxy benzaldehyde (I) andchrysophanol-8-O-β-D-galactopyranoside.

References : Chhetri (2005), Dash *et al.* (2003), Rastogi & Mehrotra (1999a, b), Sood & Thakur (2004), Watt (1889).

Saccharum spontaneum L.

Syn: *Imperata spontanea* Beauv.; *Saccharum canaliculatum* Roxb.; *S. semidecumbens* Roxb.

Family: Poaceae.

English, Hindi & Sanskrit Names:

Eng. : Thatch Grass, Wild Sugar Cane;
Hindi : Kagara, Kans, Kas, Khasi, Kosa, Kuggara;
Sans. : Ikshugandha, Kahsa, Kasalu, Kasha, Khaggara.

Distribution : Found Throughout the Country Upto 1,800 m in the Himalaya; Open or Disturbed Habitats Near Water.

Part Used : Root.

Utilization : About 20 ml extr. given only once in the morning to kill abdominal worms in men. Decoction prescribed in urinary system troubles, kidney stones and its paste with 21 black peppers (*P. nigrum*) applied on hydrocele.

References : Chhetri (2005), Dhiman (2003), Girach (1992), Maliya (2005), Parrotta (2001), Watt (1889).

Saccolabium papillosum Lindl.

Syn: *Acampe papillosa* Lindl.

Family: Orchidaceae.

Hindi & Sanskrit Names:

Hindi : Gandhata;
Sans. : Gandhata, Nakuli, Rasana.

Distribution : U. P. to E. India, Assam to S. India.

Part Used : Root.

Utilization : Considered for rheumatism. In the Konkan, it is used as cooling remedy.

References : Agarwal (2003), Ambasta (1986), Caius (2003), Chopra *et al.* (1956).

Salacia chinensis Linn.

Syn: *S. latifoia* Wall. *ex* M. Laws; *S. prinoides* DC.
Family: Hippocrateaceae.

Hindi & Sanskrit Names:.

Hindi : Madhuphal, Saptrangi;
Sans. : Saptachakra, Swarnmuula.

Distribution : Throughout India, Including Andaman Islands (Seashore & River Banks, as well as in Forests Upto 750 m).

Part Used : Root.

Utilization : Credited with antidiabetic and abortifacient properties. A decoction, however, used in amenorrhoea, dysmenorrhoea and venereal diseases.

Active Constituents : Comprised of manomeric leucopelargonidin, its monomer, dimer, tetramer, mangiferine, phlobatannin, glycosidal tannins, triterpene ether, friedalan-1, 3-dione, friedelan-1, 3-dione -24-ol, friedelan-1, 3-dione-24-oic acid, 24,25-oxido friedelan-1, 3-dione and 7,24-oxido-friedelan-1,3-dione.

Biological Activity : Leuco anthocyanidin obtained from the roots produced indifferent changes in blood sugar.

References : Agarwal (2003), Ambasta (1986), Khare (2004), Rastogi & Mehrotra (1999b).

Salacia fruticosa Heyne *ex* Laws.

Family: Hippocrateaceae.

Distribution : W. Peninsula.

Part Used : Root.

Utilization : Decoction given orally to diabetic patients.

Active Constituents : Friedelan-3-one-29-al and fridelan-3-one-29-ol- have been isolated.

References : Augustine & Sivadasan (2004), Rastogi & Mehrotra (1999b).

Salacia oblonga Wall. *ex* Wight & Arn.

Family: Hippocrateaceae.

Distribution : W. Peninsula.

Part Used : Root bark.

Utilization : Known for rheumatism, gonorrhoea, itches, skin diseases and ear troubles (decoction or as powder or boiled in milk).

References : Ambasta (1986), Chopra *et al.* (1956).

Salacia reticulata Wight.

Family: Hipporateaceae.

Hindi Name:

Hindi : Anukudu.

Distribution : Grows in Orissa & Kerala.

Parts Used : Root. Root bark.

Utilization : Root bark beneficial in rheumatism, itches, gonorrhoea and skin diseases while root extr. useful in child birth.

References : Agarwal (2003), Ambasta (1986), Chopra *et al.* (1956), Kirtikar & Basu (1935).

Salvadora oleoids Decne.

Syn: *S. indica* Royle; *S. persica* L.; *S. stockii* Wight.

Family: Salvadoraceae.

Hindi Names:

Hindi : Bahapilu, Chota Pilu, Jhal, Pilu.

Distribution : Arid Regions of Punjab & W. India.

Part Used : Root.

Utilization : Applied locally as analgesic in dental diseases like pyorrhoea.

References : Agarwal (2003), Ambasta (1986), Singh & Pandey (1998), Trivedi (2002), Watt (1889).

Salvadora persica L.

Syn: *Cissus arborea* Forssk.; *Savadora indica* Wight.; *S. wightiana* Planch.
Family: Lamiaceae.

English & Hindi Names:

Eng. : Tooth Brush Tree;
Hindi : Jal, Jhak, Kharjal, Pilu.

Distribution : Dry, Arid Regions of India on Saline Sands.

Parts Used : Root. Root bark.

Utilization : Fresh bark used as vesicant and as an ingredient of snuff; decoction used as tonic, febrifuge and emmenagogue. Patelia tribes, however, apply its paste with mustard oil on the blisters, also considered good in gonorrhoea, catarrh and spleen troubles.

Active Constituents : Reported to contain β-sitosterol, elemental-γ-monoclinic.

References : Agarwal (2003), Ambasta (1986), Rastogi & Mehrotra (1999b, 2001), Singh & Pandey (1998), Watt (1889).

Salvia moorcroftiana Wall.

Family: Lamiaceae.

Part Used : Root.

Utilization : Fresh/dried root extr. used for cough, asthma and skin infection.

Reference : Singh (1996).

Salvinia cucullata Roxb.

Family: Salviniaceae.

Hindi & Sanskrit Names:

Hindi : Bhopli;
Sans. : Akhukarni.

Distribution : Throughout India in Ponds & Rivers.

Part Used : Root.

Utilization : Reported so far as digestive, febrifuge, anthelmintic, diuretic; beneficial in the treatment of gonorrhoea and skin diseases.

Reference : Chatterjee & Pakrashi (1997).

Sambucus ebulus Linn.

Family: Caprifoliaceae.

English Names:

Eng. : Danseblood, Dansweed, Danewort, Walewort.

Distribution : Dry Inner Valleys of the Himalaya (Kagan valley), Kashmir, Pangi & Chamba 2,000 to 3,300 m.

Part Used : Root.

Utilization : Possesses purgative properties; also used in dropsy.

References : Chopra *et al.* (1956), Kirtikar & Basu (1984), Watt (1889).

Sambucus javanica Reinw. *ex* Blume; C.B. Clark in part

Syn: *S. hookeri* Rehder; *S. rubra* Ham.; *S. thunbergiana* Miq.; *S. wightianus* Wall.

Family: Caprifoliaceae.

English Name:

Eng. : Himalayan Elder.

Distribution : E. Himalaya: 1,300 to 2,600 m & Khasia Hills.

Part Used : Root.

Utilization : Decoction taken in urinary troubles.

References : Ambasta (1986), Chhetri (2005), Watt (1889).

Sambucus wightiana Wall. *ex* Wt. & Arn.

Syn: *S. ebulus* Hook.f. (non L.)

Family: Sambucaceae.

Distribution : Very Common on Heavily Grazed Slopes, Forest Floor, Waysides.

Part Used : Root.

Utilization : Decoction reported as an effective diuretic.

Active Constituents : Stigmasterol and inositol characterised from roots.

References : Rastogi & Mehrotra (1999a), Sharma (2003).

Sansevieria roxburghiana Schult & Schult.f.

Syn: *S. zeylanica* Willd.

Family: Liliaceae

English, Hindi & Sanskrit Names:

Eng. : Bow String Hemp, Indian Bowstring Hemp;
Hindi : Marul, Murva;
Sans. : Muruva.

Distribution : Coromandel of India.

Part Used : Root.

Utilization : Paste minimizes the excess of heat and used in snake bites, haemorrhoids, consumptive diseases and chronic coughs.

References : Agarwal (2003), Ambasta (1986), Chopra *et al.* (1956), Reddi *et al.* (2005).

Sapindus emarginatus Vahl.

Syn: *S. laurifolius* Vahl. var. *emarginatus* (Vahl) Cooke; *S. trifoliatus* L.

Family: Sapindaceae.

English, Hindi & Sanskrit Names:

Eng. : Soap Nut Tree;

Hindi : Bara Reetha, Ritha;

Sans. : Arishta, Phenila, Rithakaranja.

Distribution : Cultivated in Villages of W.Bengal & S. India.

Part Used : Root.

Utilization : Known in India as demulcent, expectorant; also beneficial in gout, rheumatism and paralysis.

References : Ambasta (1986), Chatterjee & Pakrashi (1997).

Sarcococca saligna (Don) Muell.

Syn: *Buxus saligna* Don; *Sarcococca pruniformis* Lindl.; *S. salicifolia* Bril.; *S. sumatrana* Blume.

Family: Buxaceae.

Hindi Name:

Hindi : Geru.

Distribution : Himalaya Upto 3,000 m.

Part Used : Root.

Utilization : Paste cures skin eruptions and its decoction described to be useful in phosphoteuria and gonorrhoea.

References : Agarwal (2003), Sharma (2003), Sood & Thakur (2004).

Sarcostemma secamone (L.) Bennet.

Syn: *Oxystelma esculentum* R.Br. *ex* Schult; *O. secamone* (L.) Karst.; *Periploca esculenta* L.f.; *P. secamone* L.; *Sarcostemma esculentum* (L.f.) Holm.

Family: Asclepiadaceae.

Hindi & Sanskrit Names:

Hindi : Dudhialata;

Sans. : Dugdhika.

Distribution : Throughout the Plains & Lower Hills of India including Paddy Fields & Semi-Marshy places.

Part Used : Root.

Utilization : Efficacious in jaundice.

Active Constituents : These are calogenin-3-o-β-D-oleandropyranosyl (1→4)-O-β-D-thevetopyranosyl (1→4)-O-β-D-cymaropyranosyl (1→4)- O-β-D- O- digitoxop yranoside (oxysine), a triglycoside sarcogenin-3-O-β-D-thevetopyranosyl-(1→4)-O-β-D-cymaro-pyranosyl-(1→4)-O-β-D-oleandropyranoside (esculetin), and cardenolide tetraglycosides.

Reference : Chatterjee & Pakrashi (1997).

Satyrium nepalensis Don

Family: Orchidaceae.

English Name:

Eng. : Satyrion.

Distribution : Kashmir, Khasia Hills, Darjeeling, Manipur, Hilly Districts of S. India: 2,700 to 3,000 m.

Part Used : Root.

Utilization : Paste of about 20-25g freshroot cooked like 'chapati' and eaten to increase body strength particularly at old age. Also, roots given to treat malaria and dysentery.

References : Arya & Prakash (2001), Bhattacharjee (2001), Sharma (2003).

Saurauia nepalensis DC.

Family: Dilleniaceae.

Hindi Names:

Hindi : Goganda, Gogina, Pangara.

Distribution : Temperate Himalaya (Bhutan & Sikkim): 1,600-2,300 m, Garwhal (800-1,600 m), Khasia Mountains, 1,600 m, Mishmi Hills.

Part Used : Root bark.

Utilization : Consumed for stomach pain and its decoction given as purgative.

References : Ambasta (1986), Hooker (1874), Jamir & Rao (1990).

Saussurea albescens (DC.) Sch.

Family: Asterceae.

Part Used : Root.

Utilization : Powdered dried roots mixed with mustard oil applied topically on skin diseases.

Reference : Singh (1996).

Saussurea lappa (Decne.) Sch.

Syn: *Aplotaxis lappa* Decne.; *Saussurea costus* (Falc.) Lipsch.

Family: Asteraceae.

English, Hindi & Sanskrit Names:

Eng. : Costus;

Hindi : Kot, Kur, Kust, Kuth;

Sans. : Agada, Amaya, Bhasura, Dushta, Gadakhya, Gadavha, Kakla, Kuthika, Kushtha, Kutsita, Niruja, Paklam, Paribhavya, Roga, Rogavhaya, Utpala, Vaniraja.

Distribution : Cultivated in Lahul & Spiti District & Along Indo-Tibetan Border: 2,500-3,500 m.

Part Used : Root.

Utilization : Extr. obtained by braying its pieces in mustard oil for 30 min used as ear drops for relieving pain. Contents of crushed roots applied as poultice over painful joints and boils. Root of this along with that of *Glycyrrhiza glabra* Linn. and fruit of *Hippophae rhamnoides* used for the treatment of inflammation of lungs associated with irritating cough accompanied by frothy blood stained phlegm and burning sensation in the chest. Also, roots considered tonic, stomachic, aphrodisiac, antiseptic, astringent, sedative, carminative, stimulant; used as a spasmodic in asthma, cough, rheumatism, leucoderma, itching, ringworm, erysipelas, diseases of the blood, vomiting, scabies, diseases of the liver and kidney, deafness, paralysis, ophthalmia, old fevers, chronic skin diseases, advanced stages of typhous fever, dysentery, ulcer, chronic bronchitis, colic, cough, dental trouble, flatulence, diarrhoea, fever, headache, hiccups, hysteria, heart troubles, menstruation trouble, and promoting urination.

Active Constituents : Found to contain sesquiterpene lactones, β-elemene, β-ionene, β-selenene, aplotaxen, an essential oil 1.5 %, saussurine, 0.05 %, resin 6.0 %, inulin, potassium nitrate, sugars, kushtin, (E) 6, 10-dimethyl-9-methylideneundec-5-en-2-one, 22-dihydrostigmasterol, costunolide, dehydrocos tuslactone, costic acid, palmitic and linoleic acids, beta-sitosterol, alpha-cyclocostunolide, costunolide, dehydrocostuslactone, alpha-cyclocostunolide, alantolactone, beta –cyclocostunolide, iso-alantolactone, (E) - 9-isopropyl-6-methyldeca-5, 9-diene-2-one, acetic acid, 3-methylbutyric acid, hexanoic acid, heptanoic acid, octanoic acid, 7-octanoic acid, 3-isopropylpenthanaoic acid, 4-ethyloctanoic acid, alpha-amorphenic acid, sesquiterpenes-(-) α-costol, (+) γ-costol, (-) elema-1, 3, 11 (13)-trien-12 ol, (+) γ-costal isolated from root oil in addition to (-) α-selinene, (+) selina-4, 11-diene, (-) α-trans-bergamotene, (-)α-costal, (+) β-cotol, (-)elema-1,3,11 (13)-trien-12-al, (-) E trans-bergamota-2, 12-diene-14al, and (-) caryophyllene-oxide.

Biological Activity : Saussurine has a depressant action on the vagus centre in the medulla as well as on the involuntary muscle fibres of the bronchioles and gastrointestinal tract. It produces slight but persistant rise of blood pressure and increases the force of contraction and amplitude of ventricles. Also, it exhibits a smooth muscle relaxant activity.

References : Agarwal (2003), Ambasta (1986), Aswal & Mehrotra (1994), Beigh *et al.* (2003), Bhattacharjee (2001), Caius (2003), Chatterjee & Pakrashi (1997), Chauhan (1999), Chopra *et al.* (1956), Jain (1968), Kala & Manjrekar (2000), Khare (2004), Kirtikar & Basu (1984), Nadeem *et al.* (2001), Rastogi & Mehrotra (1999a, 2002), Sarin (1990), Sharma (2003), Sharma & Sood (1997), Singh (1999), Singh & Kumar

(2000b), Sinha & Sinha (2001), Sood *et al.* (2001), Sood *et al.* (2005c), Thakur *et al.* (1989), Watt (1889).

Saussurea obvallata Wall.

Syn: *Aplotaxis obvallata* DC.; *Carduus obvallatus* Wall.; *C. tectus* Wall.

Family: Asteraceae.

Hindi Name:

Hindi : Brahma Kamal.

Distribution : High altitudes; Rohal Kanda, Chansal Peak, Chandramohan (Shimla District), Sangla & Chitkul Passes (Kinnour District) of H.P.

Part Used : Root.

Utilization : So far useful as nervine tonic, powerful antiseptic, and applied on wounds, bruises and cuts

References : Agarwal (2003), Ambasta (1986), Sharma (2003), Sharma & Rana (2000), Watt (1889).

Schefflera hypoleuca (Kurz.) Harms.

Syn: *Heptapleurum hypoleucum* Kurz.

Family: Araliaceae.

Part Used : Root.

Utilization : Decoction given as a tonic after child birth.

References : Ambasta (1986), Chhetri (1994).

Schebera swietenioides Roxb.

Family: Oleaceae.

English, Hindi & Sanskrit Names:

Eng. : Beam Tree, Weaver's Beam Tree;

Hindi : Banpalas, Gay Ka Lundi, Ghant, Moka;

Sans. : Muskaka.

Distribution : Tropical & Sub-Tropical Himalaya including (Kumaon & Terai), S. & C. India, (Rajasthan to W.Bengal).

Parts Used : Root. Root bark.

Utilization : Root useful in leprosy and its bark decoction enters in the preparation of oil applied to burns and boils.

References : Agarwal (2003), Ambasta (1986), Chatterjee & Pakrashi (1997), Chopra *et al.* (1956).

Scindapsus officinalis (Roxb.) Scott.

Family: Araceae.

Hindi & Sanskrit Names:

Hindi : Bari-Pipli, Gajapipal, Pippal Jhauca;
Sans. : Gaja-Pipali, Kapiballi, Kari-Pippali.

Distribution : Distributed in E. India, A.P. & Andamans.

Part Used : Root.

Utilization : Crushed root of the young plant (10 g) given to relieve stomachache, also prescribed in cough and fever. A preparation of the root consisting of 52 ingredients administered in cough and tuberculosis.

References : Agarwal (2003), Ambasta (1986), Sarkar *et al.* (2000), Saxena *et al.* (1997).

Scirpus grossus Linn.f.

Syn: *S. kysoor* Roxb.
Family: Cyperaceae.

Hindi & Sanskrit Names:

Hindi : Kaseru, Kasurn, Kesur;
Sans. : Gundakanda, Kaseru.

Distribution : Practically Throughout India in Swamps.

Part Used : Root.

Utilization : Credited with astringent properties and also given to check diarrhoea and vomiting.

References : Agarwal (2003), Ambasta (1986), Caius (2003), Watt (1889).

Scleria levis Retz.

Syn: *S. hebecarpa* Nees.
Family: Cyperaceae.

Part Used : Root.

Utilization : Its paste diluted in 100 ml of cow's milk, administered orally 3-4 times for dysentery.

References : Ambasta (1986), Rajendran & Henry (1994).

Scoparia dulcis Linn.

Syn: *Gratiola micrantha* Nutt.; *Scoparia grandiflora* Nash.
Family: Scrophulariaceae.

English & Hindi Names:

Eng. : Sweet Broomweed;
Hindi : Chini Buti, Ghoda Tulsi.

Distribution : Cultivated Fields & Roadsides in Lowland Areas of W. Bengal, Tamil Nadu & S. India.

Part Used : Root.

Utilization : Extr. taken thrice daily for five days to cure stomachache. Root paste with sugar used for diarrhoea and fever; also known in India as febrifuge and for the removal of placenta, dysentery, warts and excessive menstruation.

Active Constituents : Properties attributed to α-amyrin, betulinic acid, dulcioic and ifflaionic acids, friedelin, coixol i.e. 6-methoxybenzoxazolinone, glutirol, D-mannitol, hexacosanol and β-sitosterol.

References : Agarwal (2003), Bora (2000), Chatterjee & Pakrashi (1997), Jain *et al* (1995), Parrotta (2001), Rao & Henry (1996), Rastogi & Mehrotra (1999b), Reddi *et al.* (2005), Srivastava *et al.* (2003), Varghese (1996).

Secamone emetica (Retz.) R.Br. *ex* Schult.

Family: Asclepiadaceae.

English & Hindi Names:

Eng. : Chow-Chow;
Hindi : Squash Kadu.

Distribution : Sub-Tropical Regions of India, Nilgiris.

Part Used : Root.

Utilization : Paste mixed with rice gruel given to cattle to induce easy delivery; also as acrid and emetic.

References : Agarwal (2003), Ambasta (1986), Binu *et al.* (2003), Watt (1889).

Securinega virosa (Roxb. *ex* Willd.) Pax. & Hoffm.

Syn: *Flueggea microcarpa* Blume; *F. virosa* (Roxb. *ex* Willd.) Baillon; *Securinega obvata* Muell.-Arg.
Family: Euphorbiaceae.

Hindi Names:

Hindi : Bakarcha, Dalme, Patala, Rithoul.

Distribution : Throughout India.

Part Used : Root.

Utilization : Paste of root with onion externally applied on burns. Also, known to be analgesic and aphrodisiac.

Active Constituents : Contains norbergenin and securinne.

References : Agarwal (2003), Ambasta (1986), Balasubramanian & Prasad (1996), Rastogi & Mehrotra (1998).

Selinum vaginatum (Edgew.) Clarke

Family: Apiaceae.

Hindi & Sanskrit Names:

Hindi : Pushwari;
Sans. : Bhootkeshi.

Distribution : Kashmir to Kumaon Hills Upto 4,000 m.

Part Used : Root.

Utilization : Credited with hypotensive, sedative, analgesic and curative skin properties.

Active Constituents : So far known to contain selinidin, vaginidin, angelicin, oroselol, lomatin, selinone, vaginatin, 45.5 % alphapinene, 25.3 % limonene, 5.7 % camphene, 5.2 % beta-phellanetrine, 1.2 % alphathujene, 3.2 % fenchyl alcohol, 3.8 % terpeneol, anomalin, isopteryxin, n-triacontane, dotriacontane, triacontanol, 12-tricosanol, sucrose, D-mannitol and elemol (26.0 %).

References : Agarwal (2003), Ambasta (1986), Chauhan (1999), Kala & Manjrekar (2000), Rastogi & Mehrotra (1999a), Sharma & Sood (1997).

Selinum wallichianum (DC.) Raizada & Saxena *ex* Raizada

Syn: *Pencedanum wallichianum* DC.; *Selinum candolli* Edgew.; *S. elatum* (Edgew.) Hiroe; *S. filicifolium* (Edge) Nasir; *S. tenuifoium* Wall. *ex* C.B. Clarke

Family: Apiaceae.

Distribution : Himalaya (Kashmir to Bhutan): 2,700-4,000 m.

Part Used : Root.

Utilization : Regarded as a nervine sedative and taken for cough, asthma, piles and hysteria.

Active Constituents : Bergapten, heraclenin, heraclenol, angelicin and xanthotoxol have been characterised.

Biological Activity : Oil from root has hypotensive, sedative and analgesic properties.

References : Chauhan (1999), Negi *et al.* (1993), Sharma (2003), Singh (1996, 2000), Sood *et al.* (2001).

Semecarpus anacardium L.f.

Family: Anacardiaceae.

English, Hindi & Sanskrit Names:

Eng. : Marking Nut Tree, Oriental Cashew;
Hindi : Bhela, Bhelwa, Bhilawa;
Sans. : Arshagni, Bhallatakah, Bhallika.

Distribution : Common Throughout Sub-Himalayan & Tropical Regions of India, Particularly in Deciduous Forests of Peninsular India, Often in Association with Sal (*Shorea robusta*) in the Foot Hills of the Himalaya to an Elevation of 1,000 m.

Part Used : Root.

Utilization : 100 ml decoction of 10g dried powdered root boiled in 200 ml water given once every morning for 5 days to cure eczema. Roots cooked in sour rice water reported to cause sterility in women when eaten.

References : Girach (1992), Jain (1991), Parrotta (2001), Patil (2000), Reddi *et al.* (2005).

Senecio jacquemontianus (Decne.) Benth. *ex* Hook.f.

Syn: *Senecilluis jacquemontiana* Decne.

Family: Asteraceae.

Distribution : W. Himalaya: 3,000-4,000 m.

Part Used : Root.

Utilization : Fresh root extr. used as nervine tonic and dried once in joint pains. Its poultice applied to external boils or swellings for quick relief and in rheumatic pains, cutaneous troubles and as an adulterant of 'kuth' (*Saussurea lappa* C.B. Clarke).

References : Ambasta (1986), Chopra *et al.* (1956), Kaul (1997), Sharma (2003), Singh (1996), Sinha & Sinha (2001), Watt (1889).

Sesamum oriantale Linn.

Syn: *S. indicum* L.; *S. luteum* Retz.; *S. occidentale* Heer & Regel.

Family: Pedaliaceae.

English, Hindi & Sanskrit Names:

Eng. : Gingely, Sesame;
Hindi : Til;
Sans. : Snehaphala, Tila, Tilah.

Part Used : Root.

Utilization : Decoction used variously to treat cough, asthma and stomach pains.

References : Duhoon & Tripathi (2003), Parrotta (2001), Reddi *et al.* (2005), Sur & Halder (2002), Watt (1889).

Sesbania grandiflora (L.) Poir.

Syn: *Aeschynomene grandiflora* (L.) L.; *Agati grandiflora* (L.) Desv.; *Coronilla grandiflora* (L.) Willd.; *Robinia grandiflora* L.

Family: Fabaceae.

English, Hindi & Sanskrit Names:

Eng. : Agathi, Sesban, Swamp Pea;
Hindi : Agasti, Bak, Basna, Natiya;
Sans. : Agasthya, Agasti, Agastyan, Manitaru, Vaka, Vangasena.

Distribution : Cultivated in Many Parts of India.

Part Used : Root.

Utilization : Used for rheumatism, painful swellings (extr.), and as expectorant (juice with honey).

References : Ambasta (1986), Bhattacharjee (2001), Bondya *et al.* (2006), Chatterjee & Pakrashi (1997), Deshmukh *et al.* (2000), Dhiman (2003), Kurian (1995), Parrotta (2001), Reddi *et al.* (2005), Watt (1889).

Sida acuta Burm.f.

Syn: *S. acuta* var. *carpnifolia* (L.f.) Shumann; *S. carpinifolia* L.f.; *S. scoparia* Lour.; *S. stauntoniana* DC.; *S. stipulata* Car.

Family: Malvaceae.

Hindi & Sanskrit Names:

Hindi : Bariara, Kharenti, Kungyi, Pahari Bariara;
Sans. : Bala, Bribnunagabala, Pata.

Distribution : Found Throughout Tropical & Sub-Tropical India in Fields (Along Roads & Forest Edges).

Part Used : Root.

Utilization : Described so far as bitter, tonic, astringent, cooling, stomachic, vermifuge, febrifuge, aphrodisiac, antipyretic, digestive, diuretic, blood purifier, diaphoretic; and used in chronic bowel complaints, blood and pile disorders, general debility, nervous and urinary diseases. Its decoction (10 ml) given twice daily to get relief in physical weakness, venereal diseases and colic pain.

Active Constituents : Oxalic acid, α-amyrin, ecdysterone, ephedrine and cryptolepine are the main constituents.

References : Agarwal (2003), Ambasta (1986), Caius (2003), Chopra *et al.* (1956), Dastur (1970), Gogoi & Das (2003), Jain *et al.* (1995), Kirtikar & Basu (1935), Maheshwari *et al.* (1997), Nayak *et al.* (2003), Parrotta (2001), Pawar & Patil (2004), Rastogi & Mehrotra (2001, 2002), Shukla & Verma (1996), Subramaniam (2003), Varghese (1996), Watt (1889).

Sida cordata (Burm.f.) Borss.

Syn: *Melochia cordata* Burm.f.; *Sida humilis* Cav.; *S. humilis* var. *varonicaefolius* (Lam.) Masters; *S. multicaulis* Cav.; *S. veronicifolia* Lam.

Family: Malvaceae.

Hindi & Sanskrit Names:

Hindi : Bananiyar, Baria, Barial, Kharenti, Khoroeti, Kungyi;

Sans. : Badiyalaka, Bala, Bhadra, Bhumibala, Brela, Jayanti, Motapati, Odani, Odanika, Raktatanula, Shitapaki, Variga, Vatyali, Vilala.

Distribution : Distributed Throughout the Hotter Parts of India Upto 1,500 m.

Parts Used : Root. Root bark.

Utilization : Root paste applied locally to drive out pus from boils and in piles (applied along with cow's milk). Fresh root juice with 2-3 powdered fruits of *Piper nigrum* L. given twice or thrice daily as aphrodisiac and its decoction with ginger given in intermittent fever. Pounded roots of this plant with those of *Celastrus paniculatus* Willd. (2:1) taken twice a day for one month for amenorrhoea, and about 10 g of its powder given with 100 ml water orally in case of snake bite and to cure the effect of sun stroke. The powder of root bark given with milk and sugar for relief of frequent micturition, gonorrhoea, leucorrhoea, and also in combination with sesamum oil and milk it is very efficacious in curing cases of facial paralysis and sciatica when caused by the inflammation of the nerves concerned.

References : Agarwal (2003), Caius (2003), Chatterjee & Pakrashi (1997), Khanna *et al.* (1996), Maliya & Singh (2003), Panthi & Chaudhary (2003), Parrotta (2001), Sahoo & Mudgal (1995), Singh (2003), Singh & Pandey (1998).

Sida cordifolia L.

Syn: *S. althaeifolia* Sw.; *S. herbacea* Micans; *S. rotundifolia* Cav.

Family: Malvaceae.

English, Hindi & Sanskrit Names:

Eng. : Bala, Mahot Balais;

Hindi : Barial, Khareti, Kungyi;

Sans. : Badiyalaka, Bala, Baladhya, Brela, Jayanti, Kanaka, Krura, Nilaya, Suvarna, Svetberela, Vatyapushpi.

Distribution : Throughout the Tropical & Sub-Tropical India.

Part Used : Root. Root bark.

Utilization : Decoction of roots taken internally as astringent, cooling, diuretic, tonic (10 ml, twice daily) and also good for fever (with ginger), leucorrhoea (with sweet milk), dysuria, urinary troubles (2 teaspoons twice a day for 3-5 days). In infective viral hepatitis, its decoction (20 g) with 1-2 drops of shoot juice of *Scoparia dulcis* given twice daily for a week. A piece of root (±3) also given to chew and suck before bed time to adult patients during the course of treatment. Infusion of roots given in nervous and urinary diseases, disorders of the blood and bile, bleeding piles, asthma, colic, strangury, sciatica, gonorrhoea, cystitis, leucorrhoea, fevers and burning micturition. The root of white flowered plant pounded with cow's milk said to be good for promoting fertility. Garasia tribe use the root in urinary and nervous diseases;

particularly to cure facial paralytic attack and hemiplegia. Decoction of root-bark given with milk and sugar for giving relief in micturition and leucorrhoea. Its extr. in 'til' oil applied externally for facial paralysis and sciatica.

Active Constituents : Therapeutic properties attributed to ephedrine, si-ephedrine, beta-phenylamine, carboxylated tryptamines, the quinazoline alkaloids vasicinone, vasicine, vasicinol, choline and betaine.

References : Ambasta(1986), Bhatt *et al.* (2002), Borthakur *et al.* (2004), Bhattacharjee (2001), Chatterjee & Pakrashi (1997), Dastur (1970), Dwivedi (1993), Jain & Patole (2002), Khare (2004), Kumar & Lal (2003), Meena *et al.* (2003), Reddi *et al.* (2005), Sarin (1996), Sarma *et al.* (2001), Singh & Pandey (1998), Sood & Thakur (2004), Upadhyay & Singh (2005), Watt (1889).

Sida rombifolia Linn. ssp. *rhombifolia* Borssum

Syn: *S. canariensis* Willd.; *S. compressa* Wall.; *S. orientalis* Cav.; *S. rhombifolia* (L.) var. *obovata* Wall. *ex* Mastera; *S. rhombifolia* var. *rhomboidea* (DC.) Masters.

Family: Malvaceae.

English, Hindi & Sanskrit Names:

Eng. : Broomjute Sida, Paddy's Lucerne, Queens Land Hemp, Sida Hemp;

Hindi : Atibala, Bariara, Bhinunli, Pitabala, Sahadeva, Swetabarela;

Sans. : Ahikhanda, Atibala, Bala, Devaarha, Gandhavalli, Kesarika, Mahabala, Sarini.

Distribution : Open Degraded Sites Besides Rivers & Streams; Cultivated Throughout India, Particularly in Moist Regions Upto 1,800 m in the Himalaya.

Part Used : Root.

Utilization : Known so far in rheumatism, child birth (taken internally), leucorrhoea (root paste with milk on oral administration with honey relieves pain), and gargles for toothache (decoction).

References : Caius (2003), Chatterjee & Pakrashi (1997), Chopra *et al.* (1956), Kurian (1995), Parrotta (2001), Watt (1889), Yoganarasimhan *et al.* (1981).

Sida spinosa Linn.

Syn: *S. alba* L.
Family: Malvaceae.

English, Hindi & Sanskrit Names:

Eng. : Prickly Sida;
Hindi : Bariar, Gangeran, Gulasakari, Janglimethi, Khareti;
Sans. : Arishta, Bhadrandahi, Chatupala, Devadanda, Ghanta, Kharagandhini, Kharyashthika, Mahagandha, Mahodaya, Nagabala, Vishvadeva.

Distribution : Throughout Hotter Parts of India from N. W. India to Sri Lanka.

Parts Used : Root. Root bark.

Utilization : The root acts as a gentle tonic, diaphoretic and employed in mild cases of debility and fever; also useful in the treatment of some forms of cattle diseases. The decoction of root bark and root used as a demulcent in irritability of the bladder and gonorrhoea. Powdered root bark mixed with sugar taken orally regularly with milk as a sexual tonic.

Biological Activity : Ethnolic extr. of root hypoglycaemic.

References : Ambasta (1986), Caius (2003), Chatterjee & Pakrashi (1997), Chopra *et al.* (1956), Kirtikar & Basu (1935), Singh & Pandey (1998), Srivastava *et al.* (2003).

Siegesbeckia orientalis Linn.

Syn: *Miryranthes hetrophylla* Turcz.; *S. brachiata* Roxb.; *S. glutinosa* Wall.
Family: Asteraceae.

Hindi & Sanskrit Names:

Hindi : Lichkura;
Sans. : Likura.

Distribution : Common Throughout India.

Part Used : Root.

Utilization : Decoction useful in leucorrhoea (1 teaspoon in one cup of water twice daily for 15 days) and to trat urinary troubles.

References : Agarwal (2003), Reddi *et al.* (2005), Sood & Thakur (2004).

Smilacina fusca Hk.f.

Family: Liliaceae.

Distribution : All over N.E. Region & Sikkim: 3,000-5,000 m.

Part Used : Root.

Utilization : Powder given in malaria and dysentery.

Reference : Kumar (2002).

Smilax macrophylla Roxb.

Syn: *S. grandis* Wall.; *S. grangifolia* Voigt.; *S. ovalifolia* Roxb.; *S. prolifera* Wall.; *S. retusa* Roxb.; *S. roxburghii* Kunth.; *S. zeylanica* Linn.

Family: Liliaceae.

Hindi Names:

Hindi : Jangliaushbah, Chob-Chini.

Distribution : E. Bengal, Assam, Chutia Nagpur, W. Peninsula, Madras. Occurs Plentifully on the Lower Himalaya from Kumaon Eastwards, Ascending to about 2,000 m.

Part Used : Root.

Utilization : 10g powder given daily after meals in piles. Also, used as a substitute for Sarsaparilla (*Hemidesmus indicus* Linn.) for venereal diseases, rheumatism, urinary complaints and dysentery.

References : Ambasta (1986), Chopra *et al.* (1956), Jha & Verma (1996), Watt (1889).

Smilax orthoptra A. DC.

Family: Liliaceae.

Part Used : Root.

Utilization : Paste along with honey (about 2g) taken orally in venereal diseases and urinary problems.

Reference : Panthi & Chaudhary (2003).

Smilax ovalifolia Roxb. *ex* D. Don

Family: Liliaceae.

English & Hindi Names:

Eng. : Green Beaiar, Saraspatrilla Plant;

Hindi : Jangliaushbah.

Distribution : Throughout Tropical Parts of India.

Part Used : Root.

Utilization : Decoction taken twice a day in spermatorrhoea, jaundice, weakness, and to clear urine after high fever and considered very effective to cure venereal diseases, rheumatic swellings, dysentery, and urinary complaints. Its paste with milk prescribed for fever and heat in stomach: (50g root ground in 200 ml milk; 50 ml twice daily for 3 days). Roots with those of *Bredelia tomentosa* ground to paste and applied on rheumatic pain and gonorrhoea.

References : Agarwal (2003), Bhattacharjee (2001), Lalramnghinglova & Jha (1999), Varghese (2003).

Smilax parviflora Wall.

Syn: *S. glaucophylla* Klotzsch; *S. longibracteolata* Hook.f.

Family: Liliaceae.

Distribution : Himalaya, Nagaland & Meghalaya: 1,500-2,400 m.

Part Used : Root.

Utilization : Decoction of fresh roots used in venereal diseases and sores.

Active Constituents : Diosgenin and its 3-O-β-D-glucoside isolated from roots.

References : Ambasta (1986), Kapur (1996b), Rastogi & Mehrotra (2001).

Smilax prolifera Roxb.

Syn: *S. perfoliata* Lour.

Family: Liliaceae.

Distribution : Tropical W. Himalaya, Kumaon, Nepal, Sylhet, Bengal, Bihar & Deccan Peninsula.

Part Used : Root.

Utilization : One teaspoonful of crushed roots taken with a cup of water twice a day in sunstroke. Root brayed in coagulated milk and drunk against blood dysentery.

References : Ambasta (1986), Bhardwaj & Gakhar (2003), Chopra *et al.* (1956), Tirkey (2004), Tiwari & Tiwari (1996).

Smilax zeylanica L.

Family: Liliaceae.

English, Hindi & Sanskrit Names:

Eng. : Sarsaparilla;
Hindi : Chob Chini, Jangliaushbah, Kumarika, Ramdatun;
Sans. : Vanamadhusnahi.

Distribution : Hilly Areas Throughout Tropical India From Himalaya to Kerala about 2,000 m.

Part Used : Root.

Utilization : Its juice mixed with root paste of *Asparagus racemosus* Willd. given to cure spermatorrhoea. Decoction beneficial to cure anaemia, leucorrhoea, antifertility, paralysis, venereal diseases, skin diseases, haematuria, rheumatism, sores, swellings, abscess, boils, dysentery, weakness and as cooling. Root paste also taken orally for skin diseases.

Active Constituent : Diosgenin identified in roots.

References : Agarwal (2003), Ambasta (1986), Chhetri (2005), Dash & Mishra (2000), Islam (1996), Islam & Jha (2003), Panda & Das (2000), Parrotta (2001), Rastogi & Mehrotra (2001), Reddi *et al.* (2005).

Solanum anguivi Lam.

Syn: *S. indicum* auct. non Linn.;
S. sodomeum L. nom. ambig; *S. violaceum* Ortega.
Family: Solanaceae.

English, Hindi & Sanskrit Names:

Eng. : Indian Night Shade, Poison Berry;

Hindi : Barhanta, Bhirhatta;

Sans. : Brhati, Brihatee.

Distribution : Found Throughout Tropical India, Ascending to 1,600 m.

Part Used : Root.

Utilization : Considered carminative, anthelmintic, diaphoretic, stimulant, expectorant and used in asthma, cough, fever, colic, dysuria, inchuria, nasal ulcers, bronchitis, catarrhal affections, dropsy, toothache, difficult parturition and worm complaints. Extr. of its paste given orally to relieve from spermatorrhoea.

References : Ambasta (1986), Balasubramanian & Prasad (1996), Chatterjee & Pakrashi (2003), Gogoi *et al.* (2003b), Kumar & Lal (2003), Reddi *et al.* (2005), Singh *et al.* (2003), Subramaniam (2000), Watt (1889).

Solanum erianthum D. Don

Syn: *S. verbascifolium* auct. non L.
Family: Solanaceae.

English, Hindi & Sanskrit Names:

Eng. : Common Night Shade, Potato Tree;

Hindi : Asheta;

Sans. : Gandisa, Pathi, Priyamkari, Vidari.

Distribution : Throughout Tropical & Sub-Tropical Parts of India, Andaman Islands; Cultivated in S. India.

Part Used : Root.

Utilization : Decoction good against vertigo, urinary troubles, shooting pain all over the body and discomfort after meals. Root paste applied all over the body of 2-3 month's child to cure fever.

Active Constituents : Solanine, solasonine, solamargine and rolanidine are the pharmacologically active ingredients of root.

References : Agarwal (2003), Ambasta (1986), Hosagoudar & Henry (1996b), Parrotta (2001), Rastogi & Mehrotra (2001).

Solanum ferox L.

Syn: *S. indicum* L.
Family: Solanaceae.

English, Hindi & Sanskrit Names:

Eng. : Poison Berry;

Hindi : Barhanta, Bhatakataiya, Birhatta, Kattarha, Vadikadheri;

Sans. : Brhati, Brhatikaa, Hinguli, Kateri, Kshudrabhantaki, Prasah, Sinhi, Vartaki.

Distribution : Throughout Tropical & Sub-Tropical India Upto 1,500 m.

Part Used : Root.

Utilization : Decoction regarded cure for cough, asthma, colic, difficult parturition and nasal ulcers. Also, recorded as pungent, bitter, heating, expectorant, digestive, antibacterial, astringent to the bowels, anthelmintic, mouth freshner, and beneficial for cardiac troubles, leucoderma, fever, asthma, pain, bronchitis, vomiting, toothache, cough, catarrhal affections, dysuria and inchuria (half teaspoonful twice daily). Its poultice applied on itches, boils, burns, cuts while extr. considered digestive, and for reducing labour pains.

References : Agarwal (2003), Ambasta (1986), Bhatt *et al.* (1999), Khare (2004), Kirtikar & Basu (1984), Kumar (2002), Parrotta (2001), Shukla *et al.* (2003).

Solanum incanum L.

Family: Solanaceae.

Distribution : Punjab, Mewara & Peninsula.

Part Used : Root.

Utilization : Paste prepared in butter or coconut oil applied over wounds for relief and cure. Fresh roots chewed for stomach complaints.

References : Balasubramanian & Prasad (1996), Chopra *et al.* (1956), Reddy *et al.* (2003).

Solanum melanogena L.

Family: Solanaceae.

English, Hindi & Sanskrit Names:

Eng. : Brinjal, Egg Plant;

Hindi : Badanjan, Baigun, Bhanta;

Sans. : Bhantaki, Hingoli, Vartakee, Vatigana.

Distribution : Cultivated Throughout India.

Part Used : Root.

Utilization : Root antasthmatic, stimulant; pounded and applied on ulcers, nose bleeding, otitis and toothache.

Active Constituents : Found to contain p-aminobenzyldehyde, ethyl-caffeate, trans-ferulic acid, N-trans-feruloyltyramine, N-trans-p-coumaroyltyramine, vanillin and isoscopoletin.

References : Agarwal (2003), Ambasta (1986), Chatterjee & Pakrashi (1956).

Solanum nigrum L.

Syn: *S. erythraeum* Dunal.; *S. fistulosum* Rich.; *S. incertum* Dunal; *S. miniatum* Bernh.; *S. nodiflorum* Jacq.; *S. paludosum* Dunal; *S. roxburghii* Dunal; *S. rubrum* Mill.; *S. triangulare* Lamk.; *S. ulginosum* Dunal.

Family: Solanaceae.

English, Hindi & Sanskrit Names:

Eng. : Black Night Shade, Common Night Shade, Garden Night Shade, Hound's Berry, Morelle, Petty Morel;

Hindi : Gurkamal, Kabaiya, Mekoi;

Sans. : Bahuphala, Bahutikta, Dhvankshamachi, Ghanaghanam, Guchhaphala, Kaka, Rasyani, Sarvatikta, Tiktika, Vayasavha.

Distribution : All Parts of India Upto 3,000 m.

Parts Used : Root. Root bark.

Utilization : Root extr. (about twospoons) mixed with a pinch of sugar taken orally by men in the morning for a month as an aphrodisiac. Bark considered laxative; useful in diseases of the ear, eye, nose, ulcers on the neck, burning of the throat, inflammation of the liver, chronic fever and gripping (not to be given to pregnant woman).

Active Constituents : Properties attributed to uttronin A, furostanol glycosides, uttrosides A & B, glucose, fructose, caffeoyl glucose, caffeic, chlorogenic, iso- and neo-chlorogenic acids, solasodine, tomatidenol, tigogenin and uttronin B.

References : Chatterjee & Pakrashi (1997), Khare (2004), Kirtikar & Basu (1984), Pawar & Patil (2004), Rastogi & Mehrotra (2001), Sood & Thakur (2004).

Solanum surattense Burm.f.

Syn: *S. armatum* Br.; *S. difusum* Roxb.; *S. jacquinii* Willd.; *S. maccani* Sant.; *S. virginiaum* Jacq.; *S. xanthocarpum* Schrad.; *S. virginiaum* Jacq.

Family: Solanaceae.

English, Hindi & Sanskrit Names:

Eng. : Dwarf Lid Brinjal, Wild Egg Plant, Yellow Berried Nightshade;

Hindi : Katai, Kateli, Ringni, Satyanashi;
Sans. : Brhati, Kantakari, Nidigadhika, Vyaghri.

Distribution : Distributed Throughout India, Particularly in Deccan, Kerala & Punjab.

Parts Used : Root. Root bark.

Utilization : About 10-50 ml root decoction given twice a day against respiratory diseases, sun stroke, malarial and other fevers. Also, prescribed with the addition of long pepper and honey for cough, catarrh and spasmodic cough (with rock salt and asafoetida). Root considered pungent, bitter, heating, aphrodisiac, appetiser, laxative, stomachic, anthelmintic, diuretic, emetic, expectorant, febrifuge; useful in bronchitis, cough, asthma, malaria, fever, T.B., ozaena, strangury, lumbago, enlargement of liver and spleen, dropsy, dysuria, cold, scorpion bite, pains, piles, thirst, urinary concretions and diseases of the heart. Poultice of crushed bark applied locally for the treatment of hernia and said to be absolutely effective for young patients.

Active Constituents : Contains carpestrol and solasodine.

References : Agarwal (2003), Ambasta (1986), Beg *et al.* (2005), Bhatt *et al.* (1999), Billor *et al.* (1998), Chatterjee & Pakrashi (1997), Dastur (1970), Dwivedi (1993), Jain (1968), Khare (2004), Kirtikar & Basu (1984), Kurian (1995), Parrotta (2001), Reddi *et al.* (2005), Sarin (1996), Sharma (2002), Singh (1999), Shukla & Verma (1996), Sood & Thakur (2004), Upadhyay & Singh (2005), Varghese (1996).

Solanum trilobatum L.

Family: Solanaceae.

Hindi & Sanskrit Names:

Hindi : Alarka;
Sans. : Alarka.

Distribution : Hedges, Wasteplaces, Swampy Areas Throughout Tropical Parts of India.

Part Used : Root.

Utilization : Root bitter and decoction efficacious in tuberculosis and other lung diseases.

References : Agarwal (2003), Ambasta (1986), Chatterjee & Pakrashi (1997).

Solena amplexicaulis (Lam.) Gandhi

Syn: *Melothria hetrophylla* (Lour.) Cogn.; *Zehneria umbellata* (Klein *ex* Willd.) Thw.; *Z. umbellate* var. *napalensis* (Seringe) Cl.

Family: Cucurbitaceae.

Hindi & Sanskrit Names:

Hindi : Tarali;
Sans. : Sukakamdah.

Distribution : Common Throughout India Upto 2,100 m.

Part Used : Root.

Utilization : Powder given in the dose of 2g twice a day with milk in seminal debility and spermatorrhoea, and its decoction for burning micturition. Root paste used on rheumatism and 15-20 ml of its extr. given thrice a day for 3 days to induce menstruation.

References : Reddi *et al.* (2005), Sharma (2003), Sharma & Singh (2000), Sood & Prakash (2007), Varghese (1996).

Solidago virga-aurea L.

Family: Asteraceae.

Part Used : Roots.

Utilization : Decoction given for urinary trouble and considered antiseptic. Dried root powder used for diarrhoea in infants.

Reference : Singh & Pandey (1998).

Sonchus arvensis Linn.

Syn: *S. longifolius* Wall.; *S. orixensis* Roxb.; *S. wightiana* DC.
Family: Asteraceae.

English & Hindi Names:

Eng. : Corn Sow Thistle;
Hindi : Sadhi, Sahadevibari.

Distribution : Wasteplaces, Hedges, Gardens, Margins of Cultivated Lands & Almost Throughout India.

Part Used : Root.

Utilization : Beneficial in jaundice and diseases of respiratory tract, cough, bronchitis, asthma and pertusis.

References : Agarwal (2003), Ambasta (1986), Caius (2003), Chatterjee & Pakrashi (1997), Chopra *et al.* (1956), Watt (1889).

Sonchus asper (L.) Hill.

Syn: *S. ferox* Wall.; *S. olearaceus* Wall.
Family: Asteraceae.

English & Hindi Names:

Eng. : Milk Thistle, Spiny Leaved Sow Thistle;
Hindi : Didhi, Dodak.

Distribution : Throughout India, Including Himalaya to 4,000 m.

Part Used : Root.

Utilization : Milky latex pounded with roots applied on boils for suppuration. Roots used as vermifuge; ointment made from its extr. beneficial for ulcers and wounds.

References : Agarwal (2003), Chhetri (1994), Sood & Thakur (2004).

Sonchus wightianus DC.

Family: Asteraceae.

Part Used : Root.

Utilization : Given in jaundice.

Reference : Yoganarasimhan *et al.* (1981).

Sonerila maculata Roxb.

Family: Melastomaceae.

Distribution : Throughout N.E. Region Common in Damp Localities Upto 2,000 m.

Part Used : Root.

Utilization : Boiled and consumed for stomachic ailments.

References : Kharkongor & Joseph (1997), Kumar (2002).

Sonerila tenera Royle

Family: Melastomataceae.

Hindi Name:

Hindi : Huring.

Distribution : Found in E. India.

Part Used : Root.

Utilization : Fresh roots used for treating diseases of cattle.

References : Agarwal (2003), Ambasta (1986).

Sorbaria tomentosa Lindl.

Syn: *Spiraea lindleyana* Wall. *ex* Lindl.; *S. sorbifolia* Hook.f.in part, non Linn.

Family: Rosaceae.

English Name:

Eng. : Hardhack Steeple Bush.

Distribution : W. Temperate Himalaya (Kunawur to Kumaon): 2,300-3,300 m.

Part Used : Root.

Utilization : Decoction good for bilious fever.

References : Ambasta (1986), Hooker (1879), Nautiyal *et al.* (2003).

Soymida febrifuga (Roxb.) A Juss.

Syn: *Swietenia febrifuga* Roxb.; *S. rubra* Rottl.; *S. soymida* Duncan.
Family: Meliaceae.

English, Hindi & Sanskrit Names:

Eng. : Bastard Cedar, Indian Red Wood;
Hindi : Rakat, Rohan, Rohunna;
Sans. : Patranga.

Distribution : Practically Throughout India Except W. Maharashtra.

Part Used : Root.

Utilization : Useful in cough, cold and as abortifacient.

References : Agarwal (2003), Ambasta (1986), Khanna (2002b), Reddi *et al.* (2005), Trivedi (2002), Watt (1889).

Spermadictyon sauveolens Roxb.

Syn: *Hamiltonia mysorensis* W. & A.; *H. propinqua* Decne.; *H. scabra* Don; *H. suaveolens* Roxb.; *Spermadictyon suaveolens* Roxb.
Family: Rubiaceae.

Distribution : Tropical & Sub-Tropical Himalaya: 660-1,700 m (Kashmir to Bhutan), C. India & W. Peninsula, Ascending to 1,300 m.

Part Used : Root.

Utilization : To prevent conception, paste of its root (25 g) and 4 black peppers ground in cow's milk given to woman on empty stomach once daily for 7 days after menstruation. The root paste along with mustard oil applied externally on body for relieving fever and piles.

References : Brahmam & Saxena (1990), Hooker (1886), Kothari & Moorthy (1996), Singh & Kumar (2000b).

Sphaeranthus hirtus Burm.

Family: Asteraceae.

Distribution : Peninsula. Banks of Rice Fields.

Parts Used : Root. Root bark.

Utilization : Root powder stomachic while the bark considered a valuable remedy for piles.

Reference : Drury (1978).

Sphaeranthus indicus L.

Syn: *S. senegalensis* L.

Family: Asteraceae.

English, Hindi & Sanskrit Names:

Eng. : East Indian Gole-Thistle;
Hindi : Gorakh Mundi, Mundi;
Sans. : Mundi, Mahamundi, Śravani.

Distribution : Rice Fields, Dry Wasteplaces, Cultivated Lands in Tropical India.

Parts Used : Root. Root bark.

Utilization : Finely powdered root bark given with butter milk in bleeding piles. Root considered anthelmintic, stomachic; its extr. efficacious in chest pains, cough, and bowel diseases and juice employed to cure epilepsy, impotency, bleeding and diarrhoea. Dried root powder with 5 black peppers (*P. nigrum*) given 3g per dose twice a day for 3 days to cure blood dysentery

References : Agarwal (2003), Ambasta (1986), Chatterjee & Pakrashi (1997), Dastur (1970), Girach (1992), Reddi *et al.* (2005).

Stachyphrynium spicatum (Roxb.) Scum.

Family: Marantaceae.

Part Used : Roots.

Utilization : Paste administered orally for stomachache.

Reference : Rajendran & Henry (1994).

Stauranthera umbrosa Clarke

Family: Gesneriaceae.

Distribution : Upper Assam.

Part Used : Root.

Utilization : Juice used against boils.

Reference : Kumar (2002).

Stemona tuberosa Lour.

Family: Stemonaceae.

Hindi Name:

Hindi : Ijedigadda.

Distribution : Arunachal Pradesh, Coastal Regions of A.P. & Tamil Nadu.

Part Used : Root.

Utilization : So far known to be as antibacterial, antiseptic and used in pthisis, cough, and soothing of respiratory tract. Its decoction good for tuberculosis and fever (10 ml twice daily).

Active Constituents : Tuberostemonine, stemonine, isostemotinine, stemonidin, stemonine LG, stemoninoamide and tuberostemoninol have been characterised form roots.

References : Agarwal (2003), Ambasta (1986), Lalramnghinglova & Jha (1997), Rastogi & Mehrotra (1998, 1999b, 2002).

Stephania elegans Hook. & Thomb.

Family: Menispermaceae.

Distribution : Himalaya (Garhwal to Sikkim) & Khasia Hills: 1,000-2,100 m.

Part Used : Root.

Utilization : Powdered root good against piles (½ -1g twice a day for 15 days).

Active Constituents : Active principles are due to alkaloiisosino acutine, (-) sinoacutine, hentriacontane, β-sitosterol, sinaacutine, tetrahydropalmatine, cyclanoline, magnoflorine, epihernandolinol, N-methyl corydalmine, hasubanonine, aknadinine, cyclanoline, magnoflorine, isotetrandrine, isochondrodendrine and cycleanine.

References : Rastogi & Mehrotra (2001), Sood & Thakur (2004).

Stephania glabra (Roxb.) Miers.

Syn: *Cissampelos glabra* Roxb.; *Clypea wightii* Arn.; *Cocculus finlaysoniana* Wall.; *C. roxburghianus* Wall.; *Stephania rotunda* Hook.

Family: Menispermaceae.

Hindi & Sanskrit Names:

Hindi : Gindaru;
Sans. : Nimuka.

Distribution : Throughout Temperate & Sub-Tropical Himalaya.

Part Used : Root.

Utilization : Tuberous root given to treat asthma, dysentery, tuberculosis, fever, diarrhoea, dyspepsia and urinary complaints. Fresh root mixed with rock salt cures flatulence in cattle. Extr. obtained by boiling pieces of its root in mustard oil useful against inflamed parts.

References : Agarwal 92003), Chopra *et al.* (1956), Jain *et al.* (2003), Sood & Thakur (2004).

Stephania japonica (Thunb.) Miers.

Syn: *S. hernandifolia* (Willd.) Walp.

Family: Menispermaceae.

English, Hindi & Sanskrit Names:

Eng. : Tape-Vine;
Hindi : Akanadi, Par;
Sans. : Ambashtha, Patha, Vanatiktika.

Distribution : Punjab, Dehradun, W. Bengal, Bihar, Assam, E. & W. Coast & S. India.

Part Used : Root.

Utilization : Its infusion mixed with powdered dried ginger effective in piles, constipation (powdered root with curdled milk), obsterical practice and heart trouble (in Meghalaya). Also, reported to be useful in fever, diarrhoea, urinary diseases, dyspepsia, stomachache, diabetes, indigestion, deworming, urinary diseases, spermatorrhoea and arthritis.

References : Ambasta (1986), Caius (2003), Chatterjee & Pakrashi (1997), Gogoi *et al.* (2003a), Jain & Saklani (1992), Reddi *et al.* (2005), Saklani & Jain (1994), Srivastava *et al.* (2003), Yoganarasimhan *et al.* (1981).

Sterculia urens Roxb.

Family: Sterculiaceae.

English, Hindi & Sanskrit Names:

Eng. : Kuteera Gum Tree, Sterculia;
Hindi : Gular, Gulu, Karrai, Katira, Kulu;
Sans. : Arjuna, Balika.

Distribution : Exposed Hill Slopes in the Sub-Himalayan Hills from the Ganges Eastwards, Gujarat, Rajasthan & in C. & S. India to N. Karnataka; Also Reported in Mizoram & Sri Lanka.

Parts Used : Root. Root bark.

Utilization : Pounded bark made into a poultice and applied externally over wounds, fractures and inflamed testes. Root extr. good against scorpion sting and taken with black pepper to relieve fever.

Active Constituent : 7-hydroxy-6-methoxycoumarin.

References : Kurian (1995), Parrotta (2001), Rastogi & Mehrotra (1998), Reddi *et al.* (2005), Trivedi (2002).

Stereospermum colais (Dillw.) Mabberley

Syn: *Bignonia colais* Buch-Ham. *ex* Dillw.; *Stereospermum chelonoides* sensu Cl.; *S. personatum* (Hassk.) Chatt.; *S. tetragonum* DC.

Family: Bignoniaceae.

English, Hindi & Sanskrit Names:

Eng. : Trumpet Flower, Yellow Snake Tree;
Hindi : Padri, Parral;
Sans. : Patali.

Distribution : Throughout Moist Parts of India.

Part Used : Root.

Utilization : Decoction used in brain affections, inflammatory chest, intermittent and perpetual fever, asthma, cough, excessive thirst and considered antiasthmatic and antiemetic.

Active Constituents : Contains ceryl alcohol, oleic, palmitic and stearic acids and lapachol.

References : Agarwal (2003), Ambasta (1986), Chatterjee & Pakrashi (1997), Rastogi & Mehrotra (2001).

Stereospermum suaveolens Roxb.

Syn: *Bignonia chelonoides* L.f.; *B. suaveolens* Roxb.; *Heterophragma suaveolens* Dalz & Gibs.; *Stereospermum chelonoides* DC.; *Tecoma suaveolens* G. Don

Family: Bignoniaceae.

Hindi & Sanskrit Names:

Hindi : Padal, Padaria, Padiala, Pandri, Paral, Pur, Purula;

Sans. : Abhipriya, Alivallabha, Ambuvasani, Amova, Kalavrinla, Kuberakshi, Kumbi, Patali, Sthali, Toyadhivasini, Toyapushpi, Vasantaduti.

Distribution : Throughout N.E. Region Upto a height of 1,500 m.

Part Used : Root.

Utilization : Decoction used in intermittent and perpetual fevers, inflammatory affections within the chest and brain. Also, its decoction along with the root of *Curcuma longa* and jaggery (2 teaspoons twice a day for 8 days) given for the treatment of fever accompanying child birth. To clear deposited blood in the uterus after delivery, its extr. along with that of *Sida rhombifolia* (1 teaspoon twice a day) is given.

Active Constituents : Reported to contain lapachol, dehydro-α-lapachone and dehydrotectol.

References : Agarwal (2003), Ambasta (1986), Chopra *et al.* (1956), Dhiman (2003), Kumar (2002), Parrotta (2001), Rastogi & Mehrotra (1999a, b), Singh & Prakash (1994), Watt (1889).

Strychnos bourdilloni Brandis

Syn: *S. cinnamomifolia* Thw.

Family: Loganiaceae.

Distribution : Deccan, W.Ghats, S. Kanara to Travancore in Evergreen Forests Upto 1,000 m.

Part Used : Root.

Utilization : Decoction beneficial in rheumatism, ulcers, elephantiasis, fever and epilepsy.

References : Chopra *et al.* (1956), Kirtikar & Basu (1984).

Strychnos nux-vomica Linn.

Syn: *S. columbine*Wight.

Family: Loganiaceae.

English, Hindi & Sanskrit Names:

Eng. : Nux-Vomica, Snake Wood, Strychnine Tree;

Hindi : Bailewa, Chibbige, Kaira, Kajra, Kuchila, Kuchla;

Sans. : Garadruma, Karaskarah, Kulaka, Kupilla, Vishamushti.

Distribution : Throughout Tropical India in Semi-Evergreen, Moist & Dry Deciduous Forests to 1,300 m from U. P. to W. Bengal Southwards to Karnataka, Kerala & Tamil Nadu.

Parts Used : Root. Root bark.

Utilization : Root bark mixed with lime juice made into pills for cholera. The bitter root and root bark used as a febrifuge in intermittent fever. Root tincture considered tonic and used in cardiac failure to improve pulse, and regarded useful in fever, indigestion, constipation, dysentery dyspepsia and antidote to chloroform poisoning.

Active Constituents : These are 4-hydroxy-3 methoxy-strychnine, normacusine B, protostrychnine, brucine, strychnine, vomicine, methoxystrychinine (root bark), while C-mavacurine (root).

References : Chatterjee & Pakrashi (1997), Chopra *et al.* (1956), Goud & Pullaiah (1996), Parrotta (2001), Rastogi & Mehrotra (1999a, b), Reddi *et al.* (2005), Sharma (2003), Subramaniam (2000).

Sutherlandia frutescens R.Br.

Syn: *S. microphylla* Burchardt *ex* DC.

Family: Fabaceae.

English Names:

Eng. : Cancer Wort, Cape Balloon Pea.

Distribution : Grown in Indian Gardens.

Part Used : Root.

Utilization : Extr. anticancerous and good against eye diseases.

References : Agarwal (2003), Ambasta (1986).

Swertia chirata Buch.-ham.

Syn: *S. chirayita* (Roxb. *ex* Flem.) Karst.

Family: Gentianaceae.

English, Hindi & Sanskrit Names:

Eng. : Chirata;
Hindi : Chiraita, Chirayita;
Sans. : Bhuunimba, Katutikta, Kiraatatikata, Trnani.

Distribution : Temperate Himalaya: 1,200-1,300 m (Kashmir to Bhutan), Khasia Hills & Meghalaya: 1,200-1,500 m.

Part Used : Root.

Utilization : Used for influenza, leprosy, leucoderma and intermittent fever.

Active Constituents : Nine tetraoxygenated xanthones-1-hydroxy-3, 5, 8-trihydroxy-3, 7-dimethoxyxanthone III, 1, 8-

dihydroxy-, 5-dimethoxyxanthone, 1, 3, 8-trihydroxy-5-methoxyxanthone, 1, 5, 8-trihydroxy-3-mwthoxyxanthone, 1, 3, 5, 8 tetrahydroxy xanthone, 1, 3,7, 8-tetrahydroxyxanthone and 1, 3, 6, 7-tetrahydroxyxanthone-C_2-β-D-glucoside (mangiferin) are the active constituents.

Biological Activity : Choloform extr. exhibited neuromuscular blocking activity on rat phrenic nerve diaphragm.

References : Agarwal (2003), Ambasta (1986), Chhetri (2005), Jain (1991), Khare (2004), Rastogi & Mehrotra (1999a, b), Saklani & Jain (1994).

Swertia ciliata (G. Don) Burtt.

Syn: *S. purpurascens* Wall. *ex* C. B. Clarke.

Family: Gentianaceae.

Distribution : Throughout Temperate & Alpine Slopes of C. Himalaya Near Forest Edges & Moist Slopes Upto 4,000 m.

Part Used : Root.

Utilization : Decoction given to bring down fever rapidly. Chewing of its roots valued as a vermifuge.

References : Ambasta (1986), Negi *et al.* (1999), Sharma (2003).

Swertia corymbosa (Griseb.) Wt. *ex* C.B. Clarke

Family: Gentianaceae.

Distribution : Mountains of the S. Deccan Peninsula: 300- 2,300 m.

Part Used : Root.

Utilization : Roots made into a paste in cow's milk and given orally in diarrhoea.

References : Augustine & Sivadasan (2004), Chopra *et al.* (1956).

Swertia dilata Wall.

Family: Gentianaceae.

Distribution : Meghalaya Upto 1,000 m.

Part Used : Root.

Utilization : Powdered root administered in dysentery, malaria and syphilis.

Reference : Kumar (2002).

Symphorema polyandrum Wight.

Syn: *S. involucratum* Kew.
Family: Verbenaceae.

Distribution : S. Deccan Peninsula in the Hills.

Part Used : Root.

Utilization : Paste of 10g of its root and 7 seeds with 7 black peppers given in single dose as an antidote for snake, cat, mad dog and mosquito bites; also reported to be good for suppuration of boils (500g of its root paste taken with honey once daily on empty stomach).

References : Brahmam & Saxena (1990), Hooker (1885), Sen & Behera (2003).

Tabernaemontana divaricata (L.) R.Br. *ex* Roem. & Schults.

Syn: *Ervatamia coronaria* (Jacq.) Stapf.; *Nerium coronarium* Jacq.; *N. divaricatum* L.; *Tabernaemontana coronaria* (Jacq.)Willd.
Family: Apocynaceae.

English, Hindi & Sanskrit Names:

Eng. : East India Rosebay;
Hindi : Chandni, Tagar;
Sans. : Nandivriksha, Nandyavartah, Tagar, Tagara.

Distribution : Forests & Hedges Throughout the Sub-Himalayan Hills From Garhwal Eastwards to Assam. Cultivated in Gardens For Ornamental & Fragrant Flowers.

Part Used : Root.

Utilization : Root chewed thrice a day in toothache and corneal diseases. Mixture of its root powder with that of *Achyranthes aspera* used as tooth powder for dental caries and pyorrhoea. Decoction of 100-200g of its roots taken three times to cure jaundice. Rubbed into a thin paste with water, the root bark taken internally as a vermicide.

Active Constituents : These are tabernaemontanine, dregamine, vobasine, voacamne, voacangine (roots), 5-hydroxy-6-oxocoronaridine, 5-oxocoronaridine, 6-coronaridine, 19-hydroxy coronaridine, α-amyrin, lupeol, β-sitosterol, cycloartenol, campesterol, bezoic acid, aurantiamide acetate, coronaridine, ibogamine, voacamine, coronaridine hydroxyindolenine and pseudovobparicine (root bark).

References : Anonymous (1968), Chatterjee & Pakrashi (1997), Dhiman (2003), Pandey & Verma (2002), Pandey *et al.* (1996), Parrotta (2001), Rao *et al.* (1996), Rastogi & Mehrotra (2001, 2002), Reddi *et al.* (2005), Saklani & Jain (1994).

Tacca leontopetaloides (L.) O. Kuntze.

Syn: *T. pinnatifida* Forst. & Forst.f.;
T. pinnatifolia Gaertn. *T. sitorea* Rumph.
Family: Taccaceae.

English & Hindi Names:

Eng. : Fijji Arrowroot, Indian Arrowroot, Tahiti Arrowroot;
Hindi : Diva, Surna.

Distribution : Grown for its Ornamental Foliage in M. P., Bihar & Deccan.

Part Used : Rootstock.

Utilization : Rootstock used to cure dysentery and piles.

References : Agarwal (2003), Reddi *et al.* (2005).

Tagetes erecta L.

Family: Asteraceae.

English, Hindi & Sanskrit Names:

Eng. : African or Aztec Marigold;
Hindi : Genda, Gultera;
Sans. : Ganduga, Sandu, Sthulapushpa.

Distribution : Gardens in Various Parts of India.

Part Used : Root.

Utilization : Regarded as laxative.

Active Constituents : Contains α-terthienyl.

References : Agarwal (2003), Ambasta (1986).

Tamarix ericoides Rottl.

Syn: *T. mucronata* Smith; *T. tenacissima* Ham.; *Trichaurus ericoides* W.& A.

Family: Tamaricaceae.

Hindi Names:

Hindi : Gajrijhaui, Jhau.

Distribution : River Beds in S. India & Chotanagpur.

Part Used : Root.

Utilization : Paste applied externally to cure alti (reddish small boil-like structures).

References : Agarwal (2003), Ambasta (1986), Sen & Behera (2003), Watt (1889).

Tamarix gallica auct. non Linn.

Syn: *T. troupii* Hole

Family: Tamaricaceae.

English, Hindi & Sanskrit Names:

Eng. : The Tamarisk;
Hindi : Jhau, Jhav, Jhavuka;
Sans. : Jhavuka, Shavak.

Distribution : Throughout N. India on Saline Soils & Sandy regions.

Part Used : Root.

Utilization : Decoction of root nodule given in spruce and diarrhoea.

References : Agarwal (2003), Ambasta (1986), Jadeja (1999).

Tanacetum dolichophyllum Kitam.

Syn: *T. himachalensis* Aswal & Mehrotra; *T. longifolium* Wallich *ex* DC.
Family: Asteraceae.

Distribution : Himalaya: (Kashmir to Nepal): 3,000-4,200 m.

Part Used : Root.

Utilization : Useful against gastroenteritis and colic pain.

Active Constituents : Contains terpinen-4-ol (25.8), sabinene (23.4) and p-cymene (12.3%).

References : Negi *et al.* (1993), Rastogi & Mehrotra (1998), Sharma (2003).

Tanacetum umbelliferum Boiss.

Family: Asteraceae.

English Name:

Eng. : Sweet Pellitory.

Part Used : Root.

Utilization : Known to be useful as aphrodisiac, tonic and deobstruent.

Reference : Ambasta (1986).

Taraxacum officinale Weber *ex* Wiggers (Pl. 8C)

Syn: *Leontodon taraxacum* Linn.; *Taraxacum dens-leonic* Desf.
Family: Asteraceae.

English & Hindi Names:

Eng. : Bitter Wort, Blowball, Dandelion, Fortune Teller, Milk Gowan;

Hindi : Barau, Dulal, Kanphul.

Distribution : Throughout Temperate Himalaya.

Part Used : Roots.

Utilization : Aq. extr. given as diuretic, blood purifier, tonic, laxative, cholagogue, antirheumatic, aperient; also useful in India for dysuria, bowel complaints, chronic disorders of kidney, fever, liver gall stones, gastric ulcer, piles and warts. Powdered root in doses of 600-900 mg considered efficacious in chronic visceral diseases, chronic skin diseases and as a hepatic stimulant.

Active Constituents : Reported to contain resin, essential oil, enzyme, fatty acids including melissic acid, p-hydroxyphenyl acetic acid, β sitosterol, stigmasterol, taraxsterol, homotaraxasterol, taraxacoside and characterised as β-O-[4′-O-(hydroxyphenylacetyl)-β-D-glucopyranosyl-β-hydroxy-γ-butyrolactone, tetrahydroridentine B, a eudesmanolide-β-D-glucopyranoside, 4-α-15, 11-β-13-tetrahydrolacetate, β-D-glucopyranoside, taraxacolide-1′-O-β-D-glucopyranoside, taraxacolide-1′-O-β-D-glucopyranoside, 11, 13-dihydrotaraxinacid-1′-O-β-D-glucopyranoside and taraxin acid-1′-O-β-D-glucopyranoside.

Biological Activity : Extr. of root when injected subcutaneously produced slight and irregular diuretic action.

References : Agarwal (2003), Caius (2003), Chauhan (1999), Chaurasia *et al.* (2000a), Chhetri (2005), Chopra *et al.* (1956), Jain & Goel (2005), Kaul (1997), Khare (2004), Nautiyal *et al.* (2003), Negi *et al.* (1993), Rastogi & Mehrotra (2002), Singh & Kumar (2000b), Sood & Thakur (2004), Thakur *et al.* (1989).

Tarenna odorata (Roxb.) Robbins

Family: Rubiaceae.

Part Used : Root.

Utilization : Paste with water consumed for snake bite.

Reference : Lalramnghinglova & Jha (1999).

Tephrosia pumila (Lamk.) Pers.

Syn: *T. procumbens* Buch.-Ham.;
T. pupurea (L.) Pers. var. *pumila* (Lamk.) Baker

Family: Fabaceae.

Distribution : Throughout Plains of India.

Part Used : Root.

Utilization : Roots with curd useful for piles and smoked to check cough.

Active Constituents : These are praecansone B, rotenone, sumatrol, obovatin, praecansone A, 7-ethoxy, 3-3′, 4′-trihydroxy flavone, fisetin-7-ethyl ether and 7, 4′-dihydroxy-3′-methoxyisoflavone.

References : Deshmukh *et al.* (2000), Rastogi & Mehrotra (2002).

Tephrosia purpurea (Linn.) Pers.

Syn: *Cracca purpurea* L.; *Galega lanceaefolia* Roxb.; *G. purpurea* L.; *Tephrosia hamiltonii* Drumm. *ex* Gamble; *T. lanceolata* R. Grah.; *T. maxima* Pers.

Family: Fabaceae.

English, Hindi & Sanskrit Names:

Eng. : Purple Tephrosia, Wild Indigo;
Hindi : Dhamasia, Sarpanka, Sarphonka;
Sans. : Banah, Sharpukha, Vishika-Punkhaa.

Distribution : Upper Gangetic Plains & W. Himalaya. Commonly Grown as a Green Manure in Paddy Fields.

Parts Used : Root. Root bark.

Utilization : Powdered bark along with black pepper given as an antidote for snake bite, pulmonary disease and impotency (once a day for 3 days). Lukewarm root paste applied on hydrocele, arthritis, rheumatism, piles, enlarged cervical glands and to check stomachache and toothache. Decoction of root along with that of *Stereospermum suaveolens* given after menstruation to check conception. Decoction of roots mixed with garlic used as ear drops to treat otalgia; when mixed with black pepper given two times a day for three days in the treatment of fever and obstinate colic. Also, roots used to cure spleen enlargement, diarrhoea, liver diseases, conjunctivitis, diabetes in children, giddiness, earache, paralysis, snake bite, arthritis, rheumatism, fever, indigestion, colic pain, impotency, spermatorrhoea, scorpion sting, premature ejaculation, tympanitis, dyspepsia, urinary disorders, chronic diarrhoea, stomachache (juice of the root mixed with molasses and taken orally), elephantiasis (liniment prepared from roots), asthma and cough relief (pulverised roots smoked), and as diuretic, tonic, blood purifier and fertility drug to barren women.

Active Constituents : Found to contain (-) isolonchocarpine, pongamol, lanceolatin A, lanceolatin B, karanjin, kanjone, β-sitosterol, a new β-hydroxy chalcone-purpurinenone, apollinine, semiglabrin, semiglâbrinol, tephroglabrin, tepurindiol, pongamol, iso-lonchocarpin, O-methyl-pongamol, lanceolatins A & B, β-hydroxychalcone-purpurenone, (+) purprinm dehydroisoderricin, 2-methoxy-3, 9-dihydroxycoumestone & (-) maackian, and maxima substance A-C (roots).

References : Agarwal (2003), Ambasta (1986), Chatterjee & Pakrashi (1997), Chopra *et al.* (1956, 1969), Dash & Mishra (2000), Dwivedi (2003), Girach (1992), Gogoi & Das (2003), Khare (2004), Kirtikar & Basu (1935), Maya *et al.* (2003), Parrotta (2001), Patil (2000), Ram *et al.* (2002), Rao & Henry (1996), Rastogi & Mehrotra (1998, 2001, 2002), Reddi *et al.* (2005), Saxena (1997), Singh & Kumar (2000a), Singh & Pandey (1998), Singh & Prakash (1994), Srivastava *et al.* (2003), Varghese (1996), Watt (1889).

Tephrosia spinosa Pers. Baker in Part

Family: Fabaceae.

Hindi Name:

Hindi : Mukkavala.

Distribution : S. India.

Part Used : Root bark.

Utilization : Extr. given to cure rheumatism, indigestion, diarrhoea, fever, excessive thirst and swelling of the joints.

Active Constituents : Spinochalcones A, B, and femistrictin A characterised the roots.

References : Agarwal (2003), Ambasta (1986), Rastogi & Mehrotra (1998).

Terminalia chebula Retz.

Syn: *Embryogonia arborea* Teys & Binn., *Myrobalanus chebula* (Retz.) Gaertn.; *Terminalia aruta* Ham.; *T. reticulata* Roth.

Family: Combretaceae.

English, Hindi & Sanskrit Names:

Eng. : Black Myrobalan, Chebulic Myrobalan, Gall Nut;

Hindi : Harar, Pilzari;

Sans. : Abhayaa, Chetaki, Haritaki, Jivanti, Kaayasthaa, Puutanaa, Shreyasi, Vijayaa.

Distribution : Wild in the Sub-Himalayan Tracts Upto 1,000 m & Deciduous Forests.

Part Used : Root

Utilization : Paste good for conjunctivitis.

References : Chhetri (1994), Khare (2004), Parrotta (2001), Reddi *et al.* (2005), Thakur *et al.* (1989), Watt (1889).

Thalictrum alpinum L.

Family: Ranunculaceae.

English Name:

Eng. : Meadow Rue.

Distribution : Occurs at Donpur of Uttar-Kashi: 2,500-3,000 m.

Part Used : Root.

Utilization : Paste applied around the forehead in headache and hotness. Also, considered beneficial as purgative, diuretic and a good remedy in dyspepsia, convalescence after acute diseases and ophthalmia.

Active Constituents : Properties attributed to taiksimine, thalpindione, N-desmethylthalrugosaminine, neothalibrine, Omethylisoboldine, isoboldine, N-methyl-6, 7-dimethoxyisoquinolone, oxyberberine, noroxydrastinine, palmitine, berberine, columbamine, jatrorrhizin, thalifendine and magnoflorine.

References : Ambasta (1986), Bhattacharjee (2001), Kapur (1996b), Rastogi & Mehrotra (2001), Srivastava *et al.* (2003).

Thalictrum cultratum Wall.

Family: Ranunculaceae.

Distribution : Himalaya (Kashmir to Bhutan): 2,400- 4,200 m.

Part Used : Root.

Utilization : Decoction used for convalescence from the disease. A glass root extr. taken orally 2-3 times in a day against urinary irritation and food poisoning.

References : Kaul (1997), Maity *et al.* (2003), Rastogi & Mehrotra (2002), Srivastava *et al.* (2003).

Thalictrum foliosum DC.

Family: Ranunculaceae.

English, Hindi & Sanskrit Names:

Eng. : Gold Thread;
Hindi : Mamira, Pilijari, Pinjari, Shuprak;
Sans. : Mamira, Pitmula, Pitranga.

Distribution : Throughout the Himalaya (1,600-2,600 m) & Khasia Hills 1,300- 2,000 m.

Part Used : Root.

Utilization : Regarded as a purgative, diuretic, antiperiodic, aspirant, tonic, blood purifier, febrifuge; and for abdominal pain, earache, eczema, eye diseases, fever, snake bite, jaundice and cataract. Its tonic and aperient properties found to be useful in convalescence after acute diseases, in mild forms of intermittent fever and atonic dyspepsia. Its ecoction with black pepper given 3 times a day for 3 days to cure malarial fever whereas the paste along with *Allium cepa* (50g once a day) given orally for treating fever and shuffle in cattle.

Active Constituents : Root owes its active properties to thalrugosidine, thalrugosaminine, thalisopine, thalirugidine, oxyberberine (berlambine) noroxyhydrastinine, thalifendine, palmatine, berberine, columbamine, jatrorrhizine, thalidatine, dehydrodiscretamine, rugosinone, tembetarine, xanthoplanine and magnoflorine.

References : Agarwal (2003), Ambasta (1986), Caius (2003), Chopra *et al.* (1956), Drury (1978), Jain (1991), Jain *et al.* (1991), Kala & Rawat (2001), Kirtikar & Basu (1935), Nadeem *et al.* (2001), Negi *et al.* (1999), Rana *et al.* (1996), Rastogi & Mehrotra (2001), Samant *et al.* (2001), Singh (1999), Singh & Kumar (2000b), Singh *et al.* (2003), Verma & Chauhan (2006), Watt (1889).

Thalictrum javanicum Bl.

Family: Ranunculaceae.

Distribution : Himalaya (Kashmir to Sikkim) 1,800-3,600 m, Khasia Hills (1,500-1,800 m), Parasnath in Bihar 1,200 m, Kodaikanal & Nilgiris (1,500- 2,200 m).

Part Used : Root.

Utilization : Much valued for ophthalmia and gastric problems, and as antiperiodic, diuretic, aperient, purgative and bitter tonic during convalescence.

Active Constituents : Pharmacologically active properties attributed to demethyleneberberine, palmatine, berberine, columbamine, jatrorrhizine, magnoflorine, oxyberberin, rugosinone, thalifendine, thalisopine and thalrugosaminine.

References : Jain (1991), Kumar (2002), Rastogi & Mehrotra (2001, 2002), Singh (2000).

Thalictrum minus L.

Syn: *T. thunbergi* DC.
Family: Ranunculaceae.

Distribution : Upto 2,700-3,600 m in N.E. Region.

Part Used : Root.

Utilization : Used in fever, eye diseases (ophthalmia) and abdominal pain.

Active Constituents : Reported to possess berberine, magnoflorine, thalictrimine (β-allocryptofine), thalicrine, homothalicrine, thalicmidine, thalicmine, thalcminine, thaliadanine, thaliadine, adiantifoline, berberine, magnoflorine, thalmirabine, thalistine, O-methylthalibrine and N-methylcorydaldine, thalrugosine, thalphenine, columbamine, thalifendine, and jatrorrhizine.

References : Ambasta (1986), Bhattacharjee (2001), Jain (1991), Kumar (2002), Rastogi & Mehrotra (1999a, b, 2001).

Thalictrum veniforme Wale

Syn: *T. neurocarpum* Royle
Family: Ranunculaceae.

Distribution : Temperate Himalaya (Kulu to Sikkim): 2,600-3,300 m.

Part Used : Root.

Utilization : Juice used for the treatment of cataract.

References : Ambasta (1986), Hooker (1872).

Thespepsia villosa L. Pers.

Syn: *Galega argentea* lamk.; *G. barba-javis* Burm.; *Thespepsia argentea* Pers.,
Family: Fabaceae.

Hindi Name:

Hindi : Vaykkavali.

Distribution : Punjab, Rajasthan, Gujarat, M.P., U.P., W. Bengal.

Part Used : Roots.

Utilization : Root paste brushed to cure toothache.

Active Constituents : Contains rotenoids, tephrosol, rotenone, 6a, 129-dehydrorotenone and stigmasterol.

Biological Activity : Fresh roots considered hypoglycaemic.

References : Agarwal (2003), Ambasta (1986), Rajendran *et al.* (2001), Rastogi & Malhotra (2001), Watt (1889).

Thespepsia virginia L.

Family: Fabaceae.

English Name:

Eng. : Virginia Thespepsia.

Part Used : Root.

Utilization : Decoction efficacious against chronic coughing, bladder trouble and intestinal worms.

Reference : Bhattacharjee (2001).

Thespesia lampas (Cav.) Dalz. & Gibs.

Syn: *Azanza lampas* (Cav.) Alef.

Family: Malvaceae.

Hindi & Sanskrit Names:

Hindi : Jangli Bhindi, Ran-Bendy;

Sans. : Karpasi.

Distribution : Commonly in India.

Parts Used : Root. Root bark.

Utilization : Powdered root bark given (5 g twice a day) with a pinch of salt for malarial fever whereas root extr. good for heart stroke, malaria and typhoid. Root paste applied on boils, and on mixing with goat's milk it is administered orally for jaundice. Also, found to be useful in diabetes, gonorrhoea, syphilis, stomachache, sun-stroke, urinary complaint and veterinary eye diseases.

References : Agarwal (2003), Aminuddin & Girach (2005), Balasubramanian & Prasad (1996), Bhatt *et al.* (2003), Chopra *et al.* (1956), Reddi *et al.* (2005), Varghese (1996).

Thevetia peruviana (Pers.) Merr.

Syn: *Cerbera thevatia* L.; *Thevetia neriifolia* Juss *ex* Steud.
Family: Apocynaceae.

English, Hindi & Sanskrit Names:

Eng. : Lucky Nut Tree, Yellow Oleander;

Hindi : Kaner, Kulkephul, Pila Kaner, Zard Kunel;

Sans. : Ashvaghna, Ashvaha, Ashvamaraka, Ashvantaka, Karavirah, Pachchaialari, Pitapushpa, Shatakunda.

Distribution : Plains Throughout the Country.

Part Used : Root.

Utilization : Paste plastered to relieve tumours, and also rubbed on penis to cure impotence.

References : Agarwal (2003), Ambasta (1986), Bhattacharjee (2001), Chatterjee & Pakrashi (1997), Parrotta (2001), Singh & Pandey (1996).

Thottea siliquosa (Lam.) Ding Hou

Family: Aristolochiaceae.

Part Used : Root.

Utilization : Paste prepared from fresh roots applied externally for headache, and also for spider, scorpion and snake poison. Decoction administered internally for chest pain and cough.

Reference : Radhakrishnan *et al.* (1996).

Thunbergia fragrans Roxb. var. *laevis*

Syn: *T. angustifolia* Ham.; *T. javanica* Gaertn.; *T. roxburghia* Nees; *T. scandens* Pers.; *T. subsagittata* Blanco.

Family: Acanthaceae.

Distribution : Throughout India (300-1,300 m): Himalayan Terai to Sri Lanka & Tenasserim, ascending to 1,300-2,300 m.

Part Used : Root.

Utilization : Fresh root juice (1-2 drops) used as eye drops to treat redness of eyes.

References : Girach (2001), Hooker (1885).

Thysanolaena maxima (Roxb.) Kuntze

Syn: *Agrostis maxima* Roxb.; *Thysanolaena acarifera* Arn. & Nees; *T. agrostis* Nees.

Family: Poaceae.

English & Hindi Names:

Eng. : Bouquet-Grass, Broom Grass, Tiger-Grass;

Hindi : Phuljanta.

Distribution : Throughout India.

Part Used : Root.

Utilization : Extr. taken as an abortifacient and mouthwash. Root paste applied to check boils.

References : Agarwal (2003), Ambasta (1986), Chhetri (2005), Dash *et al.* (2003), Sharma (2000).

Tiliacora acuminata (Lam.) Miers.

Syn: *T. acuminata* Hook.f. & Thoms.; *T. racemosa* ,

Family: Menispermaceae.

Hindi Names:

Hindi : Bagamushada, Ramsarbel.

Distribution : Throughout Tropical India.

Parts Used : Root. Root bark.

Utilization : Root bark antiseptic. Fresh root juice drunk on empty stomach in snake bite.

Active Constituents : These are tiliacorine, tiliarine, tiliacorinine, tiliacoridine, nor-tiliacorinine A, nor-tiliacorinine B, mosine, tiliacine, corine, mohine, (+)-N-methyltiliamosine, d-quercitol, fumaric acid and glucose (roor bark) and magnoflorine, lotusine and magnocurarine chloride (roots).

Biological Activity : Magnoflorine, lotusine and magnocurarine chloride showed curarizing action and hypotensive activity.

References : Agarwal (2003), Ambasta (1986), Chatterjee & Pakrashi (1997), Chopra *et al.* (1956), Rastogi & Mehrotra (1999a, 2001), Reddy *et al.* (1996).

Tinospora cordifolia (Willd.) Hook.f. & Thoms.

Syn: *Cocculus convolvulaceous* DC.; *C. cordifolius* DC.; *C. verrucosus* Wall.; *Menispermum cordifolium* Willd.; *Tinospora palmnervis* Miers.

Family: Menispermaceae.

English, Hindi & Sanskrit Names:

Eng. : Heart-Leaved Mvonseed;

Hindi : Amrita, Giloe, Gulancha, Gulel, Gurach, Jiwantika;

Sans. : Amritaa, Amritalata, Bhishakapriya, Chakrangi, Chandrahasa, Goraksha, Madhuparni, Nirjara, Pittagni, Surakrita, Tantrika, Vishalya.

Distribution : Throughout Tropical India, Ascending 300 m.

Part Used : Root.

Utilization : Its paste mixed with 'til'oil applied to cure pimple. Starch from roots used in chronic diarrhoea and dysentery. Powder of root consumed (1 teaspoonful twice a day) to cure breathing problems, piles, ulcer diseases, cough, chronic fevers, leprosy, blood pressure, snake bite, headache, hiccups, heart stroke, skin diseases, whooping cough, splenomegaly, general debility, dyspepsia, dysentery, fever, urinary diseases, diarrhoea, diabetes, visceral obstruction and as tonic and emetic.

References : Ambasta (1986), Bhattacharjee (2001), Caius (2003), Chauhan (1999), Dhiman (2003), Dwivedi (1993), Jain *et al.* (2003), Khare (2004), Kothari & Rao (2000), Nadeem *et al.* (2001), Rao & Henry (1996), Reddi *et al.* (2005), Sen & Behera (2003), Singh (2002), Sood & Thakur (2004), Varghese (1996), Watt (1889).

Toddalia asiatica (L.) Lamk.

Syn: *Paulinia asiatica* L.; *Scopolia aculeata* Sm.; *Toddalia aculeata* Pers.; *T. angustifolia* Miq.; *T. floribunda* Wall.; *T. nitida* Lam.; *T. rubricaulis* Willd.

Family: Rutaceae.

English, Hindi & Sanskrit Names:

Eng. : Forest Pepper, Lopez Tree, Wild Orange Tree;

Hindi : Dahan, Janglikalimirch, Kanj;

Sans. : Dahana, Kaccana, Kanchana, Tiksnaksah.

Distribution : Throughout India to 2,500 m. Very Common in the Nilgiris, Palni Hills & Forests of Orissa.

Part Used : Root bark.

Utilization : Bark considered aromatic, bitter, pungent, tonic, antiperiodic, analgesic, diaphoretic, stomachic, antipyretic, also used in cases of diarrhoea, debility during convalescence after febrile and other exhausting diseases, amenorrhoea, gonorrhoea, cough, mouth ulcers, malarial and other periodic fevers.

Active Constituents : Contains toddanol, toddanone, toddasin, coumarins I & II, toddalosn, aculeatin, and toddalactone.

References : Agarwal (2003), Ambasta (1986), Chatterjee & Pakrashi (1997), Chhetri (1994), Chopra *et al.* (1956), Dey (1973), Parrotta (2001), Rao & Henry (1996), Rastogi & Mehrotra (1998, 2001), Reddi *et al.* (2005), Watt (1889).

Tragia involucrata L.

Syn: *T. cannabina* Linn.f.

Family: Euphorbiaceae.

English, Hindi & Sanskrit Names:

Eng. : Indian Stinging Nettle;

Hindi : Baraanta;

Sans. : Durabhigra, Duralambha, Durlaba, Dusparsa, Grahini, Kachchura, Kasagni, Samudranta, Virupa, Vrishchikali.

Distribution : Throughout India from Punjab & Lower Himalaya to Assam & Meghalaya to 750 m & Southwards to Kerala.

Part Used : Root.

Utilization : 10g fresh root mixed with black pepper given twice a day for treatment of epilepsy and at an interval of 1 hour in snake bite. Also, considered diaphoretic, alterative, blood purifier; and in ardent fever and itching of the skin (infusion given). Paste of root used to aid the extraction of guinea-worm, and forms the basis of an external applications in leprosy and to cure body pain; and also useful in the treatment of gout, rheumatic pains, urination, bowel movement, limb pains, skin eruptions, venereal diseases, bronchitis, dysentery and lactation problems.

References : Agarwal (2003), Ambasta (1986), Caius (2003), Chatterjee & Pakrashi (1997), Chopra *et al.* (1956), Dastur (1970), Drury (1978), Jain *et al.* (1991), Jha & Verma (1996), Kirtikar & Basu (1984), Parrotta (2001), Patil & Patil (2002), Reddi *et al.* (2005), Sur & Halder (2002), Varghese (1996), Watt (1889).

Tragopogon gracillus Spreng.

Family: Asteraceae.

Part Used : Root.

Utilization : Decoction given for stomachache.

References : Lal *et al.* (1996), Singh & Kumar (2000b).

Tragopogon pratense L.

Syn: *T. dubius* Scop.; *T. major* auct. non Jacq.

Family: Asteraceae.

English Names:

Eng. : Bucks-Beard, Goat's Beard, Meadow-Salsify, Yellow Salsify.

Distribution : W. Himalaya & W. Tibet: 1,500-3,600 m.

Part Used : Root.

Utilization : Decoction given for heart burn, loss of appetite, disorders of the breast and liver.

References : Chopra *et al.* (1956), Kapur & Srivastava (1996), Sood *et al.* (2001).

Trema orientalis L. Blume

Syn: *Celtis orientalis* Linn.; *Sponia orientalis* Planch; *S. wightii* Planch; *Trema amboinensis* auct. non Blume.

Family: Ulmaceae.

English, Hindi & Sanskrit Names:

Eng. : Charcoal Tree, Indian Nettle, Nettle Tree;
Hindi : Chickan, Jiban;
Sans. : Jivanthi.

Distribution : Throughout Moist Regions of India.

Parts Used : Root. Root bark.

Utilization : Decoction of bark used in epilepsy while that of root good against haematuria, diarrhoea and passing blood in urine.

References : Agarwal (2003), Ambasta (1986), Lalramnghinglova (1996), Reddi *et al.* (2005), Watt (1889).

Trewia nudiflora Linn.

Syn: *Rottlera indica* Willd.; *Tetragastsis orsea* Gaertn.; *Trewia macrophylla* Roth.; *T. macrostachya* Klotzsch.

Family: Euphorbiaceae.

English, Hindi & Sanskrit Names:

Eng. : False White Teak;
Hindi : Bhillaura, Gambar, Khamara, Pindara, Pitoli, Tumri;
Sans. : Karahata, Karangaha, Pindara.

Distribution : Hotter Parts of India from Kumaon Southwards & Eastwards to Assam.

Part Used : Root.

Utilization : Decoction given to relieve flatulence and applied locally in gouty or rheumatic affections.

References : Caius (2003), Chatterjee & Pakrashi (1997), Chopra *et al.* (1956), Kirtikar & Basu (1984), Watt (1889).

Trianthema decandra Linn.

Syn: *Zaleya decandra* Burm.

Family: Aizoaceae.

Hindi & Sanskrit Names:

Hindi : Gadabani;

Sans. : Punarnavi.

Distribution : Practically Throughout India.

Parts Used : Root. Root bark.

Utilization : Root considered aperient and said to be useful in hepatitis, asthma and suppression of the menses. A decoction of the root bark also given as an aperiant. Root brayed in milk given internally specifically in orchitis.

References : Agarwal (2003), Ambasta (1986), Chopra *et al.* (1956), Kirtikar & Basu (1935), Watt (1889).

Trianthema portulacastrum L.

Syn: *T. monogyna* L.; *T. obcordata* Roxb.

Family: Aizoaceae.

English, Hindi & Sanskrit Names:

Eng. : Horse-Purslane;

Hindi : Santhi, Shvet-Sabuni;

Sans. : Punarnava, Shvetapunarnava.

Distribution : Naturalized Throughout India as a Weed in Cultivated Fields, River-beds, Wasteplaces.

Part Used : Root.

Utilization : The powdered, bitter and nauseous root given in combination with ginger as a cathartic, blood purifier and abortifacient. Its infusion given internally (1-2 ounces) in constipation, jaundice, stanguary, dropsy, urinary disorders, eyes cures, corneal ulcers, itching, night blindness, bowel complaints, suppression of menses, asthma, amenorrhoea and obstruction of the liver.

References : Ambasta (1986), Chatterjee & Pakrashi (1997), Jha & Kumar (2003), Kothari & Londhe (2000), Reddi *et al.* (2005), Srivastava *et al.* (2003), Watt (1889).

Trianthema triquetra Willd. *ex* Rottl.

Syn: *T. crystallina* auct. non Vahl.

Family: Aizoaceae.

Hindi Name:

Hindi : Pather-Phor.

Distribution : Widely Distributed in India.

Part Used : Root.

Utilization : So far known as cathartic, abortifacient and substitute to patharphor.

References : Agarwal (2003), Ambasta (1986).

Tribulus terrestris L.

Syn: *T. lanuginosus* L.

Family: Zygophyllaceae.

English, Hindi & Sanskrit Names:

Eng. : Puncture-Vine;

Hindi : Chhotagokhru;

Sans. : Gokshura, Gosksura.

Distribution : Occasionally Found in Wasteplaces & Dry Habitats Throughout the Warmer Regions of India Including W. Rajasthan & Gujarat.

Part Used : Root.

Utilization : Given to check sterility in males and useful for gonorrhoea, calculus affections, kidney diseases, painful micturition and urinary discharges.

Active Constituents : Neotigogenin, campesterol, β-sitosterol, stigmasterol, diosgenin, tigogenin, hecogenin and gitogenin are the main constituents of roots.

References : Agarwal (2003), Chatterjee & Pakrashi (1997), Rastogi & Mehrotra (1998, 1999b), Reddi *et al.* (2005), Sarin (1996).

Trichodesma indicum (L.) Lehm.

Syn: *Borago spinulosa* Roxb.; *B. indica* Linn.; *Trichodesma hirsutum* Edgew.; *T. perfoliatum* Wall.; *T. sedgewickianum* Bannerjee nom. nov.

Family: Boraginaceae.

Hindi & Sanskrit Names:

Hindi : Andhahuli, Chota-Kulpha, Hetamundiya, Ratmundiya;

Sans. : Adhapushpi, Adhomukha, Avakpushpi, Darvika, Dhenujivaha, Gandhapushpika, Golomi, Romalu, Sarasa.

Distribution : A Weed in Disturbed Open Habitats Throughout Most of India to About 1,500 m in the Himalaya, (Roadsides, Cultivated Fields, Dry Stony Wasteland & Black Cotton Soil in Peninsular India).

Part Used : Root.

Utilization : Known to heal superficial injuries of skin (paste applied), dysentery (a cup of decoction given for five days), rheumatism, boils (paste applied externally on affected areas), fever, sores and wounds (root powder applied).

References : Agarwal (2003), Ambasta (1986), Girach (1992), Parrotta (2001), Saini (1996b), Samvatsar & Diwanji (2000), Singh & Pandey (1998), Sood & Thakur (2004), Upadhye *et al.* (1994), Varghese (1996).

Tricholepis elongata DC.

Syn: *Carduus elongatus* Wall.; *Tricholepis lanuginosa* Klotz.

Family: Asteraceae.

Distribution : W. Himalaya.

Part Used : Root.

Utilization : Fresh roots good for loss of appetite in cattle.

Reference : Sood & Thakur (2004).

Trichosanthes anguina L.

Family: Cucurbitaceae.

English, Hindi & Sanskrit Names:

Eng. : Snakegourd;

Hindi : Chachinda, Chachinga, Chirchita;

Sans. : Chichinda, Dadhipuspee.

Distribution : Cultivated in India; (Assam, Bihar, H.P., Sikkim, Tamil Nadu, U.P. & W. Bengal).

Part Used : Root.

Utilization : 10-15 ml decoction taken to cure constipation. Root considered antidiarrhoeal and anthelmintic and efficacious against dysentery, diarrhoea and syphilis.

References : Agarwal (2003), Chatterjee & Pakrashi (1997), Reddi *et al.* (2005), Satapathy & Brahmam (2000).

Trichosanthes bracteata (Lam.) Voigt.

Family: Cucurbitaceae.

Part Used : Root.

Utilization : Efficacious in pneumonia.

Reference : Kothari & Moorthy (1996).

Trichosanthes cordata Roxb.

Syn: *T. palmata* Wall.; *T. tuberosa* Roxb.

Family: Cucurbitaceae.

Hindi & Sanskrit Names:

Hindi : Bidarikanda;

Sans. : Bhumikushmanda.

Distribution : Different Parts of India (A.P., Arunachal Pradesh, Assam, Bihar, Manipur, Meghalaya, Nagaland, Sikkim, Tamil Nadu, Tripura, U.P. & W. Bengal).

Part Used : Root.

Utilization : Fresh roots pounded and mixed with oil applied in leprous ulcers. Also, used as tonic and in the enlargement of liver and spleen.

References : Ambasta (1986), Chatterjee & Pakrashi (1997), Chopra *et al.* (1956), Watt (1889).

Trichosanthes cucumerina L.

Syn: *T. laciniosa* Klein; *T. pilosa* Wall.

Family: Cucurbitaceae.

English, Hindi & Sanskrit Names:

Eng. : Snakegourd;
Hindi : Jangli Chichonda, Jangli-Chichinga;
Sans. : Amritaphala, Patolah.

Distribution : Wild All Over the Warmer Parts of the Country.

Part Used : Root.

Utilization : Root of this plant and that of 'Tejraj' (*Cynoglossum glochidiatum*) powdered in equal proportions and given (10g) for impotency. Also considered cathartic and good in bowel complaints and bronchitis.

Active Constituents : Known to contain methylpalmitate, palmitic acid, α-pinasterol, stigmasta-7-en-3β-ol, 3-O-β-glucopyranosides and vanillic acid.

References : Ambasta (1980), Brahmam & Saxena (1990), Chatterjee & Pakrashi (1997), Chopra *et al.* (1956), Patil & Patil (2002), Srivastava *et al.* (2003), Thakur *et al.* (1989), Watt (1889).

Trichosanthes dioca Roxb.

Family: Cucurbitaceae.

English, Hindi & Sanskrit Names:

Eng.	:	Pointed Gourd;
Hindi	:	Palval, Palwal, Parvar, Parwal;
Sans.	:	Patol, Patola

Distribution : Wild in Plains of N. India from Punjab to Assam. Extensively Cultivated All Over Warmer Regions of India.

Part Used : Root.

Utilization : Known in India as antipyretic, cathartic, febrifuge, hydragogue and tonic.

Active Constituents : Active principles are attributed to trichosanthin, resembling colocynthin, phytosterol, hentriacontane, a small amount of essential oil containing terpenes, a little fixed oil, starch and reducing sugars.

References : Chatterjee & Pakrashi (1997), Khanna (2002b), Khare (2004), Watt (1889).

Trichosanthes tricuspidata Lour.

Syn: *T. palmata* Roxb.; *T. laciniosa* Wall.; *T. lepiniana* (Naudin) Cogn.; *T. palmata* Roxb.

Family: Cucurbitaceae.

Hindi & Sanskrit Names:

Hindi	:	Lal Indrayan;
Sans.	:	Mahendrabaruni.

Distribution : Prostrate Climber in Wasteplaces in Different Parts of India.

Part Used : Root.

Utilization : Paste of its root with an equal portion of *Colocynth* root applied to carbuncles. Infusion formed by combining it with three myrobalans and turmeric flavoured with honey for gonorrhoea. Roots also considered good for body pain, boils, headache and in lung diseases of cattle.

Active Constituents : Contains bryonolic acid, cucurbitacins B, D, and trichotetrol.

References : Agarwal (2003), Alagesboopathi *et al.* (2000), Chatterjee & Pakrashi (1997), Rastogi & Mehrotra (1999a, 2002), Rawat & Chowdhury (1998), Varghese (1996), Watt (1889).

Tridex procumbens L.

Family: Asteraceae.

English, Hindi & Sanskrit Names:

Eng. : Coatbuttons, Mexican Daisy;
Hindi : Khal Muriya, Tal-Muriya, Tikki-Kasa;
Sans. : Jayanti.

Distribution : Found Upto 2,500 m Throughout India.

Part Used : Root.

Utilization : Paste of powdered root in 'Jado' (*Ricinus communis*) oil in the ratio of 1:5 by weight applied for rheumatic pain, headache and stomach pain. Also, roots found to be good against fever, skin diseases, wounds and cuts.

References : Agarwal (2003), Girach (1992), Parrotta (2001), Reddi *et al.* (2005), Srivastava *et al.* (2003), Varghese (1996).

Triumfetta pilosa Roth.

Syn: *T. glandulosa* Heyne; *T. oblongata* Link.; *T. polycarpa* Wall.; *T. tomentosa* Wall.

Family: Tiliaceae.

Distribution : Throughout Tropical Parts of India From the Himalaya to Travancore & Sri Lanka.

Part Used : Root.

Utilization : Taken orally to cure acidity in digestive system.

References : Hosagoudar & Henry (1996b), Watt (1889).

Triumfetta rhomboidea Jacq.

Syn: *T. acuminata* Wall.; *T. angulata* Wall.; *T. angulata* Lam.; *T. trilocularis* Roxb.; *T. vestita* Wall.; *T. bartramia* Roxb.

Family: Tiliaceae.

English, Hindi & Sanskrit Names:

Eng. : Burbush, Burweed;
Hindi : Chikti, Chiriyari;
Sans. : Jhinjhirita, Jhinjhirta.

Distribution : Tropical & Sub-Tropical Regions of India to 1,200 m in the Himalaya.

Part Used : Root.

Utilization : Pounded roots applied over boils, ulcers, inflamed eyelids and to treat intestinal ulcers. Their hot infusion recorded beneficial to cure gonorrhoea and to facilitate child birth. Known also as bitter, tonic, styptic, diuretic, galactagogue and antidysenteric.

References : Agarwal (2003), Ambasta (1986), Chatterjee & Pakrashi (1997), Chopra *et al.* (1956), Kirtikar & Basu (1935), Parrotta (2001), Sood & Thakur (2004), Subramaniam (2000), Upadhyay & Singh (2005).

Turraea villosa Benn.

Family: Meliaceae.

Hindi Name:

Hindi : Kapur Bhendi.

Distribution : Grown in W. & S. India.

Part Used : Root.

Utilization : Used as an application to fistulas and leprosy.

References : Agarwal (2003), Ambasta (1986), Chopra *et al.* (1956), Kirtikar & Basu (1935), Kothari & Moorthy (1996).

Tylophora fasciculata Buch.-Ham.

Family: Asclepiadaceae.

Part Used : Root.

Utilization : Powdered root of this plant with that of *Piper longum* taken for stomach pains.

Reference : Goud & Pullaiah (1996).

Tylophora indica (Burm.f.) Merr.

Syn: *Asclepias asthmatica* L.f.; *A. tunicata* Hort. Calc.; *A. vomitoria* Kaln.; *Cynanchum bracteatum* Thunb.; *C. indicum* Burm.f.; *C. vividiflorum* Sims.; *C. vomitorium* Lam.; *Tylophora asthmatica* (L.f.) Wight & Arn.; *T. mitroria* Voigt.

Family: Asclepiadaceae.

English, Hindi & Sanskrit Names:

Eng. : Country Ipecacuanha, Emetic Swallow Wort, Indian Ipecacuanha;

Hindi : Antamul, Jangli-Pikvam;

Sans. : Antamuula, Antrapachaka, Arkaparni, Gandhana, Latakshiri, Muulini, Shwasaghni.

Distribution : Assam, W. Bengal, Orissa & Peninsular India, ascending to 900 m.

Part Used : Root.

Utilization : Considered substitute for ipecacuanha for the treatment of asthma, bronchitis, whooping cough, dysentery, diarrhoea, rheumatism, gouty pains, intermittent malarial fever, bringing about vomiting, and as stimulant, cathartic, expectorant, stomachic, diaphoretic and abortifacient.

Active Constituents : Chief constituents are tylophorinidine, gamma-fagarine, demethyltylophorine, demethyl-tylophorinine, tylophoricine, ceryl alcohol, glucose, quercetin, kaempferol, caoutchouc, p-methoxysalicyladehyde, resin, tannin, α-amyrin, β-amyrin, tetratriacontanol, octacosanyl octacosanoate, β-sitosterol, stigmasterol, phytosterol, 2, 3, 6, 7-tetramethoxy-phenanthro [9, 10, 6, 7] indolizidine (tylophorine) and tylophorinine.

Biological Activity : Tylophorinidine isolated from the roots of three year old plant (yields 0.009 %), and two other bases identified as (+)-septicine and (+) isotylocrebrine exhibited antitumour activity.

References : Agarwal (2003), Ambasta (1986), Bhattacharjee (2001), Chatterjee & Pakrashi (1997), Dwivedi (1992), Jain (1968), Khare (2004), Kurian (1995), Parrotta (2001), Reddi *et al.* (2005), Sharma (2003), Varghese (1996), Watt (1889), Zafar & Ahmad (2003).

Typha angustata Bory & Chaub.

Syn: *T. angustifolia* Watt.non L.; *T. australis* Schum. & Thonn.; *T. elephantina* Roxb.

Family: Typhaceae.

English, Hindi & Sanskrit Names:

Eng. : Bulrush, Elephant Grass, Lesser Indian Reed Mace, Reed Mace;

Hindi : Hogla, Mothitrina, Patera;

Sans. : Eraka, Gundra.

Distribution : Marshy Habitats Throughout India Upto 1,700 m.

Parts Used : Rootstock. Root.

Utilization : Rootstock considered astringent and diuretic. Decoction of root given in strangury, calculus, dysuria, burning micturition, dysentery, gonorrhoea and measles. Also, its 10g paste per day used by 'Lodhas' to cure dropsy.

References : Agarwal (2003), Ambasta (1986), Bhatt *et al.* (2002), Parrotta (2001), Samanta & Das (2003), Srivastava *et al.* (2003), Subramaniam (2000).

Typhonium trilobatum (Linn.) Schott.

Syn: *Arum trilobatum* Linn.; *A. orixenese* Roxb.; *Typhonium orixenese* Schott.; *T. roxburghii* Saunders.

Family: Araceae.

Hindi Name:

Hindi : Gherkochoo.

Distribution : W. Himalaya to E. India & Deccan Peninsula.

Part Used : Root.

Utilization : Its paste used as an external application in the stings of bees, wasps, scorpions, and venomous snakes. Also, recorded so far as a most powerful stimulant, relaxing bowels and to relieve haemorrhoids.

References : Agarwal (2003), Caius (2003), Chopra *et al.* (1956), Watt (1889).

Uraria alopecuroides (Roxb.) Wt.

Syn: *Hedysarum alopecuroides* (Roxb.) Wt.; *Uraria hamasa* Wall.; *U. lagopus* DC.; *U. repanda* Wall. *ex* Benth.

Family: Fabaceae.

Distribution : Punjab, along the Himalaya to Assam & Ava, Ascending to 2,000 m in Shimla.

Part Used : Root.

Utilization : Used as an antidote to snake bite.

References : Ambasta (1986), Hooker (1879).

Uraria lagopodioides Desv. *ex* DC.

Syn: *Doodia lagopioides* Roxb.; *Hedysarum lagopioides* Burm.; *Uraria hamasa* Wall.; *U. lagopoides* DC.;

Family: Fabaceae.

Hindi & Sanskrit Names:

Hindi : Pithvan;

Sans. : Prisniparni.

Distribution : Found in the Grasslands in W. Bengal, Bihar, Orissa as well as in the Palni Hills in the W. Ghats.

Part Used : Root.

Utilization : Decoction beneficial in rheumatism, fever, catarrh, bleeding piles, stomach troubles (50 ml given three times daily for one week) and as antidote against snake bite.

References : Chatterjee & Pakrashi (1997), Dhiman (2003), Maheshwari *et al.* (1997), Sarkar *et al.* (2000), Watt (1889).

Uraria picta (Jacq.) Desv.

Syn : *Doodia picta* Roxb.; *Hedysarum pictum* Jacq.; *Uraria linearis* Hassk.

Family : Fabaceae.

Hindi & Sanskrit Names:

Hindi : Dabra, Pithvan, Shankaraja, Vishnu-Jadi;
Sans. : Prishniparni, Prithakparni, Sinhapushpi.

Distribution : Dry Grasslands, Wasteplaces & Open Forests in Sub-Himalayan Tract from Kashmir to W. Bengal & Assam Upto 1,800 m.

Part Used : Root.

Utilization : Credited with aphrodisiac properties; its decoction prescribed for cough, chills, fevers, malaria, gonorrhoea, heart ailments and as an antidote for snake bite.

References : Agarwal (2003), Ambasta (1986), Chatterjee & Pakrashi (1997), Jain & Singh (1997), Khare (2004), Kumar & Lal (2003), Parrotta (2001), Reddi *et al.* (2005), Sharma (2003), Watt (1889).

Urena lobata L. (Pl. 8D)

Syn : *Urena cana* Wall.; *U. palmata* Roxb.; *U. trilobata* Vell.

Family : Malvaceae.

English, Hindi & Sanskrit Names:

Eng. : Congujute;
Hindi : Bachata, Bachita, Brachta, Kunjuya, Lapetua, Pithia, Unga;
Sans. : Atibala, Nagabala, Vanabhenda.

Distribution : All Over N.E. Region.

Part Used : Root.

Utilization : Extr. of 100-200g of its pounded roots boiled in water with *Piper longum* L. (once daily for 20 days) cures jaundice. A poultice prepared from its roots and leaves used as an emollient. Also, roots considered diuretic, abortifacient, antirheumatic, anthelmintic, tonic and useful in windy colic (poultice), hyperacidity, dog bite, and dysentery (2-3g powder taken with rice water or plain water thrice a day). Decoction of root also said to be beneficial in magico-religious treatment of fever, syphilis and urinary troubles.

References : Agarwal (2003), Ambasta (1986), Caius (2003), Chatterjee & Pakrashi (1997), Chopra *et al.* (1956), Gogoi & Das (2003), Jain *et al.* (1995), Kumar (2002), Maheshwari *et al.* (1997), Pandey *et al.* (1996), Parrotta (2001), Rawat & Chowdhury (1998), Reddi *et al.* (2005), Shukla & Verma (1996), Singh & Srivastava (2000), Sood & Thakur (2004), Watt (1889).

Urena sinuata Linn.

Syn : *U. happago* DC.; *U. heterophylla* Smith; *U. lobata* ssp. *sinuata* (L.) Borssum Walkes; *U. morifolia* DC.; *U. muricata* DC.; *U. tomentosa* Wall.

Family : Malvaceae.

Hindi & Sanskrit Names:

Hindi : Hotloti, Kunguya, Kunjuya;

Sans. : Bala.

Distribution : Weed on Degraded Sites & Forest Clearings Throughout Hotter Parts of the Country.

Part Used : Root.

Utilization : Used as an external application for lumbago, and considered emollient and cooling.

References : Ambasta (1986), Caius (2003), Chopra *et al.* (1956), Parrotta (2001), Watt (1889).

Urtica dioica Linn.

Syn : *U. pilulifera* Linn.

Family : Urticaceae.

English, Hindi & Sanskrit Names:

Eng. : Nettle, Roman Nettle, Stinging Nettle;
Hindi : Bichhu Booti;
Sans. : Bichhuaa, Chichru.

Distribution : N.W. Himalaya from Kashmir to Shimla: 2,400-3,600 m.

Part Used : Root.

Utilization : Cold infusion of fresh root given to treat burning urinary tract and chewed to relieve pain in toothache. Paste of root considered diuretic, purgative, astringent, and its extr. given for easy delivery and treatment of jaundice.

Active Constituents : Reported to contain homovanillyl alcohol, its 4'-monoglucoside, neoolevil-4-O-β-D-glucoside, 9, 9'-bisacetyl-neoolevil and its 4-O- β-D-glucoside, sitosterol, 7α-and 7β-hydroxy-sitosterols, sitosterol-3-O-β-D-glucoside and 24 (R) ethyl-5α-cholestane-3β, 6α-diol, homovanillyl alcohol, vanillin, vanillic acid, (+) isolariciresinol, (-) secoisolariciresinol, neoolivil-six and 12 (Z)-octadecadienoic acid.

References : Bhattacharjee (2001), Chopra *et al.* (1956), Ganai & Nawchoo (2003), Khare (2004), Negi *et al.* (1999), Panthi & Choudhary (1998, 2003), Rastogi & Mehrotra (2002), Sood & Thakur (2004), Watt (1889).

Vaccaria pyramidata (L.) Medic.

Syn : *Saponaria vaccaria* L.
Family : Caryophyllaceae.

English & Hindi Names:

Eng. : Perfoliate-Soapwort;
Hindi : Musna.

Distribution : Weed Cultivation in Fields of Barley & Wheat: Throughout India.

Part Used : Root.

Utilization : Employed for asthma, jaundice, cough, liver and spleen diseases.

Active Constituents : Saponins (10.7%) have been obtained from roots.

References : Chatterjee & Pakrashi (1997), Rastogi & Mehrotra (1998).

Valeriana hardwickii Wall. *ex* Roxb.

Syn : *V. acuminata* Royle; *V. elata* Don; *V. javanica* Blume; *V. tenera* Wall.;

Family : Valerianaceae.

English Names:

Eng. : Asarum, Shumeo, Tagger.

Distribution : Kashmir to Bhutan & Khasia Hills Upto 4,000 m.

Part Used : Root.

Utilization : Paste used in epilepsy, hysteria, neurosis, stomach and mental disorders and for skin diseases.

Active Constituents : Reported to possess a crystalline acid, α-eudesmol, edemol and angelicin.

References : Agarwal (2003), Chauhan (1999), Kumar (2002), Sharma (2000), Singh (2000), Watt (1889).

Valeriana jatamansii Jones

Syn : *V. wallichii* DC.

Family : Valerianaceae.

English, Hindi & Sanskrit Names :

Eng. : Indian Valerian;

Hindi : Sugandha Bala, Tagar;

Sans. : Tagara.

Distribution : Temperate Himalaya Upto 3,000 m & in Khasia & Jaintia Hills: 1,500-1,800 m.

Parts Used : Root. Rootstock.

Utilization : 5g root powder with milk taken twice a day in insomnia (sleeplessness), hysteria, nervous unrest, hypochondriasis and as nervine tonic. Rootstock considered aromatic, bitter, antiseptic, stimulant, analgesic, carminative, nervine, antispasmodic, and also useful in hysteria, gleet, epilepsy, palpitation of the heart, muscular pains, toothache, diseases of the eye and blood, pains in the joints, diseases of the liver, spleen, kidney and cholera.

Active Constituents : These are valepotriates, the triesters of a polyhydroxy alcohol derived from terpenoid 4, 8-dimethyl cyclopenta (C)-pyran with isovaleric, acetic, isocproic, β-acetoxyvaleric acids, 2.4 % of valepotriates, α-fenchene-sesquifenchene, cyclopentapyrans, accacetin-7-O-rutinosides, valtrate, didorovaltrate, linarin, iso-valerinate, valeopotriates, valcrosidatum, calarene, β-bergamotene, valeranone, arcurcumene, maalioxide, maaliol, isovaleric acid & (+)-β-methyl-valeric-acid, formic, propionic, butyric, palmitic and stearic acids and isovalerryl ester of D- (-) alpha-hydroxyvaleric acid, β-sitosterol, α-, β- & γ-patchoulene and maaliol.

Biological Activity : Valepotriates exhibited sedative, spasmolytic, CNS depressant and antitumour properties.

References : Agarwal (2003), Bhattacharjee (2001), Dhiman (2003), Dwivedi (1993), Kaul (1997), Khare (2004), Kirtikar & Basu (1984), Panthi & Chaudhary (2003), Reddi *et al.* (2005), Sharma & Rana (2000), Sharma & Sood (1997), Singh (1996), Sinha & Sinha (2001), Thakur *et al.* (1989), Watt (1889).

Vanda tessellata (Roxb.) Hook. *ex* G. Don

Syn : *Aerides tessellata* Wight; *Cymbidium alagnata* Hamilton *ex* Wall.; *C. tessellatum* Sw.; *C. tesselloids* Roxb.; *Epidendrum tessellatum* Roxb.; *E. tesselloides* Steud.; *Vanda tesselloides* Reichb.f.; *V. unicolor* Steud.

Family : Orchidaceae.

Hindi & Sanskrit Names:

Hindi : Banda, Nai, Perasara, Persara, Vanda;

Sans. : Atirasa, Bhujankhi, Chhatraki, Elaparni, Muktarasa, Nakuli, Palankapa, Rasadya, Rasana, Rasya, Sugandha, Surasa, Vandaka, Yuktarasa.

Distribution : U. P. to W. Bengal, Bihar, M.P. & W. Peninsular India to Kerala & Tamil Nadu to 650 m.

Part Used : Root.

Utilization : Juice given for normal child birth and considered as fragrant, bitter, laxative, tonic to the liver and brain, antibacterial, antifertility agent, antitubercular and also useful in bronchitis, dyspepsia and abscesses. Mixed with Sesamum oil, it is applied externally in nervous disorders and rheumatic pains.

Active Constituents : Yields tetracosylferyllate and β-sitosterol glucoside.

References : Agarwal (2003), Ambasta (1986), Caius (2003), Kirtikar & Basu (1984), Kumar & Lal (2003), Parrotta (2001), Rastogi & Mehrotra (1998), Subramaniam (2000), Sur & Halder (2002), Upadhyay & Singh (2005), Varghese (1996).

Ventilago denticulata Willd.

Syn : *V. calyculata* Tul.; *V. madraspatana* Roxb.; *V. macrantha* Tulasne; *V. silhetiana* Tulasne; *V. smithiana* Tulasne; *S. sulphurea* Tulasne.

Family : Rhamnaceae.

Hindi Names:

Hindi : Nallathiga, Pitti, Raidhani.

Distribution : Throughout India.

Part Used : Root bark.

Utilization : Bark juice given orally (three teaspoons twice daily) to treat stomachache, debility (15 ml decoction taken 3 times a day for a fortnight), general weakness and renal problems (half glass decoction prepared from 100g root boiled in 500 ml of water, twice daily for 7 days).

Active Constituents : Possesses a new naphthoquinine-2-methoxystypandrone, a new xanthone-calyxanthone (1-hydroxy-6-methoxy-3-methylxanthone-8-carboxylic acid), ventiloquinones-ventilatones A and B, ventileins A and B.

References : Agarwal (2003), Ambasta (1986), Bhattacharjee (2001), Kumar & Pullaiah (1998), Rastogi & Mehrotra (2002), Varghese (1996).

Ventilago madraspatana Gaertn.

Syn : *V. bracteata* Wall.
Family : Rhamnaceae.

English, Hindi & Sanskrit Names:

Eng. : Red Creeper;
Hindi : Pitti, Raktavalli;
Sans. : Raktavalli.

Distribution : A Woody Climber Grown Throughout India.

Part Used : Root bark.

Utilization : Powdered bark carminative, stomachic, tonic and stimulant; useful in atonic dyspepsia, debility, slight cases of fever and in skin diseases.

Active Constituents : Contains ventilones A, B, C, D, E characterised as 4, 9-dihydro-5, 8-dihydroxy-6, 7-methyllenedioxy-1-methylnaphtho-[-1-2, 3-C] furan-4,9-dione, its 5-methylether, 1, 3, 4, 9-tetrahydro-8-hydroxy-5-methoxy-6,7-methylene-dioxy-1-methylnaphtho [2, 3-C] furan-4,9-dione, its 8-methyl & 5, 8-dimethylethers, ventiloquinones A,B,C,D,E,F,G,H, ventilaginone, ventilagol, maderone, cordeauxione, isocordeauxione, vantilagol and maderone.

References : Agarwal (2003), Ambasta (1986), Chopra *et al.* (1956), Kirtikar & Basu (1935), Rao & Henry (1996), Rastogi & Mehrotra (2002), Watt (1889).

Verbascum thapsus Linn.

Syn : *V. indicum* Wall.
Family : Scrophulariaceae.

English & Hindi Names:

Eng. : Lungwort;
Hindi : Bantamaco, Gidar Tamaku, Phulla.

Distribution : Indian Gardens from Kashmir to Meghalaya.

Part Used : Root.

Utilization : Considered beneficial against cramps, fevers, migraine, boils and blisters.

References : Agarwal (2003), Ambasta (1986), Kaul (1997), Singh & Kumar (2000b), Watt (1889).

Vernonia cinerea (L.) Lees. (Pl. 9A)

Syn : *Cocalia rotundifolia* Willd.; *Conyza cenerascens* Wall.; *C. cinerea* L.; *C. heterophylla* Lamk.; *C. incana* Wall.; *C. linifolia* Wall.; *C. mollis* Wild.; *C. ovata* Wall.; *C. prolifera* Lamk.; *Serratula cinerea* Roxb.; *Verinia abbreviata* DC.; *V. albicans* auct. non DC.; *V. conyzoides* DC.; *V. laxiflora* Less.; *V. leptophylla* DC.; *V. physalifolia* DC.

Family : Asteraceae.

English, Hindi & Sanskrit Names:

Eng. : Ash Coloured Fleabane, Purple Fleabane;
Hindi : Dantoapala, Daudotpala, Sadodi, Sahadevi;
Sans. : Sahadevi.

Distribution : Found Throughout the Country Upto 2,500 m in the Himalaya.

Part Used : Root.

Utilization : Employed for malaria, filarial, helminthiasis, stomach pain, sunstroke, dropsy, boils, diarrhoea, cough, colic and leucorrhoea. A root piece tied on the head in the form of cap to get sound sleep and its decoction taken internally to relieve malarial fever, pneumonia and aphasia.

Active Constituents : Known to yield 26-methyl-heptacosanoic acid, amyrin, δ-amyrinacetate, 3β-acetoxyurs-13 (18)-ene, campesterol, 24-hydroxytaraxer-14-ene, stigmat-5, 17 (20)-diene-3β-ol, β-sitosterol, α-spinasterol, stigmasterol, β-amyrinacetate, a new triterpenoid-3β-acetoxyurs-19-ene, and lupeol acetate.

References : Agarwal (2003), Ambasta (1986), Barua *et al.* (2000), Chatterjee & Pakrashi (1997), Dastur (1970), Goud *et al.* (2000), Khanna (2002b), Parrotta (2001),

Punjani (2002), Rastogi & Mehrotra (1998, 2002), Reddi *et al.* (2005), Singh & Pandey (1998), Varghese (1996), Watt (1889).

Vernonia divergens Benth.

Syn : *Decaneuron divergens* DC.; *Veronia multiflora* DC.

Family : Asteraceae.

Distribution : Pavsunath (1,300 m), C. India & From Concan to the Travancore Mountains.

Part Used : Root.

Utilization : Paste mixed in hot water, prescribed orally to cure cold and earache.

References : Hooker (1886), Hosagoudar & Henry (1996b).

Vetiveria zizanioides L. Nash.

Syn : *Anatherum zizanioides* (Linn.) Hitchcock & Chase; *Andropogon muricatus* Hook.; *A. squarrosus* Hook.f. non Linn.f.

Family : Poaceae.

English, Hindi & Sanskrit Names:

Eng. : Cuscus, Khuskhus, Koosa, Vetiver;

Hindi : Bala, Bena, Ganrar, Khaskhas, Onei, Panni;

Sans. : Bahu-Muulaka, Indragupta, Jalavasa, Laamajjaka, Nalada.

Distribution : Wild Throughout the Plains of India & at Lower Hills, Upto 1,200 m.

Part Used : Root.

Utilization : 10 ml filtrate of (100 g) roots boiled in 500 ml water for 30 minutes given with little sugar to cure dysuria. The aromatic roots considered a refrigerant and a decoction of these prescribed as a 'cooling medicine' in high fever, severe headache, inflammation and sexually transmitted diseases. Pounded roots of this plant and that of *Cyperus alopecuroides* Rottb. and seeds of *Coriandrum sativum* L. in equal proportions efficacious (two or three times a day) in rheumatism; root grounded and mixed with breast milk said to be good in abdominal pain for children. Its aq. extr.

(thrice a day for three days) removes constipation and paste beneficial to cure headache, malaria, toothache, burning sensation, bilious fevers, sweats, foul breath, thirst, strangury, ulcers, diseases of the blood, hysteria, asthma, gout, headache, cholera and also known to be diaphoretic, stimulant, refrigerant, febrifuge, and emmenagogue.

Active Constituents : Zizanal, epizizanal, α- and β-vetivones, khusimone and bicyclodecenone have been found to characterised the roots.

Biological Activity : Vetiver oil possesses antimicrobial and antifungal activities.

References : Ambasta (1986), Beg *et al.* (2005), Caius (2003), Chauhan (1999), Chopra *et al.* (1956), Khanna (2002b), Khare (2004), Kirtikar & Basu (1984), Kumar & Chauhan (2005), Patil & Patil (2002), Reddi *et al.* (2005), Samvatsar & Diwanji (2000), Satapathy & Brahmam (2000), Sharma (2003), Shukla & Verma (1996), Singh & Kumar (2000a), Singh & Prakash (1996), Thakur *et al.* (1989), Thomas & Britto (2003), Srivastava *et al.* (2003), Varghese (1996).

Viburnum continifolium Don

Syn : *V. polycarpum* Wall.

Family : Caprifoliaceae.

Distribution : Himalaya (Kashmir to Bhutan): 1,200-3,300 m.

Part Used : Root.

Utilization : 25-40g powdered roots mixed with wheat flour soften inflamed parts in cattle. Also, useful in menorrhagia and metrorrhagia.

References : Sood & Thakur (2004), Uniyal (1968).

Vicoa indica (L.) DC.

Syn : *Inula indica* L.; *Pentanema indcium* (L.) Ling.; *Vicoa auriculata* Cass.

Family : Asteraceae.

Hindi Name:

Hindi : Sonasali.

Distribution : Throughout Dry Regions of India.

Part Used : Root.

Utilization : Decoction (2 teaspoonfuls) given once a day to cure fever among children; also fresh root chewed to relieve abdominal pain.

References : Agarwal (2003), Ambasta (1986), Aminuddin & Girach (1991), Rastogi & Mehrotra (2001).

Viola odorata Linn.

Family : Violaceae.

English, Hindi & Sanskrit Names:

Eng. : Apple-Leaf, Bairnwort, Banwort, Blaver, Blue Vilot, Fine-Leaf, March Vilot, Sweet Vilot;

Hindi : Banafshah;

Sans. : Jvarapaha, Nilapushpa, Sukshmapatra, Vanapatra.

Distribution : Kashmir & Other Parts of W. Himalaya: 1,500-1,800 m.

Part Used : Root.

Utilization : Dried powder with hot water given at bed time as laxative. A dose from forty to fifty grains of the powdered roots acts powerfully inciting nausea, vomiting and nervous affection. Also, considered a powerful emetic and frequently used to adulterate ipecac.

Active Constituents : Contains saponins (0.1-2.5 %), a glycoside of methyl salicylate, an essential oil (approx 0.04 %) closely resembling that of flowers, 2-nitropropionic acid (0.0001-0.0002 %) and odoratine.

Biological Activity : Odoratine exhibited marked hypoensive activity.

References : Ambasta (1986), Caius (2003), Chopra *et al.* (1956), Khare (2004), Singh (1996).

Viola pilosa Bl.

Syn : *V. serpens* Wall *ex* Ging.; *V. wightiana* var. *pubescens* Thw.
Family : Violaceae.

Hindi Names:

Hindi : Banafsha, Thungtu.

Distribution : Throughout Hilly Regions.

Part Used : Root.

Utilization : Reported to possess aperient, antipyretic, cooling, demulcent, diaphoretic, emetic, emollient, purgative, expectorant and febrifuge properties; also useful in asthma, bleeding piles, cancer of the throat, constipation, cough, fever, headache and skin diseases. Its decoction good for vaginal discharges.

Active Constituents : Contains glucoside methyl salicylate, violine, viola quercitin, which is identical with rutin and saponin.

References : Chauhan (1999), Chhetri (2005), Sharma (2003), Thakur *et al.* (1989), Watt (1889).

Vitex negundo L. (Pl. 9B)

Syn : *Nika silvestris* Hermann *ex* Moldenke; *V. arborea* Desf.; *V. bicolor* Willd.; *V. paniculata* Lamk.; *V. spicata* Lour.; *V. trifolia* Graham *ex* Chavan & Oza.
Family : Verbenaceae.

English, Hindi & Sanskrit Names:

Eng. : Chaste Tree, Five Leaved Chaste Tree, Indian Privet;

Hindi : Mewri, Ningori, Nirgondi, Nasinda, Panikisambhalu, Sanbhalu, Shiwali, Sindhuca;

Sans. : Indrani, Nilapushpa, Nirgundi, Shephali, Sindhubara.

Distribution : Throughout India.

Part Used : Root.

Utilization : Paste applied on abdomen for easy delivery. Decoction of root administered orally for 3 times a day to cure cough. Also, known for the treatment of joint pains, spermatorrhoea, rheumatism, piles, ringworm infection, dyspepsia, boils, skin diseases, fever, bladder complaints, dysentery, liver complaints, and general fitness (pills with root powder and jaggery; 1 pill twice a day for 3 days with hot water) and considered as expectorant, analgesic, anthelmintic, demulcent, diuretic, and colic.

Active Constituents : Contained hentriacontane, β-sitosterol, β-sitosterol acetate and stigmasterol.

Biological Activity : Butanol, petrol ether (PE), chloroform, n-butanol (BE) extrs. of root showed antiinflammatory, analgesic and marked antiparkinsonian effect.

References : Agarwal (2003), Ambasta (1986), Chopra *et al.* (1956), Dastur (1970), Dash & Mishra (2000), Jain (1991), Khare (2004), Kumar & Lal (2003), Kurian (1995), Parrotta (2001), Pawar & Patil (2004), Punjani (2002), Rastogi & Mehrotra (2001, 2002), Reddi *et al.* (2005), Singh (2001), Sood & Thakur (2004), Varghese (1996).

Vitis vinifera L.

Family : Vitaceae.

English, Hindi & Sanskrit Names:

Eng. : Common Grape Vine, Eurpeon Grape, Wine-Grape;

Hindi : Angoor, Dakh;

Sans. : Draksha.

Distribution : Cultivated in Many Parts of India.

Part Used : Root.

Utilization : Juice depurative, diuretic, blood purifier and used in eye diseases.

Active Constituents : Betulinic acid, lupeol, 30-norlupan-3β-ol-20-one, neptacosanol, tricontanol, tricontanyl tridecanoate and stigmasterol identified in roots.

References : Agarwal (2003), Ambasta (1986), Rastogi & Mehrotra (2002).

Wattakaka volubilis (L.f.) Stapf.

Syn : *Dragea volubilis* (Linn.f.) Benth. *ex* Hook.f.; *Marsdenia volubilis* Cooke.
Family : Asclepiadaceae.

English, Hindi & Sanskrit Names:

Eng. : Cotton Milk Plant;
Hindi : Haran Dodi, Nakchiki;
Sans. : Hemajivanti, Madhumalati, Nakchikni.

Distribution : Throughout Hotter Parts of India.

Part Used : Root.

Utilization : Beneficial in rheumatism and snake bite.

Active Constituents : α-methylpachybioside and α-methyldredehonbioside (αmethyl-[3-O-methyl-6-doxy-D-allose (1→4)-D-Olivoside]) have been isolated.

References : Agarwal (2003), Ambasta (1986), Rastogi & Mehrotra (2002), Reddi *et al.* (2005).

Wedelia biflora DC.

Family : Asteraceae.

Distribution : Found Near the Sea from Bengal & Bombay to Sri Lanka.

Part Used : Root.

Utilization : Recommended for vaginal discharges, gonorrhoea, gravel; also applied on wounds and for itch.

Reference : Chopra *et al.* (1969).

Wendlandia heynei (R. & S.) Sant & Merch.

Syn : *Rondeletia cinerea* Wall.; *R. exserta* Roxb.; *R. heynei* Roem.; *R. thrysiflora* Roth.; *R. orissensis* Roth.; *Wendlandia cinerea* Wall.; *W. exserta* (Roxb.) DC.
Family : Rubiaceae.

Hindi Names:

Hindi : Birsa, Chaulai, Chila, Chilkiya. Tila, Tilki.

Distribution : Dry Forests of Tropical Himalaya.

Part Used : Root.

Utilization : Paste of its root (15 g) and 3 black peppers given once daily for 3 days after menses for preventing conception.

References : Brahmam & Saxena (1990), Sood & Thakur (2004).

Withania somnifera (L.) Dunal (Pl. 9C)

Syn : *Physalis arborescens* Thunb.; *P. somnifera* L.; *P. tomentosa* Thunb.; *P. flexuosa* L.; *P. flexuosa* L.; *Withania ashwagandha* Kaul.

Family : Solanaceae.

English, Hindi & Sanskrit Names:

Eng. : Asvagandha, Winter Cherry;

Hindi : Asgad, Aswagandha, Punir;

Sans. : Asvagandha, Ashwakanda, Balyaa, Gandharvagandhaa, Turagagandhaa, Vaajigandhaa, Vrshaa.

Distribution : Throughout the Drier Parts of India in Wasteplaces; Also Cultivated to a Certain Extent.

Part Used : Root.

Utilization : Root extr. (about 20 ml) taken orally once a day for about 15 days to arouse sexual desire in both men and women; it favours pregnancy when mixed with an equal quantity of cow ghee and consumed on fourth day of menses. Root powder (about 10 g or 1 teaspoonful) taken with milk twice a day in cases of general debility and for cough, hiccups, rheumatism, painful swellings, gout, dropsy, insomnia, burning micturition, toning up the uterus of women, loss of appetite, ulcers, inflammations, ageing, greying of hairs, liver complaints and also as stimulant, aphrodisiac, narcotic, alterative, deobstuent, abortifacient and diuretic. Root powder with milk or jaggery or purified butter prescribed as a nourishing, rejuvenating and restorative tonic for promoting vitility in men and fertility in women. Powdered roots mixed with husk of *Plantago major* L. in 1:1 ratio given with milk (1 teaspoonful twice a day) for spermatorrhoea. Paste of root applied to penis for toning up erectile tissues; mixed with mother's milk it is applied to sagging breasts. Its decoction with long pepper, ghee and honey prescribed in scrofula, consumption and chest complaints.

Active Constituents : Contains nicotine, somnine, somniferine, somniferinine, withanine, withananine, withananinine, pseudo-withanine, tropine, pseudotropine, isopelletrieriene, anahygrine, withasomnine, visamine, withanolides, [meso-1, 3-bis (2-piperidyl)-2-propanonel], cusco-hygrine, hygrine, de-isopelletierine, solasodine, 3-tropyltigloate, ducitol, hentriacontane, 5α, 20-dihydroxyl-1-oxo-6α, 7α-epoxy-20 R, 22-R-with a-2, 24-dienolide, acylsteryl-glucosides, sitoindosdes VII-X, AC-28 steroid lactone, 5, 20α-dihydroxy-6α, 7α- epoxy-1-oxowitha-2, 24-dienolide (withanolide), sitoindoside IX and sitoindoside X.

Biological Activity : Major alkaloid fraction of root extr. from cultivated plants showed prolonged hypotensive, bradycardiac and respiratory stimulant activities in dogs.

References : Agarwal (2003), Ambasta (1986), Beg *et al.* (2005), Bhatt *et al.* (1999), Bhattacharjee (2001), Chatterjee & Pakrashi (1997), Chauhan (1999), Chopra *et al.* (1956), Dastur (1970), Drury (1978), Dwivedi (1993, 2003), Jain (1968), Khare (2004), Khare & Khare (2000), Kumar & Chauhan (2005), Mitaliya *et al.* (2003), Parrotta (2001), Patil (2000), Pawar & Patil (2004), Rana *et al.*(1996), Rastogi & Mehrotra (1999a, 2002), Reddi *et al.* (2005), Said (1997), Sarin (1996), Sen *et al.* (2002), Sharma (2003), Sharma & Sood (1997), Singh (1999), Singh & Pandey (1998), Srivastava *et al.* (2003), Watt (1889).

Woodfordia fruticosa (L.) Kurz.

Syn : *Grislea tomentosa* Roxb.; *G. punctata* Ham.; *Lythrum fruticosum* Linn.; *Woodfordia floribunda* Salisb.; *W. tomentosa* Bedd.

Family : Lythraceae.

English, Hindi & Sanskrit Names:

Eng.	:	Five-Flame Bush, Shiranjitea;
Hindi	:	Dawi, Dhai, Dhaura, Dhai-Phul, Dhataki, Thawi;
Sans.	:	Agnijvala, Dhataki, Dhaatri, Kunjaraa, Taamrapushpi.

Distribution : Wild Upto 1,500 m in the Tropical Deciduous Forests.

Part Used : Root.

Utilization : Pellets made by mixing its root extr. with jaggery, given to cure leucorrhoea. Powdered root (5g twice daily) given in cough and its poultice applied to cure wounds, boils, rheumatism and as emmenagogue. Half teaspoon root decoction given orally in early morning daily for two weeks to stop burning sensation and bleeding during urination.

References : Aminuddin & Girach (1996), Jain (1991), Khare (2004), Panda & Das (2000), Parrotta (2001), Punjani (2006), Reddi *et al.* (2005), Srivastava *et al.* (2003), Sood & Thakur (2004), Tiwari *et al.* (1996).

Wrightia arborea (Dennst.) Mab.

Syn-: *W. tomentosa* (Roxb.) Rhem. & Schult.

Family : Apocynaceae.

Distribution : Common Along Hilly Slopes of A. P.

Parts Used : Root. Root bark.

Utilization : Decoction of root prescribed internally to relieve malarial fever while pills made by pounding its bark with black pepper (*Piper nigrum*) given for menstrual disorders (white discharges) -2 pills twice a day for 10 days.

References : Goud *et al.* (2000), Jain (1991), Rao & Henry (1996), Sharma (2003).

Wrightia tinctoria (Roxb.) R.Br.

Syn : *Nerium tinctorium* Roxb. var. *rothii* G. Don

Family : Apocynaceae.

English, Hindi & Sanskrit Names:

Eng.	:	Pala Indigo Plant, Sweet Indrajao;

Hindi : Khirni, Mitha Indarjou;
Sans. : Kutaja, Sveta Kutajah.

Distribution : Widely Distributed From Rajasthan to M.P. Southwards to Tamil Nadu in Deciduous Forests on Dry, Sandy Soils, Often Along Streams.

Parts Used : Root. Root bark.

Utilization : Root used to treat epilepsy, snake bite, scorpion stings and expel threadworms. Extr. of its bark administered orally for three days as antidote to snake bite, and also as a substitute for that of *Holarrhena antidysentrica* to treat chronic dysentery, diarrhoea, piles, haemorrhage, indigestion and skin diseases.

References : Goud & Pullaiah (1996), Hosagoudar & Henry (1996a), Jain (1991), Parrotta (2001), Reddi *et al.* (2005), Watt (1889).

Xanthium strumarium L. (Pl. 9D)

Syn : *X. brevirostre* Wall.; *X. indicum* DC.; *X. orientale* Blume; *X. roxburghii* Discolor
Family : Asteraceae.

English, Hindi & Sanskrit Names:

Eng. : Bur-Weed, Cocklebur;
Hindi : Banokra, Chhotagokhru, Chotadatura, Ghagra, Okura, Shankahuli;
Sans. : Arishta, Chanda, Itara, Pitapushpi.

Distribution : Throughout the Hotter Parts of India.

Part Used : Root.

Utilization : Powder fried in 'Desi ghee' prescribed in snake bite, also considered as bitter, tonic, anticancer and good for scrofula and nerves. Extr. applied to ulcers, boils, abscesses and inflammation.

Active Constituents : Therapeutic properties credited to campesterol, β-sitosterol, stigmasterol, β-sitosterol-D-glucoside, struaroside, abscisic acid, xanthatin, a sesquiterpene lactone, 3, 4-dihydroxycinnamic acid, heptacosanol, cytokinnins, zeatin and its riboside.

Biological Activity : Antitumour and hypoglycaemic activeties positive.

References : Caius (2003), Chatterjee & Pakrashi (1997), Chopra *et al.* (1956), Kumar & Chauhan (2005), Parrotta (2001), Reddi *et al.* (2005), Sharma (2003), Shukla *et al.* (2003), Singh & Srivastava (2000), Sood & Thakur (2004), Viswanathan (2001), Watt (1889).

Xanthoxylum nitidum (Roxb.) DC.

Syn : *X. hamiltonianum* Wall. *ex* Hook.f.

Family : Rutaceae.

English Name:

Eng. : Tezamul.

Distribution : Bihar to Naga Hills.

Part Used : Root.

Utilization : Good against toothache, stomachache and boils.

Active Constituents : Found to posssess nitidene chloride, oxynitidine, dihydronitidine, 6-methoxy-5, 6-dihydrochelerythrine, α-allocryptopine, skimmianine, isofagaridine, nitidine and chelerythrine.

References : Agarwal (2003), Ambasta (1986), Rastogi & Mehrotra (1998, 2001).

Xerompsis spinosa (Thunb.) Keay

Family : Rubiaceae.

English, Hindi & Sanskrit Names:

Eng. : Common Emetic Nut;
Hindi : Mainphal;
Sans. : Madana.

Parts Used : Root. Root bark.

Utilization : Root extr. (1 teaspoon 2 times a day for 3 days) given for the treatment of typhoid fever, also used for treating gonorrhoea and diseases of urinary tract. Root bark chewed in cases of scorpion sting and also sniffed for checking half headache. Paste of the pounded bark with pith removed given with milk and sugar, three times a day, in the infection of sex organs.

Active Constituents : Bark contains triterpene-1-keto-3α-hydroxy-leanane, α-& β-amyrins, oleanolic acid and β-sitosterol.

References : Agarwal (2003), Ambasta (1986), Rastogi & Mehrotra (2002), Singh & Prakash (1994).

Ximenia americana L.

Syn : *X. spinosa* Salisb.

Family : Olacaceae.

English & Hindi Names:

Eng. : False Sandal Wood, Tallow-Wood;

Hindi : Kadarangi.

Distribution : Grown in Deccan Peninsula & Coastal Regions of Andamans.

Part Used : Root.

Utilization : So far known as colic, cathartic, anthelmintic, febrifuge, and useful in toothache, venereal diseases, jaundice and diarrhoea.

References : Agarwal (2003), Ambasta (1986), Chopra *et al.* (1956).

Yucca filamentosa L.

Family : Liliaceae.

Hindi Name:

Hindi : Dori Ghas.

Distribution : Tropical Gardens of India.

Part Used : Root.

Utilization : Root paste in water good as poultice in inflammations. Its tincture beneficial in gonorrhoea and rheumatism.

References : Agarwal (2003), Ambasta (1986).

Zaleya decandra (L.) Burm.f.

Family : Aizoaceae.

Hindi & Sanskrit Names:

Hindi : Yadabani;

Sans. : Punarnavi.

Part Used : Root.

Utilization : Three drops of the root juice applied in the eyes (once a day early in the morning for 3 days) to treat jaundice. Root also given to treat malaria. Decoction of roots given to males against syphilis and swellings of sex organs and also to regularise menstruation.

References : Goud & Pullaiah (1998), Kumar & Pullaiah (1998), Reddi *et al.* (2005), Singh & Pandey (1996, 1998).

Zingiber chrysanthum Rosc.

Family : Zingiberaceae.

Distribution : Tropical Himalaya, N.E. India, Upto 1,200 m.

Part Used : Root.

Utilization : Useful in body pain and internal haemorrhage in Meghalaya.

References : Aswal (2001), Jain (1991, 1995).

Zingiber roseum Rosc.

Family : Zingiberaceae.

Part Used : Root.

Utilization : 100-200 ml aq. extr. (two times a day for 7 days) prescribed for the treatment of cough in cattle.

Reference : Singh & Kumar (2000b).

Zizyphus mauritiana Lamk.

Syn : *Rhamnus jujuba* Linn.; *Zizyphus jujuba* Lamk.; *Z. sativa* (Gaertn.) DC.; *Z. sororia* Schult.; *Z. trinervius* Roth.; *Z. vulgaris* Lamk.

Family : Rhamnaceae.

English, Hindi & Sanskrit Names:

Eng. : Chinese Date, Indian Cherry, Indian Jujuba, Indian Plum;

Hindi : Baer, Ber, Beri;

Sans. : Ajapriya, Badarah, Badari, Dridhabija, Kola, Koli, Kuvala, Madhuraphala.

Distribution : Both Wild & Cultivated Throughout India, ascending 1,500 m in the Himalaya.

Parts Used : Root. Root bark.

Utilization : Extr. of the root (2 teaspoonful twice a day for 3 days) given in asthma, stomachache, malarial fever and to regulate fertility in woman. Paste of root and 7 black peppers (*Piper nigrum* Linn.) taken internally to cure diarrhoea and dysentery, and its poultice applied to affected part in case of internal injuries, chronic ulcers and old wounds. Powdered root bark mixed with wheat dough restores normal function of the body in cattle. Also, paste applied on cuts, and wounds in foot and mouth diseases for quick healing.

References : Chatterjee & Pakrashi (1997), Dhiman (2003), Jain (1991), Kadel & Jain (2006), Khare (2004), Lalramnghinglova (1996), Panda & Das (2000), Parrotta (2001), Reddi *et al.* (2005), Sen & Batra (1997), Shukla & Verma (1996), Singh & Kumar (2000a), Singh & Pandey (1998), Sood & Thakur (2004), Sur & Halder (2002), Thomas & Britto (2000), Tiwari & Pande (2006), Varghese (1996), Watt (1889).

Ziziphus nummularia (Burm.f.) W. & A.

Syn: *Rhamnus nummularia* Burm.f.; *Ziziphus lotus* Lamk.; *Z. microphylla* Roxb.; *Z. rotundifolia* Lam.

Family : Rhamnaceae.

English, Hindi & Sanskrit Names:

Eng. : Wild Jujube;

Hindi : Jhadiaber, Jhahrberi, Jharber;

Sans. : Bhukamtaka, Karkandhu, Sukshamaphala.

Distribution : Common in Grasslands, Semi-arid & Arid Scrub Forests in Punjab, Rajasthan & Gujarat; Peninsular

India from Maharashtra & M.P. Southwards to 900 m.

Parts Used : Root. Root bark.

Utilization : Bark given orally as abortifacient to cattle. Roots used as toothbrush for preventing decay and other diseases of the teeth by 'Bhils'. Its extr. (1 teaspoonful thrice a day for 3-4 days) given in fever and diarrhoea.

Active Constituents : In the root new cyclopeptide alkaloids-nummularin, D, E & F, fragufoline, amphibine A, intergerrenine & mauritine F are present. Root bark contains cyclopeptide alkaloid-nummularine O (N-desmethyljubanine B), jujubanines A, B and mauritine.

References : Kumar & Chauhan (2006), Parrotta (2001), Rastogi & Mehrotra (1998, 2002), Singh & Kumar (2000a), Singh & Pandey (1998).

Ziziphus xylopyrus Willd.

Syn : *Rhamnus xylopyrus* Retz.; *Ziziphus caracutta* Roxb.; *Z. cuneata* Wall.; *Z. elliptica* Roxb.; *Z. orbicularis* Schult.; *Z. rotundifolia* Roth.; *Z. ruminata* Ham.

Family : Rhamnaceae.

Hindi & Sanskrit Names:

Hindi : Bhander, Chitena, Ghatbor, Ghont, Kat-Ber, Kakor;

Sans. : Ghonta.

Distribution : Throughout India in Dry, Deciduous or Scrub Forests & Grasslands, Often Forming Dense Thickets When Young.

Parts Used : Root. Root bark.

Utilization : Aq. extr. of root bark given 3 times a day for 3 days to cure dysentery while root useful for anthrax, rheumatism, diarrhoea and dysentery.

References : Parrotta (2001), Reddi *et al.* (2005), Singh & Prakash (1994), Watt (1889).

Zornia diphylla (L.) Pers.

Syn : *Z. congugata* (Willd.) Sm.

Family : Fabaceae.

Hindi Name:

Hindi : Samrapani.

Distribution : Cultivated Fields Throughout the Plains of India, Ascending to 1,300 m in Kumaon.

Part Used : Root.

Utilization : Considered sedative and given to children to induce sleep and for spermatorrhoea.

References : Agarwal (2003), Ambasta (1986), Chopra *et al.* (1956), Deshmukh *et al.* (2000), Varghese (1996).

❐❐❐

5

Epilogue

From the database presently complied, it is estimated that in India about 917 spp. belonging to 571 genera under 133 families have been recorded as root drug plants.

Category-wise analysis reveals the usage of 11 spp. (8 genera: *Actiniopteris, Adiantum, Angiopteris, Cheilanthes, Hemionitis, Lygodium, Polypodium, Salvinia*) belonging to Pteridophytes; 1 sp. (1 genus: *Ephedra*) to Gymnosperms; and out of 905 spp. (562 genera) of Angiosperms, 788 spp. (479 genera) belong to dicotyledons and the remaining 117 spp. (83 genera: *Acorus, Agave, Agropyron, Alocasia, Aloe, Alpinia, Amorphophallus, Aneilema, Areca, Arenga, Arisaema, Asparagus, Asphodelus, Bambusa, Belamcanda, Borassus, Bromus, Caladium, Campylandra, Canna, Chlorophytum, Cocos, Colchicum, Colocasia, Commelina, Costus, Crinum, Cryptocoryne, Curculigo, Curcuma, Cyanotis, Cymbidium, Cynodon, Cyperus, Dactylorrhiza, Desmostachya, Dioscorea, Dracaena, Eichhornia, Eleusine, Eulophia, Geodorum, Globba, Gloriosa, Habenaria, Herminium, Hedychium, Heteropogon, Iris, Juncus, Kaempferia, Kyllinga, Lasia, Monochoria, Musa, Pandanus, Paris, Pentatropis, Peristylus, Phoenix, Polygonatum, Raphia, Remusata, Rhynchostylis, Roscoea, Saccharum, Saccolabium, Sansevieria, Satyrium, Scindapsus, Scirpus, Scleria, Smilacina, Smilax, Stachyphrynium, Stemona, Thysanolaena, Typha, Typhonium, Vanda, Vetiveria, Yucca, Zingiber*) to monocotyledons (Appendix-I), thereby indicating traditional wisdom for a wide array of ailments in the country.

With regard to the distribution pattern of root drug plants in India, plant spp. having scattered or throughout distribution (390 spp.) predominate in usage, followed by 166 spp. in Peninsular India (Ghats, Coastal Plains, Deccan Plateau), 136 spp. in N.E. and E. Himalaya, 97 spp. in Central Highlands and Plains of India, 89 spp. in N.W. and W. Himalaya and 39 spp. in Entire Himalaya (Fig.1; Table I). It is quite obvious that these plants are judiciously used by Indian tribals in different physiographical regions for curing various ailments affecting them.

Table I. Distribution Pattern of Root Drug Plants of India

Distribution	No. of Species
Throughout or Scattered Distribution	390
N.W. &W. Himalaya	89
N.E. &E. Himalaya	136
Peninsular India (Ghats, Coastal Plains, Deccan Plateau)	166
Central Highlands & Plains of India	97
Entire Himalaya	39

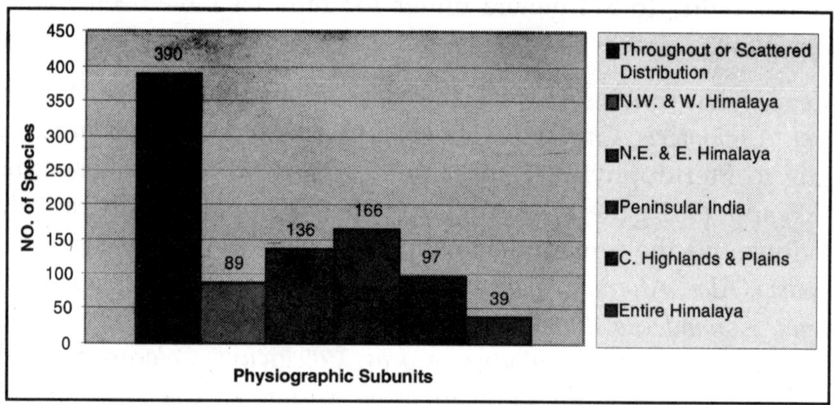

Fig. 1. Histogram Showing Distribution Patterns of Root Drug Plants of India

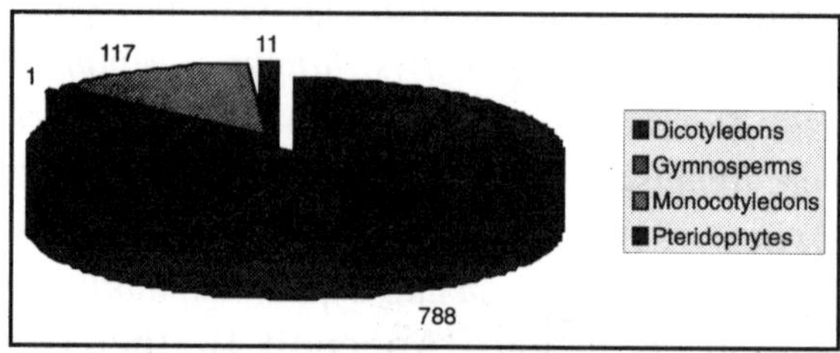

Fig. 2. Total Number of Species of Root Drug Plants Under Various Divisions of Plant Kingdom

Table II. Predominant Dicot Families of Root Drug Plants

Families	Species	%age in Taxa	Family	Species	%age in Taxa
Fabaceae	98	10.69	Gentianaceae	7	00.76
Asteraceae	70	07.33	Vitaceae	7	00.76
Ranunculaceae	33	03.60	Aizoaceae	6	00.65
Euphorbiaceae	32	03.49	Berberidaceae	6	00.65
Acanthaceae	31	03.38	Leeaceae	6	00.65
Rubiaceae	30	03.27	Melastomaceae	6	00.65
Malvaceae	27	02.94	Sterculiaceae	6	00.65
Asclepiadaceae	20	02.18	Rhamnaceae	5	00.54
Apocynaceae	18	01.96	Saxifragaceae	5	00.54
Cucurbitaceae	18	01.96	Bignoniaceae	4	00.44
Polygonaceae	17	01.85	Brassicaceae	4	00.44
Verbenaceae	17	01.85	Caprifoliaceae	4	00.44
Apiaceae	16	01.74	Celastraceae	4	00.44
Boraginaceae	15	01.63	Flacourtiaceae	4	00.44
Amarantaceae	14	01.52	Hippocrateaceae	4	00.44
Rutaceae	14	01.52	Lythraceae	4	00.44
Solanaceae	14	01.52	Meliaceae	4	00.44
Lamiaceae	13	01.42	Piperaceae	4	00.44
Convolvulaceae	12	01.30	Scrophulariaceae	4	00.44
Rosaceae	12	01.30	Bombacaceae	3	00.33
Tiliaceae	12	01.30	Geraniaceae	3	00.33
Capparidaceae	11	01.20	Nymphaeaceae	3	00.33
Moraceae	11	01.20	Ochnaceae	3	00.33
Myrsinaceae	10	01.09	Salvadoraceae	3	00.33
Menispermaceae	9	00.98	Valerianaceae	3	00.33
Urticaceae	8	00.87	Violaceae	3	00.33
Aristolochiaceae	7	00.76			

Table III. Predominant Monocot Families of Root Drug Plants

Families	Species	%age in the Taxa
Liliaceae	26	02.85
Araceae	15	01.63
Zingiberaceae	15	01.63
Orchidaceae	15	01.20
Poaceae	10	01.09
Cyperaceac	8	00.87

Contd...

Table III. Contd...

Arecaceae	7	00.76
Amaryllidaceae	4	00.44
Iridaceae	4	00.44
Commelinaceae	3	00.33
Dioscoreaceae	3	00.33

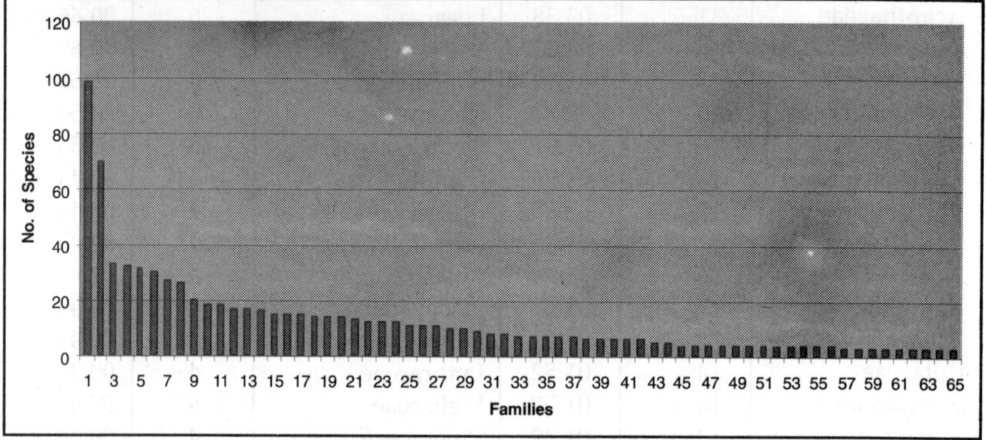

Fig. 3. Families of Root Drug Plants With Three or More Species

1. Fabaceae; 2. Asteraceae; 3. Ranunculaceae; 4. Euphorbiaceae; 5. Acanthaceae; 6. Rubiaceae; 7. Malvaceae; 8. Liliaceae; 9. Asclepiadaceae; 10. Apocynaceae; 11. Cucurbitaceae; 12. Polygonaceae; 13. Verbenaceae; 14. Apiaceae; 15. Araceae; 16. Boraginaceae; 17. Zingiberaceae; 18. Amarantaceae; 19. Rutaceae; 20. Solanaceae; 21. Lamiaceae; 22. Convolvulaceae; 23. Rosaceae; 24. Tiliaceae; 25. Capparidaceae; 26. Moraceae; 27. Orchidaceae; 28. Myrsinaceae; 29. Poaceae; 30. Menispermaceae; 31. Cyperaceae; 32. Urticaceae; 33. Arecaceae; 34. Aristolochiaceae; 35. Gentianaceae; 36. Polypodiaceae; 37. Vitaceae; 38. Aizoaceae; 39. Berberidaceae; 40. Leeaceae; 41. Melastomaceae; 42. Sterculiaceae; 43. Rhamnaceae; 44. Saxifragaceae; 45. Amaryllidaceae; 46. Bigoniaceae; 47. Brassicaceae; 48. Caprifoliaceae; 49. Celastraceae; 50. Flacourtiaceae; 51. Hippocrateaceae; 52. Iridaceae; 53. Lythraceae; 54. Meliaceae; 55. Piperaceae; 56. Scrophulariaceae; 57. Bombacaceae; 58. Commelinaceae; 59. Dioscoreaceae; 60. Geraniaceae; 61. Nymphaeaceae; 62. Ochnaceae; 63. Salvadoraceae; 64. Valerianaceae; 65. Violaceae.

Perusal of information presented in Table-II, Appendix-II, IV and Fig.3 clearly indicates that amongst the dicotyledonous families of root drug plants: Fabaceae (98 spp.) are the most dominant family followed by Asteraceae (70 spp.), Ranunculaceae (33 spp.), Euphorbiaceae (32 spp.), Acanthaceae (31 spp.), Rubiaceae (30 spp.), Malvaceae (27 spp.), Asclepiadaceae (20 spp.), Apocynaceae (18 spp.), Cucurbitaceae (18 spp.), Polygonaceae (17 spp.), Verbenaceae (17 spp.), Apiaceae (16 spp.), Boraginaceae (15 spp.), Amarantaceae (14 spp.), Rutaceae (14 spp.), Solanaceae (14 spp.), Lamiaceae (13 spp.), Convolvulaceae (12 spp.), Rosaceae (12 spp.), Tiliaceae (12 spp.), Capparidaceae (11 spp.), Moraceae (11 spp.), Myrsinaceae (10 spp.),

Menispermaceae (9 spp.), Urticaceae (8 spp.), Aristolochiaceae (7 spp.), Gentianaceae (7 spp.), Vitaceae (7 spp.), Aizoaceae (6 spp.), Berberidaceae (6 spp.), Leeaceae (6 spp.), Melastomaceae (6 spp.), Sterculiaceae (6 spp.), Rhamnaceae (5 spp.), Saxifragaceae (5 spp.), Bignoniaceae (4 spp.), Brassicaceae (4 spp.), Caprifoliaceae (4 spp.), Celastraceae (4 spp.), Flacourtiaceae (4 spp.), Hippocrateaceae (4 spp.), Lythraceae (4 spp.), Meliaceae (4 spp.), Piperaceae (4 spp.), Scrophulariaceae (4 spp.), Bombacaceae (3 spp.), Geraniaceae (3 spp.), Nymphaeaceae (3 spp.), Ochnaceae (3 spp.), Salvadoraceae (3 spp.), Valerianaceae (3 spp.), Violaceae (3 spp.), etc. Also, it is evident that the size of families naturally has a relationship with the above statistics.

So far as monocots are concerned the predominant families are those of Liliaceae (14 genera, 26 spp.), Araceae (11 genera, 15 spp.), Zingiberaceae (8 genera, 15 spp.), Orchidaceae (11 genera, 15 spp.), Poaceae (10 genera, 10 spp.), Cyperaceae (4 genera, 8 spp.), Arecaceae (6 genera, 7 spp.), Amaryllidaceae (3 genera, 4 spp.), Iridaceae (2 genera, 4 spp.), Commelinaceae (3 genera, 3 spp.), Dioscoreaceae (1 genus, 3 spp.), etc. (Table-III; Appendix-II).

The relative percentage in taxa of predominant dicotyledonous families varies from 10.69 (Fabaceae) to 00.33 (Bombacaceae, Geraniaceae, Nymphaeaceae, Ochnaceae, Salvadoraceae, Valeriaceae, Violaceae). And for monocots, it is 2.85 (Liliaceae) to 00.33 (Commelinaceae, Dioscoreaceae).

As a matter of fact, the predominantly employed dicotyledonous genera of root drug plants are: *Aconitum* (11 spp.), *Desmodium* (9 spp.), *Ficus* (8 spp.), *Cassia* (8 spp.), *Grewia* (8 spp.), *Solanum* (8 spp.), *Ardisia* (7 spp.), *Leea* (6 spp.), *Thalictrum* (6 spp.), *Trichosanthes* (6 spp.), *Argyreia* (5 spp.), *Clerodendrum* (5 spp.), *Crotalaria* (5 spp.), *Euphorbia* (5 spp.), *Flemingia* (5 spp.), *Hibiscus* (5 spp.), *Indigofera* (5 spp.), *Potentilla* (5 spp.), *Sida* (5 spp.), *Abelmoschus* (4 spp.), *Achyranthes* (4 spp.), *Amaranthus* (4 spp.), *Aristolochia* (4 spp.), *Berberis* (4 spp.), *Cirsium* (4 spp.), *Clematis* (4 spp.), *Delphinium* (4 spp.), *Mimosa* (4 spp.), *Phyllanthus* (4 spp.), *Salacia* (4 spp.), *Swertia* (4 spp.), *Acacia* (3 spp.), *Aerva* (3 spp.), *Anemone* (3 spp.), *Arnebia* (3 spp.), *Artemisia* (3 spp.), *Astragalus* (3 spp.), *Barleria* (3 spp.), *Bauhinia* (3 spp.), *Bergenia* (3 spp.), *Blumea* (3 spp.), *Casearia* (3 spp.), *Capparis* (3 spp.), *Cleome* (3 spp.), *Cynoglossum* (3 spp.), *Geranium* (3 spp.), *Inula* (3 spp.), *Jatropha* (3 spp.), *Lepidagathis* (3 spp.), *Melothria* (3 spp.), *Piper* (3 spp.), *Pogostemon* (3 spp.), *Premna* (3 spp.), *Ruellia* (3 spp.), *Rumex* (3 spp.), *Sambucus* (3 spp.), *Saussurea* (3 spp.), *Sonchus* (3 spp.), *Stephania* (3 spp.), *Tephrosia* (3 spp.), *Trianthema* (3 spp.), *Uraria* (3 spp.), *Zizyphus* (3 spp.), *Abrus* (2 spp.), *Acalypha* (2 spp.), *Alysicarpus* (2 spp.), *Ampelocissus* (2 spp.),

Andrographis (2 spp.), *Arctium* (2 spp.), *Aster* (2 spp.), *Atylosia* (2 spp.), *Blepharis* (2 spp.), *Bupleurum* (2 spp.), *Butea* (2 spp.), *Caesalpinia* (2 spp.), *Calotropis* (2 spp.), *Callicarpa* (2 spp.), *Canavalia* (2 spp.), *Carissa* (2 spp.), *Cayratia* (2 spp.), *Chasalia* (2 spp.), *Cichorium* (2 spp.), *Cissus* (2 spp.), *Citrus* (2 spp.), *Corchorus* (2 spp.), *Croton* (2 spp.), *Cyclea* (2 spp.), *Cyclista* (2 spp.), *Dalbergia* (2 spp.), *Datura* (2 spp.), *Dipteracanthus* (2 spp.), *Echinops* (2 spp.), *Ehretia* (2 spp.), *Elaeagnus* (2 spp.), *Embelia* (2 spp.), *Eranthemum* (2 spp.), *Erythrina* (2 spp.), *Ferula* (2 spp.), *Gardenia* (2 spp.), *Gentiana* (2 spp.), *Girardinia* (2 spp.), *Glycosmis* (2 spp.), *Gmelina* (2 spp.), *Gossypium* (2 spp.), *Hedyotis* (2 spp.), *Helianthus* (2 spp.), *Heliotropium* (2 spp.), *Heracleum* (2 spp.), *Heterostemma* (2 spp.), *Impatiens* (2 spp.), *Ixora* (2 spp.), *Justicia* (2 spp.), *Lagerstroemia* (2 spp.), *Launaea* (2 spp.), *Lantana* (2 spp.), *Maerua* (2 spp.), *Mallotus* (2 spp.), *Marsdenia* (2 spp.), *Momordica* (2 spp.), *Morinda* (2 spp.), *Moringa* (2 spp.), *Mucuna* (2 spp.), *Murraya* (2 spp.), *Mussaenda* (2 spp.), *Nymphaea* (2 spp.), *Osbeckia* (2 spp.), *Pentanema* (2 spp.), *Plantago* (2 spp.), *Pluchea* (2 spp.), *Plumbago* (2 spp.), *Plumeria* (2 spp.), *Podophyllum* (2 spp.), *Polygala* (2 spp.), *Rauvolfia* (2 spp.), *Rheum* (2 spp.), *Rhynchosia* (2 spp.), *Rivea* (2 spp.), *Rubus* (2 spp.), *Salvadora* (2 spp.), *Selinum* (2 spp.), *Sonerila* (2 spp.), *Sphaeranthus* (2 spp.), *Stereospermum* (2 spp.), *Strycnos* (2 spp.), *Tamarix* (2 spp.), *Tanacetum* (2 spp.), *Thespepsia* (2 spp.), *Tragopogon* (2 spp.), *Triumfetta* (2 spp.), *Tylophora* (2 spp.), *Urena* (2 spp.), *Valeriana* (2 spp.), *Ventilago* (2 spp.), *Vernonia* (2 spp.), *Viola* (2 spp.), *Wrightia* (2 spp.). Likewise, these for monocots are: *Smilax* (6 spp.), *Cyperus* (5 spp.), *Asparagus* (4 spp.), *Hedychium* (4 spp.), *Alocasia* (3 spp.), *Arisaema* (3 spp.), *Chlorophytum* (3 spp.), *Curcuma* (3 spp.), *Dioscorea* (3 spp.), *Dracaena* (3 spp.), *Iris* (3 spp.), *Curculigo* (2 spp.), *Pandanus* (2 spp.), *Phoenix* (2 spp.), *Roscoea* (2 spp.) and *Zingiber* (2 spp.) whereas these for Pteridophytes are: *Adiantum* (2 spp.), *Cheilanthes* (2 spp.) and *Lygodium* (2 spp.) (Appendix-III). All these genera hold high potential for the formulation of various drugs by pharmaceutical houses.

Thus the information complied provides enough clues to various pharmaceutical industries and research organizations engaged in biological screening programs on plant products for extracting active compounds for developing drug molecules against various ailments.

❏❏❏

6

Literature Cited

Agarwal, V.S. 1985. *Drug Plants of Indian: Root Drugs*. Kalyani Publishers, Delhi.

Agarwal, V.S. 1986. *Economic Plants of India.* Bishen Singh Mahendra Pal Singh, Dehra Dun, India.

Agarwal, V.S. 2003. *Directory of Indian Economic Plants*. Bishen Singh Mahendra Pal Singh, Dehra Dun, India.

Agharkar, S.P. 1991. *Medicinal Plants of Bombay Presidency*. Scientific Publishers, Jodhpur.

Ahuja, B.S. 1965. *Medicinal Plants of Saharanpur*. Gurukul Kangri Vishwavidyalaya, Hardwar

Ainslie, W. 1813. *Materia Medica of Hindoostan*. Neeraj Publishing House, New Delhi.

Ainslie, W. 1926. *Materia Indica*. Neeraj Publishing House, Delhi.

Alagesaboopathi, C., Dewarakan, P. & Balu, S. 2000. Plants used as medicine by tribals of Shevaroy Hills, Tamil Nadu: 391-393. In Maheshwari, J.K. (*ed.*) *Ethnobotany and Medicinal Plants of Indian Subcontinent*. Scientific Publishers, Jodhpur, India.

Ambasta, S.P. (*ed.*) 1986. *The Useful Plants of India.* C.S.I.R., New Delhi.

Ambasta, S.P., Ramachandran, K., Kashyapa, K. & Chand, R. 1992. *The Useful Plants of India.* Publications & Information Directorate, C.S.I.R., New Delhi.

Aminuddin & Girach, R.D. 1991. Ethnobotanical studies on Bondo tribe of district Koraput (Orissa). *Ethnobotany* **3** (1 & 2): 15-19.

Aminuddin & Girach, R.D. 1993. Observations on ethnobotany of the Bhunjia- a tribe of Sonabera plateau. *Ethnobotany* **5**: 83-86.

Aminuddin & Girach, R.D. 1996. Native phytotherapy Among the Paudi Bhuinya of Bohai hills. *Ethnobotany* **8**: 66-70.

Anonymous, 1948-1976. *The Wealth of India: Raw Materials*, I-XI. Publications & Information Directorate, C.S.I.R., New Delhi.

Anonymous, 1968. *Medicinal Plants of India*. Vol.- I. I.C.M.R. New Delhi.

Anonymous, 1981. *Medicinal Plants of Kerala Forest*. Kerala Forest Research Institute, Peechi, Kerala.

Anonymous, 1982. *Supplement to Cultivation and Utilization of Medicinal Plants*. R.R.L. Jammu.

Anonymous, 1990. *Use of Traditional Medicine in Primary Health Care, a Manual for Health Workers in South East Asia*. World Health Organisation, South-East Asia Regional Office, Delhi.

Anonymous, 1993. *Chemistry of Medicinal Plants Part I*. Central Council for Research in Unani Medicine, Delhi.

Anonymous, 1999. *Medicinal Plants of Andhra Pradesh*. CCRUM, New Delhi.

Ansari, A.A. 1991. Ethnobotanical notes on some plants of Khirsu, Pauri Garhwal,U.P. *Ethnobotany* **3** (1 & 2): 105-106.

Arya, K.R. & Prakash, V. 2000. Ethnobotanical study of a remote tribal area of Almora district: a survey report-Part-1: 247-251. In Maheshwari, J. K. (*ed.*) *Ethnobotany and Medicinal Plants of Indian Subcontinent*. Scientific Publishers, Jodhpur, India.

Asolkar, L.V. 1992. *Glossary of Indian Medicinal Plants with Active Principles, Second Supplement*. Council of Scientific & Industrial Research Publications & Information Directorate, Delhi.

Asolkar, L.V., Kakkar, K.K. & Chakre, O.J. 1992. *Second Supplement to Glossary of Indian Medicinal Plants with Active Principles* Part-I (A-K) (1965-1981). Publications & Information Directorate (C.S.I.R.), New Delhi.

Aswal, B.S. 1996. Conservation of ethnomedicinal plants diversity of Garhwal Himalaya in India: 133-135. In Jain, S.K. (*ed.*) *Ethnobiology in Human Welfare*. Deep Publications, New Delhi.

Aswal, B.S. 2001. Zingiberaceae of Western Himalaya: 101-118. In Pande, P.C. & Samant, S.S. (*eds.*) *Plant Diversity of the Himalaya*. Gyanodhya Prakashan, Nainital, India.

Aswal, B.S & Mehrotra, B.N. 1994. *Flora of Lahaul- Spiti (A Cold Desert in North-West Himalaya)*. Bishen Singh Mahendra Pal Singh, Dehra Dun, India.

Atal, C.K. 1982. *Cultivation and Utilization of Medicinal Plants*. Regional Research Laboratory (CSIR), Jammu, Tawi.

Atkinson, E.T. 1982. *Economic Botany of the Himalayan Region*. Cosmos Pub., New Delhi.

Augustine, Jony & Sivadasan, M. 2004. Ethnomedicinal Plants of Periyar Tiger Reserve, Kerala, India. *Ethnobotany* **16**: 44-49.

Baburaj, D., Suresh, Britto, S.J., Mathew, G. K. & Rajan, S. 2000. Cultivated medicinal plants useful in homoeopathy found in Nilgiri district, Tamil Nadu: 381-389. In Maheshwari, J.K. (*ed.*) *Ethnobotany and Medicinal Plants of Indian Subcontinent.* Scientific Publishers, Jodhpur, India.

Bal, S.N. 1940. *Outline of Pharmacopoeial Drugs of Vegetable Origin.* P. Gosh & Co., Calcutta.

Balasubramanian, P. & Prasad, S.N. 1996. Medicinal plants among the Irulars of Attapady and Boluvampatti forest in the Nilgiri biosphere reserve. *J. Econ. Taxo. Bot.* Additional Series **12**: 253-259.

Ballabh, B. & Chaurasia, O.P. 2006. Ethnobotanical studies on Boto tribe in Ladakh. *Ethnobotany* **18**: 87-95.

Banerjee, A. 2000. Ethnobotany of a few plant species in the eroded soil of Birbhum, West Bengal: 527-530. In Maheshwari, J.K. (*ed.*) *Ethnobotany and Medicinal Plants of Indian Subcontinent.* Scientific Publishers, Jodhpur, India.

Barua, K.N., Barua, I.C. & Das, M. 2000. Ethnobotany of Rajbanshis of Assam: 609-614. In Maheshwari, J.K. (*ed.*) *Ethnobotany and Medicinal Plants of Indian Subcontinent.* Scientific Publishers, Jodhpur, India.

Beg, M.J., Beg M.Z. & Ali, S.J. 2005. Folklore treatment of some common ailments in Eastern U.P: 147-177. In Khan, I.A. & Khanum, A. (*eds.*) *Role of Biotechnology in Medicinal and Aromatic Plants.* Vol.-XII. Ukaaz Publications, Hyderabad-36, India.

Behera, S. & Rout, N.C. 2003. Diversity of Zingiberaceae plants from Orissa : 546-550. In Singh, V. & Jain, A.P. (*eds.*) *Ethnobotany and Medicinal Plants of India and Nepal.* Scientific Publishers, Jodhpur, India.

Beigh, S.Y., Nawchoo, I.A. & Iqbal, M. 2003. Plants in traditional medicine of Kashmir Himalaya: 99-104. In Singh, V. & Jain, A.P. (*eds.*) Vol.-I. *Ethnobotany and Medicinal Plants of India and Nepal.* Scientific Publishers, Jodhpur, India.

Bennet, S.S.R. 1986. *Name Changes in Flowering Plants of India and Adjacent Regions.* Triseas Publishers. Dehra Dun, India.

Bhalla, N.P. & Bhalla, S. 2002. Ethnomedicinal and ecophysiological studies in Fabaceae (Leguminosae): 255-278. In Trivedi, P.C. (*ed.*) *Ethnobotany.* Aavishkar Publishers & Distributers, Jaipur (Raj.), India.

Bhandary, M.J., Chandrashekar, K.R. & Kaveriappa, K.M. 1996. Ethnobotany of Gowlis of Uttara Kannada district, Karnataka. *J.Econ. Taxo. Bot. Additional Series* **12**: 244-249.

Bhardwaj, S. & Gakhar, S.K. 2003. Ethnomedicinal plants used by the tribals of Mizoram to cure dysentery. *Ethnobotany* **15**: 51-54.

Bhatt, D.C. 2002. Studies on some ethnobotanical plants from Saurashtra: 119-127. In Trivedi, P.C. (*ed.*) *Ethnobotany*. Aavishkar Publishers & Distributers, Jaipur (Raj.), India.

Bhatt, D.C., Mehta, S.K. & Mitaliya, K.D. 1999. Ethnomedicinal plants of Shetrunjaya hill of Palitana, Gujarat. *Ethnobotany* **11**: 22-25.

Bhatt, D.C. & Mitaliya, K.D. 1999. Ethnomedicinal plants of Victoria Park (Reserve Forest) of Bhavnagar, Gujarat, India. *Ethnobotany* **11**: 81-84.

Bhatt, D.C., Mitaliya, K.D., Mehta, S.K. & Joshi, P.N. 2002. Note on some ethnomedicinal plants of Pachham Hills of Kachh. *Ethnobotany* **14**: 34-35.

Bhatt, D.C., Jadeja, B.A., Odedra, N.K. & Baxi, U.S. 2003. Enumeration of wild plants used as anti-diabetic in Barda Hills of Gujarat, India: 897-903. In Singh, V. & Jain, A.P. (eds.) Vol.-II. *Ethnobotany and Medicinal Plants of India and Nepal*. Scientific Publishers, Jodhpur, India.

Bhattacharjee, S.K. 2001. *Handbook of Medicinal Plants*. Pointer Publishers, Jaipur, India.

Bhogaonkar, P.Y. & Devarkar, V.D. 2002. Some unique ethnomedicinal plants of Korkus of Melghat Tiger Reserve (Maharashtra). *Ethnobotany* **14**: 34-35.

Billore, K.V., Joseph, T.G. & Dave, S.K. 1998. Interesting folk remedies by the 'Lok Vaidyas' of Rajasthan for 'Swas Roga'. *Ethnobotany* **10**: 42-45.

Binu, S., Shanavaskhan, A.E., Santoshkumar, E.S., & Pushpangadan, P. 2003. Plants used as medicine by the Irulas of Palghat district, Kerala, India: 808-814. In Singh, V. & Jain, A.P. (*eds.*) Vol.-II. *Ethnobotany and Medicinal Plants of India and Nepal*. Scientific Publishers, Jodhpur, India.

Bisht, A. & Jain, S.P. 2006. Review of Ethnobotanical studies of genus Rubus (Rosaceae) from North-Western Himalayas. *Ethnobotany* **18**: 127-130.

Biswas, K. 1959. *Cultivation of Medicinal Plants*. Director of Publicity. Calcutta.

Biswas, K. & Chopra, R.N. 1982. *Common Medicinal Plants of Darjeeling and the Sikkim Himalayas*. Periodical Expert Book Agency, Delhi.

Bondya, S.L., Khanna, K.K. & Singh, K.P. 2006. Ethnomedicinal uses of leafy vegetables from the tribal folk-lore of Achanakmar-Amarkantak Biosphere Reserve (Madhya Pradesh and Chhattisgarh). *Ethnobotany* **18**: 145-148.

Bondya, S.L. & Sharma, H.P. 2004. Ethnobotanical studies on plants used in diabetes (Madhumeha) under the Baragora Block of Jharkhand. *Ethnobotany* **16**: 139-140.

Borthakur, S.K. 1993. Native phytotherapy for child and women diseases from Assam. *Ethnobotany* **5**: 87-91.

Borthakur, S.K., Choudhury, B.T. & Gogoi, R. 2004. Folklore hepato-protective herbal recipes from Assam in Northeast India. *Ethnobotany* **16**: 76-82.

Bose, K.C. 1932. *Pharmacopoeia Indica, Being a Collection of Vegetable, Mineral and Animal Drugs in Common Use in India.* Bishen Singh Mahendra Pal Singh, Dehra Dun.

Brahmam, M. & Saxena, H.O. 1990. Ethnobotany of Gandhamardan hills-some noteworthy folk-medicinal uses. *Ethnobotany* **2**: 71-79.

Caius, J.F. (*ed.*) 2003. *The Medicinal and Poisonous Plants of India.* Scientific Publishers, Jodhpur, India.

Chakraborty, M.K., Bhattacharjee, A. & Pal, D.C. 2003. Ethnomedicinal uses of some exotic plants of Purulia District, West Bengal, India: 559-563. In Singh, V. & Jain, A.P. (*eds.*) Vol.-II. *Ethnobotany and Medicinal Plants of India and Nepal.* Scientific Publishers, Jodhpur, India.

Chandrasekar, K. & Srivastava, S. K. 2003. Ethnomedicinal studies in Pin valley national park, Lahaul-Spiti, Himachal Pradesh. *Ethnobotany* **15**: 44-47.

Chatterjee, A. & Pakrashi, S.C. (*ed.*) 1997. *The Treatise on Indian Medicinal Plants.* Vol.1-5. C.S.I.R., New Delhi.

Chaudhari, U.S. & Hutke, V. 2002. Ethno-medico-botanical information on some plants used by Melghat tribes of Amravati district, Maharashtra. *Ethnobotany* **14**: 100-102.

Chaudhury, D. & Neogi, B. 2000. Ethnobotany of Khasi and Chakma tribes of North East India: 583-589. In Maheshwari, J.K. (*ed.*) *Ethnobotany and Medicinal Plants of Indian Subcontinent.* Scientific Publishers, Jodhpur, India.

Chauhan, N.S. 1999. *Medicinal and Aromatic Plants of Himachal Pradesh.* Indus Publishing Company, New Delhi.

Chaurasia, O.P., Singh, B. & Sareen, S.K. 2000a. Ethnomedicinal plants of arctic desert-Ladakh used in veterinary practices: 161-162. In Maheshwari, J.K. (*ed.*) *Ethnobotany and Medicinal Plants of Indian Subcontinent.* Scientific Publishers, Jodhpur, India.

Chaurasia, O.P., Singh, B. & Sareen, S.K. 2000b. Ethno-medico-botanical survey of Nubra Valley: 167-172. In Maheshwari, J.K. (*ed.*) *Ethnobotany and Medicinal Plants of Indian Subcontinent.* Scientific Publishers, Jodhpur, India.

Chhetri, R.B. 1994. Further observations on ethnomedicobotany of Khasi hills in Meghalaya, India. *Ethnobotany* **6**: 33-36.

Chhetri, D.R. 2005. Ethnomedicinal plants of the Khangchendzonga national park, Sikkim, India. *Ethnobotany* **17** (1&2): 96-103.

Chopra, I.C. 1960. *Medicinal Plants of Arid Zone.* Today & Tomorrow's Printers & Publishers, Delhi.

Chopra, R.N. 1932. *Medicinal and Economic Aspects of Some Indian Medicinal Plants.* Patna University, Patna.

Chopra, R.N. 1933. *Indigenous Drugs of India.* The Art Press, Kolkata.

Chopra, R.N. & Chopra, I.C. 1955. *Review of Work on Indian Medicinal Plants (Including Indigenous Drugs and Poisonous Plants).* Indian Council of Medicinal Research, Delhi.

Chopra, R.N., Nayer, S.L. & Chopra, I.C. 1956. *Glossary of Indian Medicinal Plants.* Council of Scientific & Industrial Research, New Delhi.

Chopra, R.N., Chopra, I.C. & Verma, B.S. 1969. *Supplement to Glossary of Indian Medicinal Plants.* Publications & Information Directorate Hillside Road, New Delhi.

Chowdhury, J.U., Alam, M.K. & Hassan, M.A. 2001. Some traditional folk formularies against dysentery and diarrhoea in Bangladesh. *J. Econ. Taxon. Bot.* Additional Series. **12**: 20-23.

Chunckar, K.C. 1972. *Glossary of Vegetable Drugs in Brhattrayi.* Chowkhamba Sanskrit Series Office, Varanasi.

Dagar, H.S. & Dagar, J.C. 2000. Plant folk medicines for gynaecological, urino-genital and other related problems among aborigines of Andaman and Nicobar Islands: 561-567. In Maheshwari, J.K. (*ed.*) *Ethnobotany and Medicinal Plants of Indian Subcontinent.* Scientific Publishers, Jodhpur, India.

Das, A.K. & Sharma, G.D. 2003. Ethno-medicinal uses of Plants by Manipuri and Barman communities of cachar district, Assam: 421-429. In Singh, V. & Jain, A.P. (eds.) Vol.-I. *Ethnobotany and Medicinal Plants of India and Nepal.* Scientific Publishers, Jodhpur, India.

Dash, S.S., Maiti, A. & Rai, S.K. 2003. Traditional uses of plants among the urban population of Gangtok-Sikkim: 317-324. In Singh, V. & Jain, A.P. (*eds.*) Vol.- I. *Ethnobotany and Medicinal Plants of India and Nepal.* Scientific Publishers, Jodhpur, India.

Dash, S.S. & Mishra, M.K. 2000. Taxonomic survey and systematic census of economic plants of Narayanapatna hills of Koraput district, Orissa: 473-498. In Maheshwari, J.K. (ed.) *Ethnobotany and Medicinal Plants of Indian Subcontinent.* Scientific Publishers, Jodhpur, India.

Dastur, J.F. 1970. *Medicinal Plants of India and Pakistan.* D.B. Taraporevala Sons & Co., Naoroji Road, Bombay.

Datta, S.C. 1949. *Pharmacognosy of Indian Root and Rhizome Drugs.* Central Council for Research in Ayurveda & Siddha, New Delhi.

Datta, S.C. & Mukerji, B. 1952. *Pharmacognosy of Indian Leaf Drugs.* Government of India Press, Calcautta.

Deshmukh, V.R., Muratkar, G.D. & Rothe, S.P. 2000. Preliminary observation on the medicinal and economically important leguminous plant species from Amravati Tehsil: 283-289. In Maheshwari, J.K. (*ed.*) *Ethnobotany and Medicinal Plants of Indian Subcontinent.* Scientific Publishers, Jodhpur, India.

Deshmukh, V.R. & Rothe, S.P. 2003. Ethnomedicinal study from Melghat tribal regions of Amravati District, Maharashtra: 582-584. In Singh, V. & Jain, A.P. (*eds.*) Vol.-II. *Ethnobotany and Medicinal Plants of India and Nepal.* Scientific Publishers, Jodhpur, India.

Dey, K.L. 1973. *The Indigenous Drugs of India: Short Descriptive Notices of the Principle Medicinal Products Met within British India.* Pama Primlane, The Chronica Botanica, New Delhi, India.

Dhawan, B.N. 1986. *Current Research on Medicinal Plants.* Indian National Science Academy, Delhi.

Dhiman, A.K. 2003. *Sacred Plants and Their Medicinal Uses.* Daya Publishing House, Delhi.

Drury, C.H. 1978. *The Useful Plants of India: With Notices of Their Chief Value in Commerce, Medicine, and The Arts.* International Book Distributers, Dehra Dun, India.

Duhoon, S.S. & Tripathi, A.K. 2003. Medicinal and curative properties of Sesame (*Sesamum indicum* L.): 20-28. In Singh, V. & Jain, A.P. (*eds.*) Vol.-I. *Ethnobotany and Medicinal Plants of India and Nepal.* Scientific Publishers, Jodhpur, India.

Dutt, N.B. 1928. *Commercial Drugs of India.* Bishen Singh Mahendra Pal Singh, Dehra Dun.

Dutt, U.C. 1877. *Materia Medica of the Hindus, with a glossary of Indian Plants.* Sri Satguru Publications, Delhi.

Dutta, A.C. 1985. *Dictionary of Economic and Medicinal Plants.*

Dutta, M. & Nath, S.C. 2000. Ethno-medico-botany of the Tai-Ahomas of Assam. India: 591- 598. In Maheshwari, J.K. (*ed.*) *Ethnobotany and Medicinal Plants of Indian Subcontinent.* Scientific Publishers, Jodhpur, India.

Dymock, W. 1883. *Vegetable Materia Medica of Western India.* Educ. Sco. Press, Bombay.

Dymock, W., Warden, C.J.H. & Hooper, D. 1889-1893. *Pharmacographia Indica.* 3 Vols. Thakur, Spink & Co., Calcutta.

Dwivedi, A.P. 1993. *Forests: The Non-Wood Resources.* Intenational Book Distributers, Dehra Dun, India.

Dwivedi, S.N. 2003. Ethnobotanical studies and conservational strategies of wild and natural resources of Rewa district of Madhya Pradesh: 233-244. In Singh, V. & Jain, A.P. (eds.) Vol.-2. *Ethnobotany and Medicinal Plants of India and Nepal.* Scientific Publishers, Jodhpur, India.

Farnsworth, N. 1998. Screening Plants for New Medicines: 83-87. In Wilson, E.O. (*ed.*) *Biodiversity.* National Acadamy Press, Washington.

Farnsworth, N.R. & Soejarto, D.D. 1991. Global Importance of Medicinal Plants: 25-47. In Akerele, O. Heywood. V. & Synge, H. (*eds.*) *Conservation of Medicinal Plants.* Cambridge University Press, Cambridge.

Farooqi, A.A. 1993. *Production Technology of Medicinal and Aromatic Crops.* Natural Remedies Pvt. Ltd., Bangalore.

Farooqi, A.A. 2001. *Cultivation of Medicianl and Aromatic Crops.* University Press Hyderabad.

Fleming, J. 1810. *A Catalogue of Indian Medicianl Plants and Drugs.*

Ganai, K.A. & Nawchoo, A.I. 2003. Traditional treatment of toothache by the Gujjar and Bakerwal tribes of Kashmir in India: 105-107. In Singh, V. & Jain, A.P. (*eds.*) Vol.-I. *Ethnobotany and Medicinal Plants of India and Nepal.* Scientific Publishers, Jodhpur, India.

Ghosh, S. & Sensarma, P. 1997. Ethnomedicine to modern medicine: an observational study in some villages of West Bengal. *Ethnobotany* 9: 80-84.

Girach, R.D. 1992. Medicinal plants used by Kondh tribe of district Phulbani (Orissa) in Eastern India. *Ethnobotany* **4** (1 & 2): 53-66.

Girach, R.D. 2001. Ethnobotanical notes on some plants of Mahendragiri hills, Orissa. *Ethnobotany* **13**: 80-83.

Girach, R.D. & Aminuddin. 1995. Ethnomedicinal uses of plants among the tribals of Sing Bhum district, Bihar, India. *Ethnobotany* 7: 103-107.

Girach, R.D., Aminuddin, Brahmam, M. & Misra, M.K. 1997. Observations on Ethnomedicinal Plants of Bhadrak District, Orissa, India. *Ethnobotany* 9: 44-46.

Girach, R.D., Singh, S., Brahmam, M. & Misra, M.K. 2000. Traditional treatment of skin diseases in Bhadrak District, Orissa: 499-504. In Maheshwari, J.K. (*ed.*) *Ethnobotany and Medicinal Plants of Indian Subcontinent.* Scientific Publishers, Jodhpur, India.

Godbole, S.R. 1966. *Glossary of Vegetable Drugs in Vagbhata.* Indian Drugs Research Association, Pune.

Goel, A.K. & Rajendran. 2000. A cross-cultural ethnobotanical studies of Santhal Pargana (Eastern India) and Western Ghats (Southern India): 147-150. In Maheshwari, J.K. (*ed.*) *Ethnobotany and Medicinal Plants of Indian Subcontinent.* Scientific Publishers, Jodhpur, India.

Gogoi, R. & Borthakur, S.K. 2001. Notes on herbal recipes of Bodo tribe in Kamrup district, Assam. *Ethnobotany* **13**: 15-23.

Gogoi, R. & Das, M.K. 2003. Observations on some weeds of medicinal importance in the Brahmaputra valley of Assam: 434-441. In Singh, V. & Jain, A.P. (*eds.*) Vol.-1. *Ethnobotany and Medicinal Plants of India and Nepal.* Scientific Publishers, Jodhpur, India.

Gogoi, R., Bokolial, D. & Hazarika, D. 2003a. Preliminary observation on the Medicinal Plants of Chandrapur Area of Kamrup District, Assam: 410-415. In Singh, V. & Jain, A.P. (*eds.*) Vol.-I. *Ethnobotany and Medicinal Plants of India and Nepal.* Scientific Publishers, Jodhpur, India.

Gogoi, J.C., Sharma, D.K. & Borthakur, S.K. 2003b. Folklore hepatoprotective herbal recipes from Assam in Northeastern India: 1-21. In Khan, S.S. (*ed.*) *Vistas in Ethnobotany* Vol.-1. Classic Printers & Publishers, Bhopal, India.

Gopal, G.V. & Chandra, K.S.J. 2003. Ethnobotany of Soligas of Biligirirangana hills, Karnataka, India: 640-647. In Singh, V. & Jain, A.P. (*eds.*) Vol.-II. *Ethnobotany and Medicinal Plants of India and Nepal*. Scientific Publishers, Jodhpur, India.

Gosh, S. & Sensarma 1997. Ethnomedicine to modern medicine: an observational study in some villages of West Bengal. *Ethnobotany* 9: 80-84.

Goud, P.S. & Pullaiah, T. 1996. Ethnobotany of Kurnool District, Andhra Pradesh : 410-412. In Jain, S.K. (*ed.*) *Ethnobiology in Human Welfare*. Deep Publications, New Delhi.

Goud, P.S., Pullaiah, T. & Murthy, K.S.R. 2000. Native phytotherapy for fever and malaria from Kurnool district, Andhra Pradesh: 337-340. In Maheshwari, J.K. (*ed.*) *Ethnobotany and Medicinal Plants of Indian Subcontinent*. Scientific Publishers, Jodhpur, India.

Goyal, H.R. 1998. *Medicinal Plants of Kachchh (Gujarat)*. CCRAS, New Delhi.

Greenish, H.G. 1999. *Materia Medica*. Scientific Publishers, Jodhpur, India.

Gupta, O.P., Srivastava, T.N., Gupta, S.C. & Badola, D.P. 1981. Ethnobotanical and phytochemical screening of high altitude plants of Ladakh. *Bull. Medico-Ethno. Bot. Res.* 2: 67-88.

Gupta, R. 1986. Prospects of Cultivation of Plants used in Ayurvedic Medicine into agricultural and agri-forestry systems in India. Proc. Silver Jublee Celebrations of Nehru Ayurvedic Medicinal Plants Garden & Herbarium, Pune (India): 321-333.

Hembrom, P.P. 1996. Contact therapy practised by Mundas of Chotanagpur (Bihar). *Ethnobotany* 8: 36-39.

Hembrom, P.P. & Goel, A.K. 2005. Horopathy: ethnomedicine of Mundas. *Ethnobotany* 17 (1 & 2): 89-95.

Hooker, J.D. 1872-1897. *The Flora of British India*. Vol. I-VII. Bishen Singh Mahendra Pal Singh, Dehradun, India.

Hosagoudar, V.B. & Henry, A.N. 1996a. Ethnobotany of Soligas in Biligiri Rangana Betla, Karnataka, Southern India. *J.Eco. Taxon. Bot.* Additional Series 12: 228-243.

Hosagoudar, V.B. & Henry, A.N. 1996b. Ethnobotany of tribes Irular, Kurumban and Paniyan of Nilgiris in Tamil Nadu, Southern India. *J.Eco. Taxon. Bot.* Additional Series 12: 272-283.

Husain, A. 1992. *Dictionary of Indian Medicinal plants*. Central Institute of Medicinal & Aromatic Plants, Lucknow.

Huxley, A. 1984. *Green Inheritance: The World Wildlife Fund Book of Plants*. Collins, London.

Islam, M. 1996. Ethnobotany of certain underground parts of plants of North-Eastern region, India. *J. Eco.Taxon. Bot.* Additional Series **12**: 338-343.

Islam, M.A. & Jha, R.K. 2003. Ethnopharmacology of tribals of Ranchi district in Jharkhand: 300-310. In Singh, V. & Jain, A.P. (*eds.*) Vol.-I. *Ethnobotany and Medicinal Plants of India and Nepal.* Scientific Publishers, Jodhpur, India.

Jadeja, B.A. 1999. Plants used by the tribe Rabari in Barda hills of Gujarat. *Ethnobotany* **11**: 42-46.

Jadhav, D. 2006. Ethno-medicinal survey of Maalgamdi in Ujjain district, Madhya Pradesh, India. *Ethnobotany* **18**:157-159.

Jain, A.K. & Patole, S.N. 2001. Less-known medicinal uses of plants among some tribal and rural communities of Pachmashi forest (M.P.). *Ethnobotany* **13**: 96-100.

Jain, S.K. 1968 (*ed.*). *Medicinal Plants.* National Book Trust, India.

Jain. S.K. 1987. *Manual of Ethnobotany.* Scientific Publishers, Jodhpur.

Jain, S.K. 1989 (*ed.*). *Methods and Approaches in Ethnobotany.* Society of Ethnobotanist. Lucknow.

Jain, S.K. 1991. *Dictionary of Indian Folk Medicine and Ethnnobotany.* Deep Publications, New Delhi.

Jain, S.K. 1995. Ethnobotanical diversity in Zingiber of India. *Ethnobotany* **7**: 83-88.

Jain, S.K. 2002. *Bibliography of Indian Ethnobotany.* Scientific Publishers, Jodhpur.

Jain, S.K. & Mudgal, V. 1999. *A Handbook of Ethnobotany.* Bishen Singh Mahendra Pal Singh, Dehra Dun.

Jain, S.K., Sinha, B.K. and Gupta, R.C. 1991. *Notable Plants in Ethnomedicine of India.* Deep Publications, New Delhi.

Jain, S.K., Fernandes, V.F., Lata, S., & Ayub, A. 1995. Indo-amazonian ethnobotanic connections-similar uses of some common plants. *Ethnobotany* **7**: 29-37.

Jain, S.K. & Filipps, De. 1991. *Medicinal Plants of India.* Reference Publications Inc., Algonae, Michigan.

Jain, S.K. & Goel, A.K. 2005. Some Indian plants in Tibetan traditional medicine-[2]. *Ethnobotany* **17**: 127-137.

Jain, S.K., Mudgal, V., Banerjee, D.K., Guha, A. & Das, D. 1984. *Bibliography of Ethnobotany.* Bot. Surv. India, Calcutta: 1-157.

Jain, S.K. & Saklani, A. 1992. Cross-cultural ethnobotanical studies in Northeast India. *Ethnobotany* **4** (1& 2): 25-38.

Jain, S.K. & Sharma, P.P. 2000. A review of the parallel or unique indigenous uses of twenty five plants in India and some African countries. *Ethnobotany* **12**: 51-55.

Jain, S.K. & Sikarwar, R.L.S. 1998. Some Indian plants used in Latin American ethnomedicine. *Ethnobotany* **10** (1 & 2): 61-65.

Jain, S.K., Sinha, B.K. & Gupta, R.C. 1991. *Notable Plants in Ethnomedicine of India.* Deep Publications, New Delhi.

Jain, S.P. 1996. Ethno-medico-botanical survey of Chaibasa, Singhbhum District, Bihar. *J. Eco.Taxon. Bot.* Additional Series **12**: 403-407.

Jain, S.P. 2003. Comparative ethno-medico-botanical studies of tribes of Neterhat plateau, Chhotanagpur and Singhbhum district of Jharkhand, India: 295-299. In Singh, V. & Jain, A.P. (*eds.*) Vol.-I. *Ethnobotany and Medicinal Plants of India and Nepal.* Scientific Publishers, Jodhpur, India.

Jain, S.P. & Singh, S.C. 1997. An ethno-medico-botanical survey of Ambikapur district, M.P: 83-91. In Jain, S.K. (*ed.*) *Contribution to Indian Ethnobotany*, 3rd ed. Scientific Publishers, Jodhpur, India.

Jain, S.P., Singh, J. & Singh, S.C. 2003. Rare and endangered medicinal and aromatic plants of Madhya Pradesh: 925-932. In Singh, V. & Jain, A.P. (*eds.*) Vol.-I. *Ethnobotany and Medicinal Plants of India and Nepal.* Scientific Publishers, Jodhpur, India.

Jain, S.P., Singh, S.C., Singh, J. & Kumar, S. 2003. Ethno-medico-botanical survey of Raipur district, Chhatisgarh state: 266-271. In Singh, V. & Jain, A.P. (*eds.*) Vol.-I. *Ethnobotany and Medicinal Plants of India and Nepal.* Scientific Publishers, Jodhpur, India.

Jamir, N.S. 1997. Ethnobiology of Naga tribe in Nagaland: II-Medicinal Herbs. *Ethnnobotany* 9: 101-104.

Jamir, N.S. & Rao, R.R. 1990. Fifty new or interesting medicinal plants used by the Zeliangs of Nagaland (India). *Ethnobotany* **2** (1 & 2): 11-18.

Jha, B.J. & Kumar, M. 2003. Herbal remedies for various diseases: 78-87. In Singh, V. & Jain, A.P. (*eds.*) Vol.- I. *Ethnobotany and Medicinal Plants of India and Nepal.* Scientific Publishers, Jodhpur, India.

Jha, R.R. & Verma, S.K. 1996. Ethnobotany of Sauria Paharias of Santhal Pargana, Bihar: I. Medicinal Plants. *Ethnobotany* 8: 31-35.

Jones, W. 1799. *Botanical Observations on Select Indian Plants.* London.

Joshi, S.G. 2000. *Medicinal Plants.* Oxford & IBH Publishers, New Delhi.

Kadel, C. & Jain, A.K. 2006. Plants used in ethnoveterinary practices in Jhabua District, Madhya Pradesh. *Ethnobotany* 18: 149-152.

Kala, C.P. & Manjrekar, N. 2000. Ethno-medico-botany of India trans-Himalaya: A case study from Spiti: 177-183. In Maheshwari, J.K. (*ed.*) *Ethnobotany and Medicinal Plants of Indian Subcontinent.* Scientific Publishers, Jodhpur, India.

Kala, C.P. & Rawat, G.S. 2001. Human use and conservation status of wild medicinal herbs in the Bhyundar valley, Western Himalaya: 547-529. In Pande, P.C. & Samant, S.S. (*eds.*). *Plant Diversity of the Himalaya.* Gyanodaya Prakashan, Nainital, India.

Kapoor, S.L. 1979. *Herbal Drugs in Indian Pharmaceutical Industry*. National Botanical Research Institute, Lucknow.

Kapur, S.K. 1996a. Traditionally important medicinal plants of Bhaderwah hills-Jammu Province-III. *J. Econ. Taxon. Bot.* Additional Series **12**: 62-68.

Kapur, S.K. 1996b. Traditionally important medicinal plants of Bhaderwah hills-Jammu Province-IV. *J. Econ. Taxon. Bot.* Additional Series **12**: 70-74.

Kapur, S.K. & Nanda, S. 1996. Traditionally important medicinal plants of Bhaderwah hills (Jammu Province) Part-II. *J. Econ. Taxon. Bot.* Additional Series **12**: 56-61.

Kapur, S.K., Nanda, S. & Srivastava, T.N. 1996. Ethnobotanical uses of RRL-Herbarium-III. *J. Econ. Taxon. Bot.* Additional Series **12**: 50-55.

Kapur, S.K. & Singh, P. 1996. Traditionally important medicinal plants of Udjampur district (Jammu Province) Part- I. *J. Econ. Taxon. Bot.* Additional Series **12**: 75-81.

Kapur, S.K. & Srivastava, T.N. 1996. Traditionally important medicinal plants of Udhampur district (Jammu Province) Part-II. *J. Econ. Taxon. Bot.* Additional Series **12**: 82-88.

Karnick, C.R. 1994. *Pharmacopoeial Standards of Herbal Plants*. Vol.-I. Sri Satguru Publications, Delhi, India.

Katark, N.N. 1903. *Materia Medica of Indian and Their Therapeutics*. Neeraj Publishing House, Delhi.

Katewa, S.S., Chaudhary, B.L., Jain, A.P. & Thakha, H.K. 2001.Some plants in folk medicine of Rajasamand District (Rajasthan). *Ethnobotany* **13**: 129-134.

Kaul, M.K. 1997. *Medicinal Plants of Kashmir and Ladakh (Temperate & Cold Arid Himalaya)*. Indus Publishing Company, New Delhi.

Kayal, R.N., Saha, S. & Ghoshal, P.P. 2003. A floristic account of Indian medicinal plants (originated, cultivated, naturalized or introduced) used in homoeopathic drug preparation: 60-67. In Singh, V. & Jain, A.P. (*eds.*) Vol.- I. *Ethnobotany and Medicinal Plants of India and Nepal*. Scientific Publishers, Jodhpur, India.

Khan, A.A., Khan, S. & Singh, M.P. 2003. Weeds of soyabeen (*Glycine max* (L.) Merrill) and their medicinal importance in Rewa division (M.P.), India: 245-248. In Singh, V. & Jain, A.P. (*eds.*) Vol.-I. *Ethnobotany and Medicinal Plants of India and Nepal*. Scientific Publishers, Jodhpur, India.

Khanna, K.K. 2002a. Unreported ethnobomedicinal uses of plants from the tribal and rural folklores of Gonda District, Uttar Pradesh. *Ethnobotany* **14**: 52-56.

Khanna, K.K. 2002b. Ethnobotany of five districts of Terai region, Uttar Pradesh: 128-145. In Trivedi, P.C. (*ed.*) *Ethnobotany* Aavishkar Publishers & Distributers, Jaipur (Raj.), India.

Khanna, K.K. & Kumar, R. 2000. Ethnomedicinal plants used by the Gujjar tribe of Saharanpur district, Uttar Pradesh. *Ethnnobotany.* **12** (1 & 2): 17-22.

Khanna, K.K., Shukla, G. & Mudgal, V. 1996. New traditional medicinal uses of plants from Jalaun district, Uttar Pradesh: 108-111. In Maheshwari, J.K. (*ed.*) *Ethnobotany and Medicinal Plants of Indian Subcontinent.* Scientific Publishers, Jodhpur, India.

Khanna, K.K., Srivastava, P.K. & Mudgal, V. 1996. Noteworthy medicinal plant uses form rural folklore of Raebareli district, Uttar Pradesh. *J. Econ. Taxon. Bot.* Additional Series **12**: 118-122.

Khare, C.P. 2004. *Encyclopedia of Indian Medicinal Plants, Rational Western Therapy, Ayurvedic and Other Traditional Usage, Botany.* Springer-Verlag, Berlin-Heidelberg, New York.

Khare, P.K. & Khare, L.J. 2000. Plants used in rheumatism by the rural people of Chhatapur district, Madhya Pradesh, India: 301-304. In Maheshwari, J.K. (*ed.*) *Ethnobotany and Medicinal Plants of Indian Subcontinent.* Scientific Publishers, Jodhpur, India.

Kharkongor, P. & Joseph, J. 1997. Folklore medico-botany of rural Khasi and Jaintia Tribes in Meghalaya: 195-207. In Jain, S.K. (*ed.*) *Contribution to Indian Ethnobotany*, 3rd ed. Scientific Publishers, Jodhpur, India.

Khory. 1887. *Bombay Materia Medica and Their Therapeutics.* Periodical Expert Book Agency, Delhi.

Kirtikar, K.R. & Basu, B.D. 1935. *Indian Medicinal Plants.* Vols.I-V. Lalit Mohan Basu, Allahabad, India.

Kothari, M.J., Jadhar, C.R. & Singh, N.P. 2003. Ethnobotanical wealth of two sacred groves in Junnar taluka, Pune, Maharashtra: 585-591. In Singh, V. & Jain, A.P. (*eds.*) Vol.-I. *Ethnobotany and Medicinal Plants of India and Nepal.* Scientific Publishers, Jodhpur, India.

Kothari, M.J. & Londhe, A.N. 2000. Ethnobotany in human healthcare of Chikhaldara, Amaravati district in Maharashtra State, India: 273-281. In Maheshwari, J.K. (*ed.*) *Ethnobotany and Medicinal Plants of Indian Subcontinent.* Scientific Publishers, Jodhpur, India.

Kothari, M.J. & Moorthy, S. 1996. Ethnobotany in human welfare of Raigad district in Maharashtra State, India: 403-407. In Jain, S.K. (*ed.*) *Ethnobiology in Human Welfare.* Deep Publications, New Delhi.

Kothari, M.J. & Rao, K.M. 2000. Ethnobotanical studies of Thane district, Maharashtra: 265-272. In Maheshwari, J.K. (*ed.*) *Ethnobotany and Medicinal Plants of Indian Subcontinent.* Scientific Publishers, Jodhpur, India.

Krishnamurthy, S. 1953. *Horticultural and Economic Plants of the Nilgiris.* Government Botanic Gardens Centenary Committee, Udhagamandalam.

Kshirsagar, R.D. & Singh, N.P. 2000a. Less known ethnomedicinal uses of plants in Coorg district of Karnatka State, Southern India. *Ethnobotany* **12** (1 & 2): 12-16.

Kshirsagar, R.D. & Singh, N.P. 2000b. Less known ethnomedicinal uses of plants reported by Jenukuruba tribe of Mysore district, Southern India. *Ethnobotany* **12** (1 & 2): 118-122.

Kumar, A. 1996. Some ethnomedicinal plants of the Murias of the Indravati Tiger Reserve, Bastar (Madhya Pradesh): 201-205. In Maheshwari, J.K. (*ed.*) *Ethnobotany in South Asia.* Scientific Publishers, Jodhpur, India.

Kumar, K. & Goel, A.K. 1998. Little known ethnomedicinal plants of Santhal and Paharia tribes in Santhal Paragana, Bihar, India. *Ethnobotany* **10** (1& 2): 66-69.

Kumar, A. & Lal, J. 2003. Medicinal plant resources of the Kanha tiger reserve, Madhya Pradesh: 255-265. In Singh, V. & Jain, A.P. (*eds.*) Vol.-I. *Ethnobotany and Medicinal Plants of India and Nepal.* Scientific Publishers, Jodhpur, India.

Kumar, R.V. & Pullaiah, T. 1998. Medicinal plants used by the tribals of Prakasham district, A.P. *Ethnobotany* **10** (1 & 2): 97-102.

Kumar, S. 1979. *Marketing Directory of Medicinal Plants.* CIMAP, Lucknow.

Kumar, S. 1999. *The Economic Plants of North East India.* Scientific Publishers, Jodhpur.

Kumar, S. 2002. *The Medicinal Plants of North-East India.* Scientific Publishers, Jodhpur, India.

Kumar, S. & Chauhan, A.K.S. 2005. Medicinal plants used by local inhabitants in Bharatpur district, Rajasthan. *Ethnobotany* **17**: 179-183.

Kumar, S. & Chauhan, A.K.S. 2006. Less-known medicinal uses of plant species in Keoledeo National Park, Bharatpur, Rajasthan. *Ethnobotany* **18**: 153-156.

Kumar, T.D.C. & Pullaiah, T. 2000. Ethno-medicinal uses of some plants of Mahabubnagar District, Andhra Pradesh, India: 341-345. In Maheshwari, J.K. (ed.) *Ethnobotany and Medicinal Plants of Indian Subcontinent.* Scientific Publishers, Jodhpur, India.

Kumar, V. & Jain, S.K. 1998. A contribution to ethnobotany of Surguja district in Madhya Pradesh, India. *Ethnobotany* **10** (1 & 2): 89-96.

Kumar, V. & Rao, R.R, 2001. Some plant beverages used in traditional medicine. *Ethnobotany* **13**: 36-39.

Kumari, R.M. & Narasimhan, D. 2003. Abortifacient and contraceptive plants used by Malayalis of Javadhi Hills: 788-790. In Singh, V. & Jain, A.P. (*eds.*) Vol.-II. *Ethnobotany and Medicinal Plants of India and Nepal.* Scientific Publishers, Jodhpur, India.

Kurian, J.C. 1995. *Plants That Heal.* Oriental Watchman Publishing House, Pune, India.

Lal, B., Vats, S.K., Singh, R.D. & Gupta, A.K. 1996. Plants used as ethnomedicine and supplementary food by Gaddis of Himachal Pradesh, India: 384-387. In Jain, S.K. (*ed.*) *Ethnobiology in Human Welfare.* Deep Publications, New Delhi.

Lalramnghinglova, J.H. 1996. Ethnobotany of Mizoram- A preliminary survey. *J. Econ. Taxon. Bot.* Additional Series **12**: 439-459.

Lalramnghinglova, J.H. & Jha, L.K. 1997. Ethnomedicine from Mizoram-North East India. *Ethnobotany* **9**: 105-111.

Lalramnghinglova, J.H. & Jha, L.K. 1999. New records of ethnomedicinal plants from Mizoram. *Ethnobotany* **11**: 57-64.

Maheshwari, J.K. 1996 (*ed.*). *Ethnobotany in South Asia.* Scientific Publishers, Jodhpur, India.

Maheshwari, J.K. 2000. *Ethnobotany and Medicinal Plants of Indian Subcontinent.* Scientific Publishers, Jodhpur, India.

Maheshwari, J.K., Painuli, R.M. & Dwivedi, R.P. 1997. Notes on ethnobotany of the Oraon and Korwa tribes of Madhya Pradesh: 67-82. In Jain, S.K. (*ed.*) *Contribution to Ethnobotany*, 3rd ed. Scientific Publishers, Jodhpur, India.

Maheshwari, P. & Singh, U. 1965. *Dictionary of Economic Plants in India.* I.C.A.R., New Delhi.

Maiti, S. & Mishra, T.K. 2000. Antivenom drugs of Santals, Savars and Mahatos of Midnapore districts of West Bengal, India. *Ethnobotany* **12**: 77-80.

Maity, D., Chauhan, A.S. & Maiti, G.G. 2003. Ethnobotanical notes on some unexploited medicinal plants used by Lepchas and Nepalese communities of north Sikkim: 325-332. In Singh, V. & Jain, A.P. (*eds.*) Vol.-I. *Ethnobotany and Medicinal Plants of India and Nepal.* Scientific Publishers, Jodhpur, India.

Maliya, S.D. 2005. Uses of some Indian plants in folk-medicines of Nepal. *Ethnobotany* 17 (1 & 2): 202-204.

Maliya, S.D. & Singh, K.K. 2003. Some new or less known folk medicines of Bahraich district, Uttar Pradesh. *Ethnobotany* **15**: 132-135.

Maliya, S.D. & Singh, K.K. 2004. Some new or less known folk medicines of district Bahraich, Uttar Pradesh, India-II. *Ethnobotany* **16**:113-115.

Mao, A.A. 1993. A preliminary report on the folklore botany of Mao Nagas of Manipur (India). *Ethnobotany* **5**: 143-147.

Masih, S.K. 2003. Diversity of ethnomedicinal wealth in Amarkantak plateau region (India): 204-214. In Khan, S.S. (*ed.*) *Vistas in Ethnobotany.* Classic Printer and Publishers. Bhopal, India.

Maya, S., Menon, S.V. & Nair, S.K. 2003. Economic importance of river vegetation of Kerala-a case study: 790-803. In Singh, V. & Jain, A.P. (*eds.*) Vol.-II. *Ethnobotany and Medicinal Plants of India and Nepal.* Scientific Publishers, Jodhpur, India.

Maya, S., Prameela, K.S. & Menon, V.S. 2003. Ethnobotanical notes on the flora of sacred tanks of Kerala. *Ethnobotany* **15**: 55-59.

Mehrotra, S. & Mehrotra, B.N. 2005. Role of traditional and folklore herbals in the development of new drugs. *Ethnobotany* **17**: 104-111.

Mehrotra, S. Rawat, A.K.S., Singh, H.K. & Shome, U. 1995. Standardisation of popular ayurvedic adaptogenic preparation 'Chyavanprash' and ethnobotany of its ingredients. *Ethnobotany* **7**: 1-15.

Meena, S.L., Sharma, K.C. & Gopalan, R. 2003. Ethnomedicinal plants of Karauli district, Rajasthan: 177-180. In Singh, V. & Jain, A.P. (*eds.*) Vol.-II. *Ethnobotany and Medicinal Plants of India and Nepal.* Scientific Publishers, Jodhpur. India.

Mitaliya, K.D. & Bhatt, D.C. 2003. Rare and endangered medicinal plants of Gujarat: 845-850. In Singh, V. & Jain, A.P. (*eds.*) Vol. II. *Ethnobotany and Medicinal Plants of India and Nepal.* Scientific Publishers, Jodhpur, India.

Mitaliya, K.D. & Bhatt, D.C., Patel, J.N. & Patel, S.K. 2003. Observations on plants used for joint disorders by rural and tribal people of Gujarat: 212-216. In Singh, V. & Jain, A.P. (*eds.*) Vol.-I. *Ethnobotany and Medicinal Plants of India and Nepal.* Scientific Publishers, Jodhpur, India.

Mitra, R. 1985. *Bibliography on Pharmacognosy of Medicinal Plants.* National Botanical Research Institute, Lucknow.

Mohanty, R.B. & Padhy, S.N. 1996. Traditional phytotherapy for diarrhoeal diseases in Ganjam Phulbani districts of South Orissa, India. *Ethnobotany* **8**: 60-65.

Molla, H.A. & Roy, B. 1996. Some ethnobotanical claims from the Jalpaiguri district of West Bengal. *J. Econ. Taxon. Bot.* Additional Series **12**: 322-324. Scientific Publishers, Jodhpur, India.

Mughal, M.H., Saba, Srivastava, P.S. & Iqbal, M. 2000. Drumstick (*Moringa pterigosperma* Gaertn.): a unique source of food and medicine: 47-61. In Maheshwari, J.K. (*ed.*) *Ethnobotany and Medicinal Plants of Indian Subcontinent.* Scientific Publishers, Jodhpur, India.

Mukherjee, A. & Namhata, D. 1990. Medicinal plantlore of the tribals of Sundargarh district, Orissa. *Ethnobotany* **2** (1 & 2): 57-60.

Mukhopadhyaya, G. 1992. *History of Indian Medicine.* Munshiram Manoharlal Publishers Pvt. Ltd. Delhi.

Nadanakunjidam, S. 2003. Ethnomedicinal observations from Attapadi hills of Western Ghats: 732-740. In Singh, V. & Jain, A.P. (*eds.*) Vol.-II. *Ethnobotany and Medicinal Plants of India and Nepal.* Scientific Publishers, Jodhpur, India.

Nadeem, M., Kumar, A., Nandi, S.K. & Palni, L.M.S. 2001. Tissue culture of medicinal plants with particular reference to Kumaun Himalaya: 231-268. In Samant, S.S., Dhar, U. & Palni, L.M.S. (*eds.*). *Himalayan Medicinal Plants.* Gyanodaya Prakashan, Nainital, India.

Nadkarni, K.M. 1926. *Indian Materia Medica.* Popular Book Depot, Bombay.

Nambiar, V.P.1985. *Studies on the Medicinal Plants of Kerala Forest.* Kerala Forest Research Institute, Peechi.

Nautiyal, S., Maikhuri, R.K., Rao, K.S. & Saxena, K.G. 2003. Ethnobotany of the Tolchha Bhotiya Tribe of the buffer zone villages in Nanda Devi biosphere reserve, India: 119-142. In Singh, V. & Jain, A.P. (*eds.*) Vol.-I. *Ethnobotany and Medicinal Plants of India and Nepal.* Scientific Publishers. Jodhpur. India.

Nayak, P.K., Nayak, R.K. & Choudhury, B.P. 2003a. A check list of medicinal plants of Kalahandi district in Orissa: 519-532. In Singh, V. & Jain, A.P. (*eds.*) Vol.-II. *Ethnobotany and Medicinal Plants of India and Nepal*. Scientific Publishers, Jodhpur, India.

Nayak P.K., Nayak, R.K. & Choudhury, B.P. 2003b. Some medicinal weeds of Mahanadi delta: 533-538. In Singh, V. & Jain, A.P. (*eds.*) Vol.-II. *Ethnobotany and Medicinal Plants of India and Nepal*. Scientific Publishers, Jodhpur, India.

Nayar, M.P., Ramamurthy, K. & Agarwal, V.S. 1989. *Economic Plants of India*. Vols. I & II. Botanical Survey of India, Calcutta.

Negi, K.S., Gaur, R.D. & Tiwari, J.K. 1999. Ethnobotanical notes on the flora of Har-ki-Doon (District Uttarkashi) Garhwal Himalaya, Uttar Pradesh, India. *Ethnobotany* **11**: 9-17.

Negi, K.S. & Pant, K.C. 1990. Notes on ethnobotany of the Gangwal- a tribe of Garhwal Himalaya. *Ethnobotany* **2**: 81-85.

Negi, K.S., Tiwari, J.K., Gour, R.D. & Pant, K.C. 1993. Notes on ethnobotany of five districts of Garhwal Himalaya, Uttar Pradesh, India. *Ethnobotany* **5**: 73-81.

O' Shaughnessy, W. 1841. *The Bengal Dispensatory*. Thakur, Spink & Co., Calcutta.

Painuli, R.M. & Maheshwari, J.K. 1996. Some interesting ethnomedicinal plants used by Sahariya tribe of Madhya Pradesh. *J. Econ. Taxon. Bot.* Additional Series **12**: 179-185.

Panda, P.C. & Das, P. 2000. Medicinal plantlore of the tribals of Baliguda sub-division, Phulbani district, Orissa. In Maheshwari, J.K. (*ed.*) *Ethnobotany and Medicinal Plants of Indian Subcontinent*. Scientific Publishers, Jodhpur, India.

Panda, S. 1996. A recent perspective from eastern Himalayas. *J. Econ. Taxon. Bot.* Additional Series **12**: 332-337.

Pandey, A.K., Bora, H.R. & Deka, S.C. 1996. An ethno-medico-botanical study of Golaghat District, Assam: native plant remedies for jaundice. *J. Econ. Taxon. Bot.* Additional Series **12**: 334-349.

Pandey, A.K. & Rout, S.D. 2006. Ethnobotanical uses of plants of Simlipal Biosphere Reserve (Orissa). *Etthnobotany* **18**: 102-106.

Pandey, B. & Pande, P.C. 2000. Ethnobotanical studies on Gymnospermic plants of Kumaon Himalaya: 253-257. In Maheshwari, J.K. (*ed.*) *Ethnobotany and Medicinal Plants of Indian Subcontinent*. Scientific Publishers, Jodhpur, India.

Pandey, G. 1995. *Medicinal Plants of Himalaya*. Sri Satguru Publications, Delhi.

Pandey, H.P. & Verma, B.K. 2002. Plants in oral healthcare among the aborigines of Gonda and Balrampur regions, U.P., India. *Ethnobotany* **14**: 81-86.

Pandey, I.B. 2003. Some traditional herbal home remedies used in and around Kanpur City (Uttar Pradesh), India. *Ethnobotany* **15**: 129-131.

Pandey, K. & Tiwari, S.P. 2003. Studies on the ethnobotanical approach of the vanishing tribe Raji: 947-955. In Singh, V. & Jain, A.P. (*eds.*) Vol.-II. *Ethnobotany and Medicinal Plants of India and Nepal.* Scientific Publishers, Jodhpur, India.

Panthi, M.P. & Chaudhary, R.P. 2003. Ethnomedicinal plant resources of Arghakhanchi district, West Nepal. *Ethnobotany* **15**: 71-86.

Parmar, C. & Kaushal, M.K. 1982. *Wild Fruits of the Sub-Himalayan Region*. Kalyani Publishers, New Delhi.

Parrotta, J.A. 2001. *Healing Plants of Peninsular India.* CABI Publishing.

Patil, D.A. 2000. Sanskrit plant names in an ethnobotanical perspective. *Ethnobotany* **12**: 60-64.

Patil, D.A. 2004. Origin of medicines vis-à-vis Doctrine of Signatures. *Ethnobotany* **16**: 52-58.

Patil, D.A. & Patil, M.V. 2002. Sanskrit plant names in an ethnobotanical perspective:[22]. *Ethnobotany* **14**: 87-92.

Pawar, S. & Patil, D.A. 2004. Herbal folk-remedies for urinogenital complaints in Jalgaon district of Maharashtra. *Ethnobotany* **16**: 86-90.

Prajapati, N.D., Purohit, S.S., Sharma, A.K. & Kumar, T. 2006. *A Handbook of Medicinal Plants A Complete Source Book.* Agrobios (India), Jodhpur.

Prakash, V. & Mehrotra, B.N. 1991. Ethno-medicinal uses of some plants among Garos of Meghalaya. *Ethnobotany* **3**: (1& 2): 41-45.

Prasad, K.V., Rajagopal, T. & Badarinath, K.V.S. 2002. Ethnobotany of Kondareddis of Rampa agency, Eastern Ghats, Andhra Pradesh: 205-219. In Trivedi, P.C. (*ed.*) *Ethnobotany* Aavishkar Publishers & Disributers, Jaipur (Raj.), India.

Prasad, K.V., Rajagopal, T. & Badarinath, K.V.S. 2003. Notes on economic importance of wild plants of Rampa agency-East Godavari district, Andhra Pradesh, India: 603-612. In Singh, V. & Jain, A.P. (*eds.*) Vol.-II. *Ethnobotany and Medicinal Plants of India and Nepal.* Scientific Publishers, Jodhpur, India.

Prescott-Allen, R.P. 1982. *What's Wildlife Worth?* International Institute for Environment and Development, London.

Principe, P.P. 1991. Valuing the biodiversity of medicinal plants: 79-124. In Akerele, O., Heywod, V. & Synge, H. (*eds.*) *Conservation of Medicinal Plants.* Cambridge Univ. Press, Cambridge.

Punjani, B.L. 2002. Plants used in contact therapy by tribals of Sabarkantha (Gujrat). *Ethnobotany* **14**: 57-59.

Punjani, B.L. 2003. Poisonous plants in the forest areas of Vijaynagar Taluka (North Gujrat): 196-204. In Singh, V. & Jain, A.P. (*eds.*) Vol.-I. *Ethnobotany and Medicinal Plants of India and Nepal.* Scientific Publishers, Jodhpur, India.

Punjani, B.L. 2004. Ethnomedicinal uses of Vitaceae among the tribals of North Gujarat. *Ethnobotany* **16**: 83-85.

Punjani, B.L. 2006. Ethno-medicobotanical study of Kathodi tribe of Sabarkantha in Gujarat. *Ethnobotany* **18**: 135-138.

Pullaiah, T. & Kumar, T.D.C. 1996. Herbal plants in Mannanur forest Mahaboobnagar district, Andhra Pradesh. *J. Econ. Taxon. Bot.* Additional Series **12**: 218-220.

Radhakrishnan, K., Pandurangan, A.G., Pushpangadan, P. & Sasidharan, A.1996. Less known ethnomedicinal plants of Kerala State & their conservation. *Ethnobotany* **8**: 82-84.

Raghunathan, K. 1976. *Medicinal Plants Growing in the Districts of Andhra Pradesh Through Which the River Krishna Flows.* Central Council for Research in Indian Medicine & Homoeopathy, Delhi.

Ragupathy, S. & Mahadevan, A. 1991. Ethnobotany of Kodiakkarai reserve forest, Tamil Nadu, South India. *Ethnobotany* **3** (1 & 2): 79-82.

Rahman, M.A. 2000. Ethno-medico-botanical knowledge among tribals of Bangladesh: 89-93. In Maheshwari, J.K. (*ed.*) *Ethnobotany and Medicinal Plants of Indian Subcontinent.* Scientific Publishers, Jodhpur, India.

Rai, L.K. 1994. *Medicinal Plants of the Sikkim Himalaya: Status, Usage and Potential.* Bishen Singh Mahendra Pal Singh, Dehra Dun.

Rai, L.K.& Sharma, E. 1994. *Medicinal Plants of Sikkim Himalaya.* Bishen Singh Mahendra Pal Singh, Dehra Dun, India.

Rajan, S., Gupta, H.C. & Kumar, S. 2000. Exotic medicinal plants of Lucknow: 205-222. In Maheshwari, J.K. (*ed.*) *Ethnobotany and Medicinal Plants of Indian Subcontinent.* Scientific Publishers, Jodhpur, India.

Rajendran, A. & Henry, A.N. 1994. Plants used by the tribe Kadar in Anamalai hills of Tamil Nadu. *Ethnobotany* **6**: 19-24.

Rajendran, A, Rao, R.N. & Henry, A.N. 1997. Rare and noteworthy plants of the eastern ghats in Andhra Pradesh with their ethnic uses. *Ethnobotany* 9: 39-43.

Rajendran, A. & Sikarwar, R.L.S. 2003. Intra-cultural ethnobotanical studies on the tribe Kadars (Tamil Nadu) and Sahariyas (Madhya Pradesh), India: 662-664. In Singh, V. & Jain, A.P. (*eds.*) Vol.-II. *Ethnobotany and Medicinal Plants of India and Nepal.* Scientific Publishers, Jodhpur, India.

Rajendran, S.M., Sekar, K.C. & Sundaresan, V. 2001. Ethnomedicinal lore of Seithur hills-Southern Western ghats, Tamil Nadu. *Ethnobotany* **13**:101-109.

Rajore, S. & Batra, A. 2003. *Jatropha curcas*: a plant of immense potential value: 36-41. In Singh, V. & Jain, A.P. (*eds.*) Vol.-II. *Ethnobotany and medicinal plants of India and Nepal.* Scientific Publishers, Jodhpur, India.

Ram, A.J., Raja, K. Reddy, K.E. & Raju, R.R.V.2002. Medicinal plantlore of Sugalis of Gooty forests, Andhra Pradesh. *Ethnobotany* **14**: 37-42.

Rana, T.S., Datt, B. & Rao, R.R. 1996. Strategies for sustainable utilization of plant resources by the tribals of Tons Valley, Western Himalaya. *Ethnobotany* **8**: 96-104.

Rana, T.S., Singh, K.K. & Rao, R.R. 2000. Studies on indigenous herbal remedies for *Diabetes mellitus* in India: 115-120. In Maheshwari, J.K. (*ed.*) *Ethnobotany and Medicinal Plants of Indian Subcontinent.* Scientific Publishers, Jodhpur, India.

Ranjan, V. 1996. Some ethnomedicinal plants of Lalitpur district, Uttar Pradesh, India: 149-150. In Jain, S.K. (*ed.*) *Ethnobiology in Human Welfare.* Deep Publications, New Delhi.

Rao, D.M. & Pullaiah, T. 2001. Ethno-medico-botanical studies in Guntur district of Andhra Pradesh, India. *Ethnobotany* **13**: 40-44.

Rao, N.R. & Henry, A.N. 1996. *The Ethnobotany of Eastern Ghats in Andhra Pradesh, India.* Botanical Survey of India, Calcutta.

Rao, N.S.B., Rajasekhar, D., Raju, D.C. & Nagarju, N. 1996. Ethno-medicinal notes on some plants of Tirumala hills for dental disorders. *Ethnobotany* **8**: 88-91.

Rastogi, R.P. & Mehrotra, B.N. 1998. *Compendium of Indian Medicinal Plants.* Vol.-V, 1990-1994. Central Drug Research Institute, Lucknow & National Institute of Science Communication, New Delhi.

Rastogi, R.P. & Mehrotra, B.N. 1999a. *Compendium of Indian Medicinal Plants.* Vol.-I, 1960-1969. Central Drug Research Institute, Lucknow & National Institute of Science Communication, New Delhi.

Rastogi, R.P. & Mehrotra, B.N. 1999b. *Compendium of Indian Medicinal Plants.* Vol.-II, 1970-1979. Central Drug Research Institute, Lucknow & National Institute of Science Communication, New Delhi.

Rastogi, R.P. & Mehrotra, B.N. 2001. *Compendium of Indian Medicinal Plants.* Vol.-III, 1980-1984. Central Drug Research Institute, Lucknow & National Institute of Science Communication, New Delhi.

Rastogi, R.P. & Mehrotra, B.N. 2002. *Compendium of Indian Medicinal Plants.* Vol.-IV, 1985-1989. Central Drug Research Institute, Lucknow & National Institute of Science Communication, New Delhi.

Ravisankar, T. & Henry, A.N. 1992. Ethnobotany of Adilabad District, Andhra Pradesh, India. *Ethnobotany* **4**: 45-52.

Rawat, M.S. & Chowdhury. 1998. *Ethno-medico-botany of Arunachal Pradesh (Nishi and Apatani Tribes).* Bishen Singh Mahendra Pal Singh. Dehradun, India.

Rawat, R.B.S. 2002. *Cultivation Practices of Some Commercially Important Medicinal Plants.* National Medicinal Plants Board, New Delhi.

Raychaudhuri, S.P. 1991. *Recent Advances in Medi1cinal, Aromatic and Spice Crops.* Today & Tomorrow's Printers & Publishers, Delhi.

Reddi, T.V.V.S., Prasanthi, S. & Naidu, B.V.A.R. 2005. Medicinal and aromatic plants of India: 1-146. In Khan, I.A. & Khanum, A. (*eds.*). *Role of Biotechnology in Medicinal and Aromatic Plants*. Vol.-XII. Ukaaz Publications, Hyderabad, India.

Reddy, M.H., Vijayalakshmi, K. & Raju, R.R.V. 1996. Native phytotherapy for snake bite in Nallamalais, Eastern Ghats, India. *J. Econ. Taxon. Bot.* Additional Series **12**: 214-217.

Reddy, P.R., Rao, P.P. & Prabhaskar, M. 2003. Ethnomedicinal practices amongst Chenchus of Nagarjunasagar Srisailam tiger reserve (NSTR), Andhra Pradesh- plant remedies for cuts, wounds and boils. *Ethnobotany* **15**: 67-70.

Robert, K. 1932. *Vegetable Drugs of India*. New Delhi.

Roxburgh, W. 1832. *Flora Indica*. Thakur, Spink & Co., Calcutta.

Royal, J.F. 1837. *An Essay on the Antiquity of Hindu Medicine*. London.

Rubin, V. 1960. Culture, society and health. *Annals New York Acad. Sci.* **84**: 783-1060.

Sahoo, A.K. & Bahali, D.D. 2003. Plants used in less known ethnomedicines and medico-religious beliefs in Phulbani district of Orissa: 500-504. In Singh, V. & Jain, A.P. (eds.) Vol.-I. *Ethnobotany and Medicinal Plants of India and Nepal*. Scientific Publishers, Jodhpur, India.

Sahoo, A.K. & Mudgal, V. 1995. Less known ethnobotanical uses of plants of Phulbani District, Orissa, India. *Ethnobotany* 7: 63-67.

Said, H.M. 1997. *Hamdard Pharmacopoeia of Eastern Medicine*. Sri Satguru Publications, Delhi.

Saikia, N. & Nath, S.C. 2003. Indigenous practice of treating dental disorders in Assam: 496-499. In Singh, V. & Jain, A.P. (*eds.*) Vol.-I. *Ethnobotany and Medicinal Plants of India and Nepal*. Scientific Publishers, Jodhpur, India.

Saini, D.C. 1996a. Ethnobotany of Tharus of Basti district. Uttar Pradesh. *J. Econ. Taxon. Bot.* Additional Series **12**: 138-153.

Saini, V.K. 1996b. Plants in the welfare of tribal women and children in certain areas of central India: 140-144. In Jain, S.K. (*ed.*) *Ethnobiology in Human Welfare*. Deep Publications, New Delhi.

Saklani, A. & Jain, S.K. 1994. *Cross-Cultural Ethnobotany of Northeast India*. Deep publications, New Delhi.

Samant, S.S., Dhar, U. & Rawal, R.S. 2001. Diversity distribution and indigenous uses of threatened medicinal plants of Askot wildlife sanctuary in West Himalaya: conservation and management prospectives: 167-184. In Samant, S.S., Dhar, U. & Palni, L.M.S. (*eds*). *Himalayan Medicinal Plants: Potential and Prospects*. Gyanodaya Prakashan, Nainital, India.

Samanta, A.K. & Das, D.C. 2003. Ethnobotanical studies on *Typha elephantina* Roxb. (Typhaceae) in the southern parts of West Bengal, India: 576-579. In Singh, V. & Jain, A.P. (*eds.*) Vol.-II. *Ethnobotany and Medicinal Plants of India and Nepal.* Scientific Publishers, Jodhpur, India.

Samvatsar, S. & Diwanji, V.B. 1996. Plants used in snake, scorpion and insect bites/ stings by Adibasis of Jhabua (M.P.) India. *J. Econ. Taxon. Bot.* Additional Series **12**: 199-200.

Samvatsar, S. & Diwanji, V.B. 2000. Plants used for rheumatism by the tribals of Western M.P: 305-314. In Maheshwari, J.K. (*ed.*) *Ethnobotany and Medicinal Plants of Indian Subcontinent.* Scientific Publishers, Jodhpur. India.

Sanyal, D. 1924. *Vegetable Drugs of India.* Bishen Singh Mahendra Pal Singh, Dehra Dun.

Sarangi, N. & Sahu, R.K. 2003. Ethnomedicinal plants used in venereal and gynaecological disorders in Kalahandi, Orissa. *Ethnobotany* **15**: 16-20.

Saren, A.M., Sen, R. & Pal, D.C. 2000. A contribution to the ethnobotany of Bankura district, West Bengal: 545-555. In Maheshwari, J.K. (*ed.*) *Ethnobotany and Medicinal Plants of Indian Subcontinent.* Scientific Publishers, Jodhpur, India.

Sarkar, N., Rudra, S. & Basu, S.K. 2000. Ethnobotany of Bangriposi, mayurbhanj, Orissa: 509-574. In Maheshwar, J.K. (*ed.*) *Ethnobotany and Medicinal Plants of Indian Subcontinent.* Scientific Publishers, Jodhpur, India.

Sarin, Y.K. 1990. Some less known but effective folk remedies in North-West Himalayan region. *Ethnobotany* **2**: 39-43.

Sarin, Y.K. 1996. *Illustrated Manual of Herbal Drugs Used in Ayurveda.* C.S.I.R. Delhi.

Sarma, S.K., Bhattacharya, D.K. & Devi, B. 2001.Medicinal plants used by Bodo tribe of Nalbari district in Assam. *Ethnobotany* **13**: 133-139.

Sarma, S.K., Bhattacharya, D.K. & Devi, B. 2002. Traditional use of herbal medicines by Madahi tribe of Nalbari district of Assam. *Ethnobotany* **14**: 103-111.

Sasikala, K., Vajravelu, E. & Daniel, P. 2003. Medicinal aroids of India: 49-52. In Singh, V. & Jain, A.P. (*eds.*) Vol.-I. *Ethnobotany and Medicinal Plants of India and Nepal.* Scientific Publishers, Jodhpur, India.

Satapathy, K.B. & Brahmam, M. 2000. Some interesting phytotherapeutic claims of tribals of Jaipur district (Orissa), India: 467-472. In Maheshwari, J.K. (*ed.*) *Ethnobotany and Medicinal Plants of Indian Subcontinent.* Scientific Publishers, Jodhpur, India.

Satyavati, G.V., Raina, M.K. & Sharma, M. (*ed.*) 1976. *Medicinal Plants of India.* I.C.M.R., New Delhi.

Saxena, B.R., Moli, M.C. & Saxena, R.C. 2000. Preliminary ethnomedical and phytochemical study of *Cleome viscosa* L. *Ethnobotany* **12**: 47-50.

Saxena, H.O. 1971. *Medicinal Plants of Patalkot.* State Forest Research Institute, Jabalpur.

Saxena, H.O. & Dutta, P.K. 1975. Studies on the ethnobotany of Orissa. *Bull. Bot. Surv. India* **17**: 124-131

Saxena, H.O., Brahmam, M. & Dutta, P.K. 1997. Ethnobotanical studies in Orissa: 123-135. In Jain, S.K. (*ed.*) *Contribution to Indian Ethnobotany*, 3rd ed. Scientific Publishers, Jodhpur, India.

Schultes, R.E. 1960. Tapping our heritage of ethnobotanical lore. *Econ. Bot.* **8**: 224-227.

Schultes, R.E. & Raffauf, R.F. 1990. *The Healing Forest*. Dioscorides Press, Portlan Oregon.

Sen, S. & Batra, A. 1997. Ethno-medico-botany of household remedies of Phagi tehsil of Jaipur district (Rajasthan). *Ethnobotany* **9**: 122-128.

Sen, S., Batra, A. & Ali, Z.D. 2002. Some herbal home remedies from Jaipur district, Rajasthan. *Ethnobotany* **14**: 116-122.

Sen, S.K. & Behera, L.M. 2003. Ethnomedicinal plants used against skin diseases in Bargarh district in Orissa. *Ethnobotany* **15**: 90-96.

Seetharam, Y.N., Chalgeri, G., Haleshi, C. & Vijay. 1999. Folk medicine and ethnomedicine of North-Eastern Karnataka. *Ethnobotany* **11**: 32-37.

Shah, N.C. 1995. *Rauvolfia serpentina*: from folkmedicine to modern medicine-some unknown aspects. *Ethnobotany* **7**: 109-112.

Shah, N.C. & Singh, S.C. 1990. Hither to unreported phytotherapeutical uses from tribal pockets of Madhya Pradesh (India). *Ethnobotany* **2**: 91-95.

Sharma, B.D. & Rana, J.C. 2000. Traditional medicinal uses of plants of Himachal Hills: 173-176. In Maheshwari, J.K. (*ed.*) *Ethnobotany and Medicinal Plants of Indian Subcontinent*. Scientific Publishers, Jodhpur, India.

Sharma, H.M., Devi, A.R. & Sharma, B.M. 2003. Ethnomedicinal uses of monocotyledonous plants by the Meities of Manipur: 473-480. In Singh, V. & Jain, A.P. (*eds.*) Vol.-I. *Ethnobotany and Medicinal Plants of India and Nepal*. Scientific Publishers, Jodhpur, India.

Sharma, K.R. & Sood, M. 1997. *Important Medicinal Plants of Himachal Pradesh*. Y.S. Parmar University of Horticulture and Forestry, Nauni, Solan (H.P.).

Sharma, N.K. 2002. Ethno-medico-religious plants of Hadoti plateau (S.E. Rajasthan)-a preliminary survey: 394-411. In Trivedi, P.C. (*ed.*) *Ethnobotany* Aavishkar Publishers & Distributors. Jaipur (Raj.), India.

Sharma, P.C., Yelne, M.B. & Dennis, T.J. 2001. *Data Base on Medicinal Plants Used in Ayurveda*. Vols. I-II. C.C.R.A.S., New Delhi.

Sharma, P.K., Chauhan, N.S. & Lal, B. 2003. Commercially important medicinal and aromatic plants of Parvati valley, Himachal Pradesh: 937-942. In Singh, V. & Jain, A.P. (*eds.*) Vol.-I. *Ethnobotany & Medicinal Plants of India & Nepal*. Scientific Publishers. Jodhpur. India.

Sharma, P.P. & Singh, N.P. 1999. Less known ethnobotanical uses of plants in Dadra and Nagar Haveli (U.T.): Part I (A-K). *Ethnobotany* **11**: 109-114.

Sharma, P.P. & Singh, N.P. 2000. Less known ethnobotanical uses of plants in Dadra and Nagar Haveli (U.T.): Part II (L-Z). *Ethnobotany* **12**: 81-85.

Sharma, R. 2003. *Medicinal Plants of India- An Encyclopaedia.* Daya Publishing House, Delhi.

Sharma, S.K. 2004. *Ethnobotanical Studies on Trees, Shrubs and Climbers of Jogindarnagar, Mandi, (H.P.).* Ph.D Thesis. Himachal Pradesh University, Shimla.

Sharma, U.K. 2000. Folk and herbal medicine among nepalese of Assam: 599-603. In Maheshwari, J.K. (*ed.*) *Ethnobotany and Medicinal Plants of Indian Subcontinent.* Scientific Publishers, Jodhpur, India.

Shil, S. & Sharma, G.D. 2003. A few medicinal plants from Reang tribals of North Tripura: 451-456. In Singh, V. & Jain, A.P. (*eds.*) Vol.-I. *Ethnobotany and Medicinal Plants of India and Nepal.* Scientific Publishers. Jodhpur. India.

Shukla, G. & Verma, B.K. 1996. Roots-a vital plant part to cure body ailments among tribal/ rural folklore of Western Bihar. *J. Econ. Taxon. Bot.* Additional Series 12: 392-394.

Shukla, G., Verma, B.K. & Pandey, H.P. 2003. Folklore herbotherapy of Jalaun district, U.P., India: 143-147. In Singh, V. & Jain, A.P. (*eds.*) Vol.-I. *Ethnobotany and Medicinal Plants of India and Nepal.* Scientific Publishers, Jodhpur, India.

Sikarwar, R.L.S. 1996. Ethno-veterinary herbal medicines in Morena district of Madhya Pradesh, India :194-196. In Jain, S.K. (*ed.*) *Ethnobiology in Human Welfare.* Deep Publications, New Delhi.

Sikarwar, R.L.S. 2002. Ethnogynaecological uses of plants new to India. *Ethnobotany* 14: 112-115.

Singh, A.K. 2000. A contribution to the ethnobotany of Sub-Himalayan region of eastern Uttar Pradesh: 237-246. In Maheshwari, J.K. (*ed.*) *Ethnobotany and Medicinal Plants of Indian Subcontinent.* Scientific Publishers, Jodhpur, India.

Singh, G.C. 2002. Minor forest products of Sariska national park: an ethnobotanical profile: 221-238. In Trivedi, P.C. (*ed.*) *Ethnobotany.* Aavishkar Publishers & Distributors. Jaipur (Raj.), India.

Singh, G.S. 2000. Ethnobotanical study of useful plants of Kullu district in North Western Himalaya, India: 185-198. In Maheshwari, J.K. (*ed.*) *Ethnobotany and Medicinal Plants of Indian Subcontinent.* Scientific Publishers, Jodhpur, India.

Singh, G.S. 2003. Ethnobotanical inventory of Sariska national park in the Aravali Hills of Eastern Rajasthan, India: 181-195. In Singh, V. & Jain, A.P. (*eds.*) Vol.-I. *Ethnobotany and Medicinal Plants of India and Nepal.* Scientific Publishers, Jodhpur, India.

Singh, H. 1985. *Medicinal Chemistry Research in India.* Central Drug Research Institute, Lucknow.

Singh, H. 2006. Some unrecorded dye-yielding plants and their less known ethnomedicinal uses in Uttaranchal. *Ethnobotany* **18**: 119-121.

Singh, H. & Maheshwari, J.K. 1993. Phytotherapy for diphtheria by the Bhoxas of Nainital District, Uttar Pradesh, India. *Ethnobotany* **5**: 63-65.

Singh, H. & Maheshwari, J.K. 1998. Phytotherapy for diphtheria by the Bhoxas of Nainital District, Uttar Pradesh, India. *Ethnobotany* **5**: 63-65.

Singh, H.B., Hynneiwta, T.M. & Bora, P. J. 1997. Ethno-medico-botanical studies in Tripura, India. *Ethnobotany* **9**: 56-58.

Singh, J., Bhuyan, T.C. & Ahmed, A. 1996. Ethnobotanical studies on the mishing tribes Assam with reference to food and medicinal plants-I. *J. Econ. Taxon. Bot.* Additional Series **12**: 330-356.

Singh, K.K. 1996. Ethnobotanical observations on Sonbhadra district of Southern Uttar Pradesh, India: utilization and conservation. In Jain, S.K. (*ed.*) *Ethnobiology in Human Welfare.* Deep Publications, New Delhi.

Singh, K.K. & Kumar, K. 2000a. Ethnotherapeutics of some medicinal Plants used as antipyretic agents among the tribals of India: 135-141. In Maheshwari, J.K. (*ed.*) *Ethnobotany and Medicinal Plants of Indian Subcontinent.* Scientific Publishers, Jodhpur, India.

Singh, K.K. & Kumar, K. 2000b. *Ethnnobotanical Wisdom of Gaddi Tribe in Western Himalaya.* Bishen Singh Mahendra Pal Singh, Dehradun, India.

Singh, K.K. & Maheshwari, J.K. 1992. Folk medicinal uses of some plants among the Tharus of Gorakhpur district, Uttar Pradesh, India. *Ethnobotany* **4**: 39-43.

Singh, K.K. & Prakash, A. 1994. Indigenous phytotherapy among the Gond tribe of Uttar Pradesh, India. *Ethnobotany* **6**: 37-41.

Singh, K.K. & Prakash, A. 1996. Observations on Ethnobotany of the Kol tribe of Varanasi district, Uttar Pradesh, India. *J. Econ. Taxon. Bot.* Additional Series **12**: 133-137.

Singh, K.K. & Rao, R.R. 2003. Studies on some potential ethnomedicine among the tribals of Uttar Pradesh, India: 42-57. In Khan, S.S. (*ed.*) *Vistas in Ethnobotany* Vol.-I. Classic Printer and Publishers, Bhopal, India.

Singh, P.B. 1999. *Illustrated Field Guide to Commercially Important Medicinal and Aromatic Plants of Himachal Pradesh (With Special Reference to Mandi District).* Society for Herbal Medicine and Himalayan Biodiversity, Mandi, H.P.

Singh, P.B. 2001. Medicinal plants of Kangra district in North-West Himalaya: 185-196. In Samant, S.S., Dhar, U. & Palni, L.M.S. (*eds*). *Himalayan Medicinal Plants: Potential and Prospects.* Gyanodaya Prakashan, Nainital, India.

Singh, S.C. & Srivastava, G.N. 2000. Exotic medicinal plants of Lucknow district, U.P. India: 223-236. In Maheshwari, J.K. (*ed.*) *Ethnobotany & Medicinal Plants of Indian Subcontinent.* Scientific Publishers. Jodhpur. India.

Singh, S.P., Tripathi, S. & Shukla, R.S. 2003. Ethnomedicinal heritage for bioprospecting and drug development in North Eastern States of India: 384-395. In Singh, V. & Jain, A.P. (*eds.*) Vol.-I. *Ethnobotany and Medicinal Plants of India and Nepal*. Scientific Publishers, Jodhpur, India.

Singh, U. 1996. *Dictionary of Economic Plants in India*. Indian Council of Agricultural Research, Delhi.

Singh, U., Wadhwani, A.M. & John, B.M. 1983. *Dictionary of Economic Plants of India*. Indian Coun. Agric. Res., New Delhi.

Singh, V. 1996. Ethnomedicobotany of Dards tribe of Gurez valley in Kashmir Himalaya: 129-132. In Jain, S.K. (*ed.*) *Ethnobiology in Human Welfare*. Deep Publications, New Delhi.

Singh, V. & Pandey, R.P. 1996. Ethnomedicinal plants used for venereal and gynaecological diseases in Rajasthan (India). *J. Econ. Taxon. Bot.* Additional Series **12**: 154-165.

Singh, V. & Pandey, R.P. 1998. *Ethnobotany of Rajasthan, India*. Scientific Publishers, Jodhpur, India.

Singh, V.K. 1984. *Medicinal Plants of Gwalior Forest Division, Madhya Pradesh*. Central Council for Research in Unani Medicine, Delhi.

Sinha, B.K. & Dixit, R.D. 2003. Ethnomedicinal plants sold in Omkareshwar, Madhya Pradesh. *Ethnobotany* **15**: 127-128.

Sinha, R.K. & Sinha, S. 2001. *Ethnobiology (Role of Indigenous and Ethnic Societies in Biodiversity Conservation, Human Health Protection and Sustainable Development)*. Surabhi Publications, Jaipur.

Sivakumar *et al.* 2003. Ethnobotany of poliyars of Anaimalai hills, Tamil Nadu: 679-685. In Singh, V. & Jain, A.P. (*eds.*) Vol.-I. *Ethnobotany and Medicinal Plants of India and Nepal*. Scientific Publishers, Jodhpur, India.

Sood, S.K., Bhardwaj, R. & Lakhanpal, T.N. 2005a. *Ethnic Indian Plants in Cure of Diabetes*.Scientific Publishers, Jodhpur, India.

Sood, S.K., Nath, R. & Kalia, D.C. 2001. *Ethnobotany of Cold Desert Tribes of Lahoul-Spiti (N.W. Himalaya)*. Deep Publications, New Delhi.

Sood, S.K. & Thakur, S. 2004. *Ethnobotany of Rewalsar Himalaya*. Deep Publications, New Delhi.

Sood, S.K., Rana, S. & Lakhanpal, T.N. 2005b. *Ethnic Aphrodisiac Plants*. Scientific Publishers, Jodhpur, India.

Sood, S.K., Parmar, S. & Lakhanpal, T.N. 2005c. *Ethnic Plants of India Used in Cancer Cure (A Compendium)*. Bishen Singh Mahendra Pal Singh, Dehra Dun, India.

Sood, S.K., Negi, C.J. & Lakhanpal, T.N. 2001. *Orchidaceae and Mankind*. Deep Publications, New Delhi.

Sood, S.K. & Prakash, V. 2007. *Edible Roots and Underground Stems of Ethnic India.* Satish Serial Publlishing House, Delhi.

Srivastava, A.K. & Singh, M. 2003. Medicinal potential of grasses in Rajasthan: 170-176. In Singh, V. & Jain, A.P. (*eds.*) Vol.-I. *Ethnobotany and Medicinal Plants of India and Nepal.* Scientific Publishers, Jodhpur, India.

Srivastava, R.C. 1989. *Drug-Plant Resources of Central India.* Today & Tomorrow's Printers & Publishers, Delhi.

Srivastava, R.C., Singh, V.P. & Singh, M.K. 2003. Medicinal plants of Jaunpur district (U.P.): 148-159. In Singh, V. & Jain, A.P. (*eds.*) Vol.-I. *Ethnobotany and Medicinal Plants of India and Nepal.* Scientific Publishers, Jodhpur, India.

Srivastava, S.C., Mitra, S. & Bandyopadhyay, S. 2003. Some plants of ethnomedicinal importance from Darjeeling district, West Bengal: 972-977. In Singh, V. & Jain, A.P. (*eds.*) Vol.-II. *Ethnobotany and Medicinal Plants of India and Nepal.* Scientific Publishers, Jodhpur, India.

Subramaniam, A. 2000. A survey of medicinal plants from Chitheri hills in Dharampuri district, Tamil Nadu: 395- 416. In Maheshwari, J.K. (*ed.*) *Ethnobotany and Medicinal Plants of Indian Subcontinent.* Scientific Publishers, Jodhpur, India.

Subramanian, S.P. & Goraya, G.S. 2003. Some folklore medicinal plants of Kolli hills: record of a Natti Vaidyas Sammelan: 665-678. In Singh, V. & Jain, A.P. (*eds.*) Vol.-II. *Ethnobotany and Medicinal Plants of India and Nepal.* Scientific Publishers, Jodhpur, India.

Sundararaj, D.1959. *Guide to Economic Plants of South India.* Periodical Expert Book Agency, Delhi.

Sur, P.R. & Halder, A.C. 2002. Observation on the ethnobotany of Purulia district, West Bengal, India: 146-168. In Trivedi, P.C. (*ed.*) *Ethnobotany.* Aavishkar Publishers & Distributors, Jaipur (Raj.), India.

Swami, A. & Gupta, B.K. 1996. A note on some commonly occurring medicinal weeds of Udhampur district (J. & K.). *J.Econ. Tax. Bot.* Additional Series **12**: 89-91.

Tewari, K.C., Joshi, G.C. & Misra, S.S. 2001. Plants and allied materials used in Folk medicines of Asssam, Meghalaya and Arunachal Pradesh (India): 523-546. In Pande, P.C. & Samant, S.S. (*eds.*). *Plant Diversity of the Himalaya.* Gyanodaya Prakashan, Nainital.

Thakur, R.S., Puri, H.S. & Husain, A. 1989. *Major Medicinal Plants of India.* Director, Central Institute of Medicinal and Aromatic Plants. Lucknow, India.

Thomas, J. & Britto, A.J.D. 2000. Weeds of medicinal impotance in Tiruneveli district in Tamil Nadu: 363-367. In Maheshwari, J.K. (*ed.*) *Ethnobotany and Medicinal Plants of Indian Subcontinent.* Scientific Publishers, Jodhpur, India.

Thomas, J. & Britto, A.J.D. 2003. Ethnobotanical study in Wynad district in Kerala: 815-824. In Singh, V. & Jain, A.P. (*eds.*) Vol.-I. *Ethnobotany and Medicinal Plants of India and Nepal.* Scientific Publishers, Jodhpur, India.

Tigga, M., Lakra, G.S. & Rao, P.S.N. 2003. Medicinal value of latex yielding plants from Andaman and Nicobar Islands: 765-772. In Singh, V. & Jain, A.P. (*eds.*) Vol.-II. *Ethnobotany and Medicinal Plants of India and Nepal*. Scientific Publishers, Jodhpur, India.

Tirkey, A. 2004. Some ethnomedicinal plant species of Chhattisgarh State. *Ethnobotany* **16**: 118-124.

Tiwari, K.C. & Tiwari, V.P. 1996. Some important medicinal plants of the tropical, sub tropical and temperate regions of Siang, Subansiri and Tirap districts of Arunachal Pradesh. *J. Econ. Taxon. Bot.* Additional Series **12**: 359-363.

Tiwari, K.P. 1998. *Medicinal Plants of Madhya Pradesh: Distribution, Cultivation and Trade.* SFRI Jabalpur.

Tiwari, L. & Pande, P.C. 2006. Ethnoveterinary plants of Uttarkashi District, Uttaranchal, India. *Ethnobotany* **18**: 139-144.

Tiwari, V.J., Padhye, M.D. & Madke, K.H. 1996. Ethnomedicinal plants used by Kawar tribe of Gadchiroli district of Maharashtra State, India: 160-173. In Rajak, R.C. & Rai, M.K. (*eds.*). *Herbal Medicines, Biodiversity and Conservation Strategies.* International Book Distributors, Dehra Dun, India.

Trivedi, G.N., Kayal, R.N. & Chaudhury, H.N.R. 1982. Some medicinal plants of Mayurbhanj (Orissa). *Bull. Bot. Surv. India.* **24**:117-120.

Trivedi, P.C. 2002. Ethno-medicinal plants of Rajasthan State, India: 412-439. In Trivedi, P.C. (*ed.*) *Ethnobotany*. Aavishkar Publishers & Distributors, Jaipur (Raj.), India.

Uniyal, M.R. 1968. Medicinal plants of Bhagirathi valley lying in Uttarkashi forest division. *Indian Forester* **94**: 407-420.

Uniyal, M.R. 1989. *Medicinal Flora of Garhwal Himalayas.* Baidyanath Ayurved Bhawan Ltd., Nagpur.

Uniyal, M.R. 1997. *Medicinal Plants and Minerals of Uttarankhand Himalaya.* Baidyanath Ayurvedic Research Centre, Patna.

Upadhyay, R. & Chauhan, S.V.S. 2000. Ethnobotanical observations on Koya tribe of Gundala-mandal of Khammam district, Andhra Pradesh. *Ethnobotany* **12**: 93-99.

Upadhyay, R. 2003. Ethnobotanical observations on some rare and endangered herbs of Singrauli in Madhya Pradesh: 989-992. In Singh, V. & Jain, A.P. (*eds.*) Vol.-II. *Ethnobotany and Medicinal Plants of India and Nepal*. Scientific Publishers, Jodhpur, India.

Upadhyay, R. & Singh, J. 2005. Ethnomedicinal use of plants from Tikri forest of Gonda District (U.P.). *Ethnobotany* **17**: 167-170.

Upadhye, A.S. & Kulkarni, D.K. 2004. Traditional phytotherapy for insect bites among the tribals in western Maharashtra. *Ethnobotany* **16**: 59-61.

Upadhye, A.S., Vartak, V.D. & Kumbhojkar, M.S. 1994. Ethno-medico-botanical studies in Western Maharashtra, India. *Ethnobotany* 6: 25-31.

Uphof, J.C.T. 1986. *Dictionary of Economic Plants.* Second Edition. Revised and Enlarged. Stechert-Hafner Service Agency, Inc. New York.

Varghese, E. 1996. *Applied Ethnobotany A Case Study Among The Kharias of Central India.* Deep Publications, New Delhi.

Ved, D.K. 2000. *Tropical Indian Medicinal Plants Propagation Methods.* FRLHT, Bangalore.

Vedavathy, S., Mrudula, V. & Sudhakar, A. 1997. *Tribal Medicine of Chittor District A.P.* (International Development Research Centre, Canada & Sri Venkateshwara Arts College:

Venkataratnam, K. & Raju, V.R.R. 2004. Folk medicine from Gundlabrahmeswaram wild life sanctuary, Andhra Pradesh. *Ethnobotany* **15**: 33-39.

Verma, P., Khan, A.A. & Singh, K.K. 1995. Traditional phytotherapy among the Baiga tribe of Shahdol district of Madhya Pradesh, India. *Ethnobotany* **7**: 63-79.

Verma, S. & Chauhan, N.S. 2006. Studies on ethno-medico-botany of Kunihar Forest Division, district Solan (H.P.). *Ethnobotany* **18**: 160-165.

Viswanathan, M.B. 1997. Ethnobotany of the Malayalis in North Arcot District, Tamil Nadu, India. *Ethnobotany* **9**: 77-79.

Viswanathan, M.V. 2000. Edible and medicinal plants of Ladakh (J & K): 151-154. In Maheshwari, J.K. (*ed.*) *Ethnobotany and Medicinal Plants of Indian Subcontinent.* Scientific Publishers, Jodhpur, India.

Waller, G.R. & Nowacki, E.R. 1978. *Alkaloid Biology and Metabolism in Plants.* Plenum Press, New York.

Warrier, P.K., Nambiar, V.P.K. & Raman, K. 1994-1996. *Indian Medicinal Plants–A Compendium of 500 Species.* Vol.I-V. Orient Longman Limited, Hyderabad.

Watt, G. 1889-1896. *A Dictionary of Economic Products of India.* Vol.I-VI. Periodical Experts, Delhi (India).

Wolffers, I. 1990. *Role of Traditional Asian Medicine in Primary Health Care.* Free University Press, Amsterdam.

Yoganarasimhan, S.N., Subramanyam, K. & Razi, B.A. 1981. *Flora of Chikmagalur District, Karnataka, India.* International Book Distributers, Dehradun, India.

Zafar, R. & Ahmad, S. 2003. *Tylophora indica* (Burm.f.) Merill- A review: 42-48. In Singh, V. & Jain, A.P. (*eds.*) Vol.-I. *Ethnobotany and Medicinal Plants of India and Nepal.* Scientific Publishers, Jodhpur, India.

□□□

Plates

7

Plates

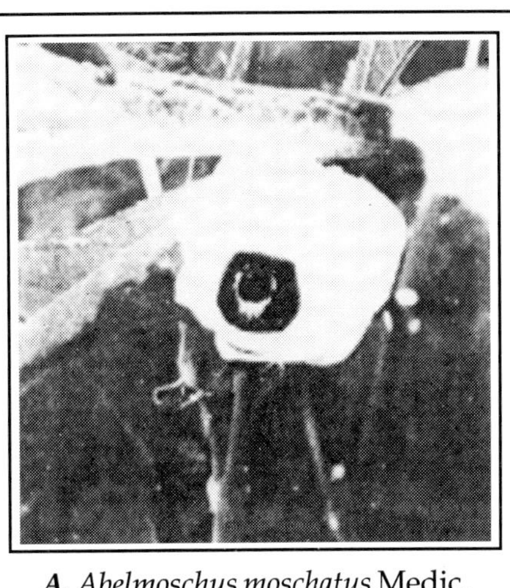

A. *Abelmoschus moschatus* Medic.

B. *Abroma augusta* Linn.

C. *Abrus precatorius* L.

D. *Acacia catechu* (Roxb.) Willd.

Pl. 1 A-D. Root Drug Plants of Ethnic India.

A. *Achyranthes aspera* L

B. *Acorus calamus* L.

C. *Aloe barbadensis* Mill.

D. *Althaea officinalis* Linn.

Pl. 2A-D. Root Drug Plants of Ethnic India.

A. *Amaranthus spinosus* L.

B. *A. viridis* L.

C. *Argemone maxicana* L.

D. *Argyreia nervosa* (Burm.f.) Boj.

Pl. 3A-D. Root Drug Plants of Ethnic India.

A. *Berginia ciliata* (Haw.) Stenb.

B. *Butea monosperma* (Lam.) Taub.

C. *Calotropis procera* (L.) Ait.

D. *Cichorium intybus* L.

Pl. 4A-D. Root Drug Plants of Ethnic India.

A. *Cissampelos prareira* L.

B. *Citrullus colocynthis* (L.) Schrad. *ex* Eckl. & Zeyh.

C. *Cleome viscosa* L.

D. *Dioscorea bulbifera* L.

Pl. 5A-D. Root Drug Plants of Ethnic India.

A. *Eclipta prostrata* L.

B. *Glycirrhiza glabra* Linn.

C. *Jatropha curcas* L.

D. *Lantana camara* Linn.

Pl. 6A-D. Root Drug Plants of Ethnic India.

A. *Mimosa pudica* L.

B. *Plantago ovata* Forssk.

C. *Plumbago zeylanica* L.

D. *Podophyllum hexandrum* Royle

Pl. 7A-D. Root Drug Plants of Ethnic India.

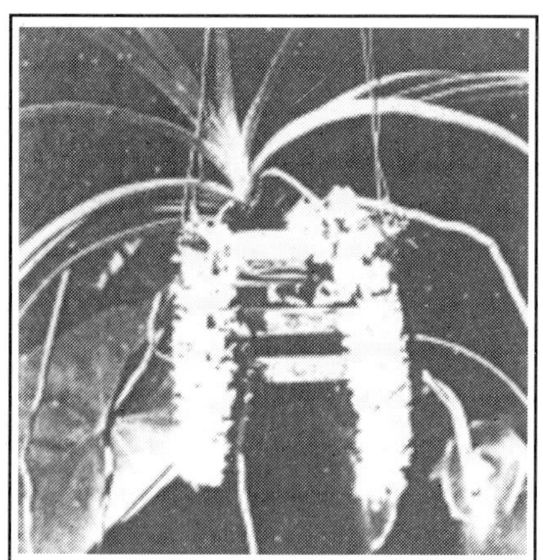

A. *Rhynchostylis retusa* (L.) Bl.

B. *Roscoea alpina* Royle

C. *Taraxacum officinale* Weber *ex* Wiggers

D. *Urena lobata* L.

Pl. 8A-D. Root Drug Plants of Ethnic India.

A. *Vernonia cinerea* (L.) Lees.

B. *Vitex negundo* L.

C. *Withania somnifera* (L.) Dunal

D. *Xanthium strumarium* L.

Pl. 9A-D. Root Drug Plants of Ethnic India.

8

Appendices

Appendix - I

Total No. of Genera and Species of Root Drug Plants Under Various Divisions of Plant Kingdom

Division	No. of Families	Genera/ Species	Total No. of Species in a Genus
Gymnosperms	1	1/1	*Ephedra* (1).
Pteridophytes	4	8/11	*Actiniopteris* (1), *Adiantum* (2), *Angiopteris* (1), *Cheilanthes* (2), *Hemionitis* (1), *Lygodium* (2), *Polypodium* (1), *Salvinia* (1).
Angiosperms	(128)	(562/905)	
Dicots	109	479/788	*Abelmoschus* (4), *Abroma* (1), *Abrus* (2), *Abutilon* (1), *Acacia* (3), *Acalypha* (2), *Acanthospermum* (1), *Acanthus* (1), *Achyranthes* (4), *Aconitum* (11), *Acrocephalus* (1), *Actinocarya* (1), *Adanosia* (1), *Adenanthera* (1), *Aegle* (1), *Aerva* (3), *Aeschynanthes* (1), *Aeschynomene* (1), *Aganosma* (1), *Ageratum* (1), *Agrimonia* (1), *Ailanthus* (1), *Ainsliaea* (1), *Ajuga* (1), *Alangium* (1), *Albizia* (1), *Alcea* (1), *Alstonia* (1), *Alternanthera* (1), *Althaea* (1), *Alysicarpus* (2), *Amaranthus* (4), *Ampelocissus* (2), *Anacyclus* (1), *Andrographis* (2), *Anemone* (3), *Angelica* (1), *Anona* (1),

Contd...

Appendix I Contd...

			Anthocephalus (1), *Antidesma* (1), *Apium* (1), *Aporusa* (1), *Aquilegia* (1), *Arctium* (2), *Ardisia* (7), *Argemone* (1), *Argyreia* (5), *Aristolochia* (4), *Armoracia* (1), *Arnebia* (3), *Artanema* (1), *Artemisia* (3), *Artocarpus* (1), *Asarum* (1), *Asclepias* (1), *Asteracantha* (1), *Aster* (2), *Astilbe* (1), *Astragalus* (3), *Atropa* (1), *Atylosia* (2), *Averrhoea* (1), *Azima* (1), *Balanites* (1), *Baliospermum* (1), *Barleria* (3), *Barringtonia* (1), *Basella* (1), *Bauhinia* (3), *Begonia* (1), *Berberis* (4), *Bergenia* (3), *Beta* (1), *Bidens* (1), *Biophytum* (1), *Bistorta* (1), *Bixa* (1), *Blepharis* (2), *Blumea* (3), *Boehmeria* (1), *Boerhaavia* (1), *Bolahaavia* (1), *Bombax* (1), *Bragantia* (1), *Brassica* (1), *Bridelia* (1), *Bupleurum* (2), *Butea* (2), *Byttneria* (1), *Cadaba* (1), *Caesalpinia* (2), *Cajanus* (1), *Callicarpa* (2), *Calotropis* (2), *Caltha* (1), *Canavalia* (2), *Cannabis* (1), *Canthium* (1), *Capparis* (3), *Capsicum* (1), *Cardiospermum* (1), *Carica* (1), *Carissa* (2), *Casearia* (3), *Cassia* (8), *Cassine* (1), *Catharanthus* (1), *Catunaregam* (1), *Cayratia* (2), *Ceiba* (1), *Celastrus* (1), *Celosia* (1), *Centaurea* (1), *Centella* (1), *Centratherum* (1), *Cephaelis* (1), *Chasalia* (2), *Chonemorpha* (1), *Chrozophora* (1), *Cichorium* (2), *Cimcifuga* (1), *Cinchona* (1), *Cinnamomum* (1), *Cirsium* (4), *Cissampelos* (1), *Cissus* (2), *Citrullus* (1), *Citrus* (2), *Clausena* (1), *Cleistanthes* (1), *Clematis* (4), *Cleome* (3), *Clerodendrum* (5), *Clitorea* (1), *Coccinia* (1), *Cocculus* (1), *Codenia* (1), *Codonopsis* (1), *Colebrookia* (1), *Coleus* (1), *Connarus* (1), *Coptis* (1), *Corallocarpus* (1), *Corbichonia* (1), *Corchorus* (2), *Cordia* (1), *Corydalis* (1), *Crataeva* (1), *Crotalaria* (5), *Croton* (2), *Cryptolepis* (1), *Cucumis* (1), *Cyananthus* (1), *Cyathula* (1), *Cyclea* (2), *Cyclista* (2), *Cynoglossum* (3), *Dalbergia* (2), *Datisca* (1), *Datura* (2), *Daucus* (1), *Delphinium* (4), *Dendrocnide* (1), *Derris* (1), *Desmodium* (9), *Desmos* (1), *Dichroa* (1), *Dichrostachya* (1), *Dictamnus* (1), *Dillenia* (1), *Diospyros* (1), *Dipsacus* (1), *Dipteracanthus* (2), *Distemon* (1), *Dregea* (1), *Dolichos* (1), *Doronicum* (1), *Drosera*

Contd...

Appendix I Contd...

			(1), *Ebermaiera* (1), *Ecbolium* (1), *Echinops* (2), *Eclipta* (1), *Ehretia* (2), *Elaeagnus* (2), *Elephantopus* (1), *Elytraria* (1), *Embelia* (2), *Emblica* (1), *Emilia* (1), *Enicostema* (1), *Eranthemum* (2), *Ervatamia* (1), *Eryngium* (1), *Erythrina* (2), *Erythroxylum* (1), *Eupatorium* (1), *Euphorbia* (5), *Evolvulus* (1), *Fagopyrum* (1), *Feronia* (1), *Ferula* (2), *Ficus* (9), *Flacourtia* (1), *Flemingia* (5), *Foeniculum* (1), *Fumaria* (1), *Gardenia* (2), *Gentiana* (2), *Geranium* (3), *Geum* (1), *Gillenia* (1), *Girardinia* (2), *Glochidion* (1), *Glossogyne* (1), *Glycosmis* (2), *Glycyrrhiza* (1), *Gmelina* (2), *Gnaphalium* (1), *Gossypium* (2), *Grangea* (1), *Grewia* (8), *Gymnema* (1), *Gymnosporia* (1), *Gynandropsis* (1), *Gynura* (1), *Hedyotis* (2), *Helenium* (1), *Helianthus* (2), *Helicteres* (1), *Heliotropium* (2), *Hemidesmus* (1), *Heracleum* (2), *Heterostemma* (1), *Hibiscus* (5), *Holarrhena* (1), *Holmskioldia* (1), *Holostemma* (1), *Homonia* (1), *Houttuynia* (1), *Humboldtia* (1), *Hybanthus* (1), *Hydrangea* (1), *Hygrophila* (1), *Hymenodictyon* (1), *Hyptis* (1), *Ichnocarpus* (1), *Impatiens* (2), *Indigofera* (5), *Inula* (3), *Ipomoea* (1), *Ixora* (2), *Janakia* (1), *Jasminum* (1), *Jatropha* (3), *Jurinea* (1), *Jurenella* (1), *Jussieua* (1), *Justicia* (2), *Knoxia* (1), *Kydia* (1), *Lablab* (1), *Lagerstroemia* (2), *Lantana* (2), *Launaea* (2), *Lavatera* (1), *Lawsonia* (1), *Leea* (6), *Lepidagathis* (3), *Leptadenia* (1), *Leucas* (2), *Litsea* (1), *Ludwigia* (1), *Luffa* (1), *Lycnis* (1), *Madhuca* (1), *Maerua* (2), *Maesa* (1), *Malachra* (1), *Mallotus* (2), *Malva* (1), *Manihot* (1), *Manilkara* (1), *Marsdenia* (2), *Martynia* (1), *Maytenus* (1), *Meconopsis* (1), *Megacarpaea* (1), *Melia* (1), *Melochia* (1), *Melothria* (3), *Memecylon* (1), *Merremia* (1), *Meyna* (1), *Mezoneuron* (1), *Michelia* (1), *Milletia* (1), *Mimosa* (4), *Mirabilis* (1), *Mitragyna* (1), *Mollugo* (1), *Momordica* (2), *Monotropa* (1), *Morina* (1), *Morinda* (2), *Moringa* (2), *Morus* (1), *Mucuna* (2), *Mukia* (1), *Mundulea* (1), *Murraya* (2), *Mussaenda* (2), *Nardostachys* (1), *Naregamia* (1), *Naringi* (1), *Nelsonia* (1), *Nelumbo* (1),

Contd...

Appendix I Contd...

			Nerium (1), *Nymphaea* (2), *Ochna* (3), *Ocimum* (1), *Olax* (1), *Onosoma*(1), *Operculina* (1), *Ophiorrhiza* (1), *Oreocnide* (1), *Oroxylum* (1), *Orthosiphon* (1), *Osbeckia* (2), *Osyris* (1), *Ougenia* (1), *Oxyspora* (1), *Paederia* (1), *Paeonia* (1), *Panax* (1), *Paracalyx* (1), *Paramignya* (1), *Parthenium* (1), *Passiflora* (1), *Pavetta* (1), *Pavonia* (1), *Pentanema* (2), *Pentapetes* (1), *Peperomia* (1), *Pergularia* (1), *Phyllanthus* (4), *Physalis* (1), *Picrorhiza* (1), *Pimpinella* (1), *Piper* (3), *Pithecellobium* (1), *Plantago* (2), *Pluchea* (2), *Plumbago* (2), *Plumeria* (2), *Podophyllum* (2), *Pogostemon* (3), *Polygala* (2), *Polygonum* (10), *Pongamia* (1), *Porana* (1), *Potentilla* (5), *Pouzolzia* (1), *Prangos* (1), *Premna* (3), *Princepia* (1), *Psidium* (1), *Pueraria* (1), *Punica* (1), *Quercus* (1), *Radermachera* (1), *Randia* (1), *Raphanus* (1), *Rauvolfia* (2), *Reinwardtia* (1), *Rheum* (2), *Rhinacanthus* (1), *Rhynchosia* (2), *Ricinus* (1), *Rivea* (2), *Rourea* (1), *Rubia* (1), *Rubus* (2), *Ruellia* (3), *Rumex* (3), *Salacia* (4), *Salvadora* (2), *Salvia* (1), *Sambucus* (3), *Sapindus* (1), *Sarcococca* (1), *Sarcostemma* (1), *Saurauia* (1), *Saussurea* (3), *Schefflera* (1), *Schrebera* (1), *Scoparia* (1), *Secamone* (1), *Securinega* (1), *Selinum* (2), *Semecarpus* (1), *Senecio* (1), *Sesamum* (1), *Sesbania* (1), *Sida* (5), *Siegesbeckia* (1), *Solanum* (8), *Solena* (1), *Solidago* (1), *Sonchus* (3), *Sonerila* (2), *Sorbaria* (1), *Soymida* (1), *Spermadictyon* (1), *Sphaeranthus* (2), *Stauranthera* (1), *Stephania* (3), *Sterculia* (1), *Stereospermum* (2), *Strycnos* (2), *Sutherlandia* (1), *Swertia* (4), *Symphorema* (1), *Tabernaemontana* (1), *Tacca* (1), *Tagetes* (1), *Tamarix* (2), *Tanacetum* (2), *Taraxacum* (1), *Tarenna* (1), *Tephrosia* (3), *Terminalia* (1), *Thalictrum* (6), *Thespepsia* (2), *Thespesia* (1), *Thevetia* (1), *Thottea* (1), *Thunbergia* (1), *Tiliacora* (1), *Tinospora* (1), *Toddalia* (1), *Tragia* (1), *Tragopogon* (2), *Trema* (1), *Trewia* (1), *Trianthema* (3), *Tribulus* (1), *Trichodesma* (1), *Tricholepis* (1), *Trichosanthes* (6), *Tridex* (1), *Triumfetta* (2), *Turraea* (1), *Tylophora* (2), *Uraria* (3), *Urena* (2),

Contd...

Appendix I Contd...

			Urtica (1), *Vaccaria* (1), *Valeriana* (2), *Ventilago* (2), *Verbascum* (1), *Vernonia* (2), *Viburnum* (1), *Vicoa* (1), *Viola* (2), *Vitex* (1), *Vitis* (1), *Wattakaka* (1), *Wedelia* (1), *Wendlandia* (1), *Withania* (1), *Woodfordia* (1), *Wrightia* (2), *Xanthium* (1), *Xanthoxylum* (1), *Xerompsis* (1), *Ximenia* (1), *Zaleya* (1), *Zizyphus* (3), *Zornia* (1).
Monocots	19	83/117	*Acorus* (1), *Agave* (1), *Agropyron* (1), *Alocasia* (3), *Aloe* (1), *Alpinia* (1), *Amorphophallus* (1), *Aneilema* (1), *Areca* (1), *Arenga* (1), *Arisaema* (3), *Asparagus* (4), *Asphodelus* (1), *Bambusa* (1), *Belamcanda* (1), *Borassus* (1), *Bromus* (1), *Caladium* (1), *Campylandra* (1), *Canna* (1), *Chlorophytum* (3), *Cocos* (1), *Colchicum* (1), *Colocasia* (1), *Commelina* (1), *Costus* (1), *Crinum* (1), *Cryptocoryne* (1), *Curculigo* (2), *Curcuma* (3), *Cyanotis* (1), *Cymbidium* (1), *Cynodon* (1), *Cyperus* (5), *Dactylorrhiza* (1), *Desmostachya* (1), *Dioscorea* (3), *Dracaena* (3), *Eichhornia* (1), *Eleusine* (1), *Eulophia* (1), *Geodorum* (1), *Globba* (1), *Gloriosa* (1), *Habenaria* (1), *Herminium* (1), *Hedychium* (4), *Heteropogon* (1), *Iris* (3), *Juncus* (1), *Kaempferia* (1), *Kyllinga* (1), *Lasia* (1), *Monochoria* (1), *Musa* (1), *Pandanus* (2), *Paris* (1), *Pentatropis* (1), *Peristylus* (1), *Phoenix* (2), *Polygonatum* (1), *Raphia* (1), *Remusata* (1), *Rhynchostylis* (1), *Roscoea* (2), *Saccharum* (1), *Saccolabium* (1), *Sansevieria* (1), *Satyrium* (1), *Scindapsus* (1), *Scirpus* (1), *Scleria* (1), *Smilacina* (1), *Smilax* (6), *Stachyphrynium* (1), *Stemona* (1), *Thysanolaena* (1), *Typha* (1), *Typhonium* (1), *Vanda* (1), *Vetiveria* (1), *Yucca* (1), *Zingiber* (2).

❑❑❑

Appendix - II

Families of Root Drug Plants

Division	Genera/ Species	Total No. of Species in a Genus
Acanthaceae	18/31	*Acanthus* (1), *Andrographis* (2), *Asteracantha* (1), *Barleria* (3), *Blepharis* (2), *Blumea* (3), *Dipteracanthus* (2), *Ebermaira* (1), *Ecbolium* (1), *Elytraria* (1), *Eranthemum* (2), *Hygrophila* (1), *Justicia* (2), *Lepidagathis* (3), *Nelsonia* (1), *Rhinacanthus* (1), *Ruellia* (3), *Thunbergia* (1).
Aizoaceae	4/6	*Corbichonia* (1), *Mollugo* (1), *Trianthema* (3), *Zaleya* (1).
Alangiaceae	1/1	*Alangium* (1).
Amarantaceae	6/14	*Achyranthes* (4), *Aerva* (3), *Alternanthera* (1), *Amaranthus* (4), *Celosia* (1), *Cyathula* (1).
Amaryllidaceae	3/4	*Agave* (1), *Crinum* (1), *Curculigo* (2).
Anacardiaceae	1/1	*Semecarpus* (1).
Anonaceae	2/2	*Anona* (1), *Desmos* (1).
Apiaceae	12/16	*Angelica* (1), *Apium* (1), *Bupleurum* (2), *Centella* (1), *Daucus* (1), *Eryngium* (1), *Ferula* (2), *Foeniculum* (1), *Heracleum* (2), *Pimpinella*, (1), *Prangos* (1), *Selinum* (2).
Apocynaceae	14/18	*Aganosma* (1), *Alstonia* (1), *Carissa* (2), *Catharanthus* (1), *Chonemorpha* (1), *Ervatamia* (1), *Holarrhena* (1), *Ichnocarpus* (1), *Nerium* (1), *Plumeria* (2), *Rauvolfia*, (2), *Tabernaemontana* (1), *Thevetia* (1), *Wrightia* (2).
Araceae	11/15	*Acorus* (1), *Alocasia* (3), *Amorphophallus* (1), *Arisaema* (3), *Caladium* (1), *Colocasia* (1), *Cryptocoryne* (1), *Lasia* (1), *Remusata* (1), *Scindapsus* (1), *Typhonium* (1).
Araliaceae	2/2	*Panax* (1), *Schefflera* (1).
Arecaceae	6/7	*Areca* (1), *Arenga* (1), *Borassus* (1), *Cocos* (1), *Phoenix* (2), *Raphia* (1).
Aristolochiaceae	4/7	*Aristolochia* (4), *Asarum* (1), *Bragantia* (1), *Thottea* (1).
Asclepiadaceae	16/20	*Asclepias* (1), *Calotropis* (2), *Cryptolepis* (1), *Dregea* (1), *Gymnema* (1), *Hemidesmus* (1), *Heterostemma* (2), *Holostemma* (1), *Janakia* (1), *Leptadenia* (1), *Marsdenia* (2), *Pergularia* (1), *Sarcostemma* (1), *Secamone* (1), *Tylophora* (2), *Wattakaka* (1).

Contd...

Appendix II Contd...

Asteraceae	47/70	*Acanthospermum* (1), *Ageratum* (1), *Ainsliaea* (1), *Anacyclus* (1), *Arctium* (2), *Artemisia* (3), *Aster* (2), *Bidens* (1), *Centaurea* (1), *Centratherum* (1), *Cichorium* (2), *Cirsium* (4), *Doronicum* (1), *Echinops* (2), *Eclipta* (1), *Elephantopus* (1), *Emilia* (1), *Eupatorium* (1), *Glossogyne* (1), *Gnaphalium* (1), *Grangea* (1), *Gynura* (1), *Helenium* (1), *Helianthus* (2), *Inula* (3), *Jurinea* (1), *Jurenella* (1), *Launaea* (2), *Parthenium* (1), *Pentanema* (2), *Pluchea* (2), *Saussurea* (3), *Senecio* (1), *Siegesbeckia* (1), *Solidago* (1), *Sonchus* (3), *Sphaeranthus* (2), *Tagetes* (1), *Tanacetum* (2), *Taraxacum* (1), *Tragopogon* (2), *Tricholepis* (1), *Tridex* (1), *Vernonia* (2), *Vicoa* (1), *Wedelia* (1), *Xanthium* (1).
Balsaminaceae	1/2	*Impatiens* (2).
Basellaceae	1/1	*Basella* (1).
Begoniaceae	1/1	*Begonia* (1).
Berberidaceae	2/6	*Berberis* (4), *Podophyllum* (2).
Bignoniaceae	3/4	*Oroxylum* (1), *Radermachera* (1), *Stereospermum* (2),
Bixaceae	1/1	*Bixa* (1).
Bombacaceae	3/3	*Adanosia* (1), *Bombax* (1), *Ceiba* (1).
Boraginaceae	9/15	*Actinocarya* (1), *Arnebia* (3), *Codenia* (1), *Cordia* (1), *Cynoglossum* (3), *Ehretia* (2), *Heliotropium* (2), *Onosoma* (1), *Trichodesma* (1).
Brassicaceae	4/4	*Armoracia* (1), *Brassica* (1), *Megacarpaea* (1), *Raphanus* (1).
Buxaceae	1/1	*Sarcococca* (1).
Cactaceae	1/1	*Datisca* (1).
Campanulaceae	2/2	*Codonopsis* (1), *Cyananthus* (1).
Cannabinaceae	1/1	*Cannabis* (1).
Cannaceae	1/1	*Canna* (1).
Capparidaceae	6/11	*Cadaba* (1), *Capparis* (3), *Cleome* (3), *Crataeva* (1), *Gynandropsis* (1), *Maerua* (2).
Caprifoliaceae	2/4	*Sambucus* (3), *Viburnum* (1).
Caricaceae	1/1	*Carica* (1).
Caryophyllaceae	2/2	*Lycnis* (1), *Vaccaria* (1).
Celastraceae	4/4	*Cassine* (1), *Celustrus* (1), *Gymnosporia* (1), *Maytenus* (1).

Contd...

Appendix II Contd...

Chenopodiaceae	1/1	*Beta* (1).
Combretaceae	1/1	*Terminalia* (1).
Commelinaceae	3/3	*Aneilema* (1), *Commelina* (1), *Cyanotis* (1).
Connaraceae	2/2	*Connarus* (1), *Rourea* (1).
Convolvulaceae	7/12	*Argyreia* (5), *Evolvulus* (1), *Ipomoea* (1), *Merremia* (1), *Operculina* (1), *Porana* (1), *Rivea* (2).
Cucurbitaceae	10/18	*Citrullus* (1), *Coccinia* (1), *Corallocarpus* (1), *Cucumis* (1), *Luffa* (1), *Melothria* (3), *Momordica* (2), *Mukia* (1), *Solena* (1), *Trichosanthes* (6).
Cyperaceae	4/8	*Cyperus* (5), *Kyllinga* (1), *Scirpus* (1), *Scleria* (1).
Dilleniaceae	2/2	*Dillenia* (1), *Saurauia* (1).
Dioscoreaceae	1/3	*Dioscorea* (3).
Dipsacaceae	2/2	*Dipsacus* (1), *Morina* (1).
Droseraceae	1/1	*Drosera* (1).
Ebenaceae	1/1	*Diospyros* (1).
Elaeagnaceae	1/2	*Elaeagnus* (2).
Erythroxylaceae	1/1	*Erythroxylum* (1).
Euphorbiaceae	20/32	*Acalypha* (2), *Antidesma* (1), *Aporusa* (1), *Baliospermum* (1), *Bridelia* (1), *Chrozophora* (1), *Cleistanthus* (1), *Croton* (2), *Emblica* (1), *Euphorbia* (5), *Glochidion* (1), *Homonia* (1), *Jatropha* (3), *Mallotus* (2), *Manihot* (1), *Phyllanthus* (4), *Ricinus* (1), *Securinega* (1), *Tragia* (1), *Trewia* (1).
Fabaceae	46/98	*Abrus* (2), *Acacia* (3), *Adenanthera* (1), *Aeschynomene* (1), *Albizia* (1), *Alysicarpus* (2), *Astragalus* (3), *Atylosia* (2), *Bauhinia* (3), *Bolahaavia* (1), *Butea* (2), *Caesalpinia* (2), *Cajanus* (1), *Canavalia* (2), *Cassia* (8), *Clitorea* (1), *Crotalaria* (5), *Cyclista* (2), *Dalbergia* (2), *Derris* (1), *Desmodium* (9), *Dichrostachys* (1), *Dolichos* (1), *Erythrina* (2), *Flemingia* (5), *Glycirrhiza* (1), *Humboldtia* (1), *Indigofera* (5), *Lablab* (1), *Mezoneuron* (1), *Millettia* (1), *Mimosa* (4), *Mucuna* (2), *Mundulea* (1), *Ougenia* (1), *Paracalyx* (1), *Pithecellobium* (1), *Pongamia* (1), *Pueraria* (1), *Rhynchosia* (2), *Sesbania* (1), *Sutherlandia* (1), *Tephrosia* (3), *Thespepsia* (2), *Uraria* (3), *Zornia* (1).
Fagaceae	1/1	*Quercus* (1).
Flacourtiaceae	2/4	*Casearia* (3), *Flacourtia* (1).

Contd...

Appendix II Contd...

Fumariaceae	2/2	*Corydalis* (1), *Fumaria* (1).
Gentianaceae	3/7	*Enicostema* (1), *Gentiana* (2), *Swertia* (4).
Geraniaceae	1/3	*Geranium* (3).
Gesneriaceae	2/2	*Aeschynanthus* (1), *Stauranthera* (1).
Gnetaceae	1/1	*Ephedra* (1).
Hippocrateaceae	1/4	*Salacia* (4).
Hydrangeaceae	1/1	*Hydrangea* (1).
Iridaceae	2/4	*Belamcanda* (1), *Iris* (3).
Juncaceae	1/1	*Juncus* (1).
Lamiaceae	10/13	*Acrocephalus* (1), *Ajuga* (1), *Colebrookia* (1), *Coleus* (1), *Hyptis* (1), *Leucas* (2), *Ocimum* (1), *Orthosiphon* (1), *Pogostemon* (3), *Salvia* (1).
Lauraceae	2/2	*Cinnamomum* (1), *Litsea* (1).
Lecythidaceae	1/1	*Barringtonia* (1).
Leeaceae	1/6	*Leea* (6).
Liliaceae	14/26	*Aloe* (1), *Asparagus* (4), *Asphodelus* (1), *Campylandra* (1), *Chlorophytum* (3), *Colchicum* (1), *Dracaena* (3), *Gloriosa* (1), *Paris* (1), *Polygonatum* (1), *Sansevieria* (1), *Smilacina* (1), *Smilax* (6), *Yucca* (1).
Linaceae	1/1	*Reinwardtia* (1).
Loganiaceae	1/1	*Strycnos* (2).
Lythraceae	3/4	*Lagerstroemia* (2), *Lawsonia* (1), *Woodfordia* (1).
Magnoliaceae	1/1	*Michelia* (1).
Malvaceae	14/27	*Abelmoschus* (4), *Abutilon* (1), *Alcea* (1), *Althaea* (1), *Gossypium* (2), *Hibiscus* (5), *Kydia* (1), *Lavatera* (1), *Malachra* (1), *Malva* (1), *Pavonia* (1), *Sida* (5), *Thespesia* (1), *Urena* (2).
Marantaceae	1/1	*Stachyphrynium* (1).
Marattiaceae	1/1	*Angiopteris* (1).
Melastomaceae	4/6	*Memecylon* (1), *Osbeckia* (2), *Oxyspora* (1), *Sonerila* (2).
Meliaceae	4/4	*Melia* (1), *Naregamia* (1), *Soymida* (1), *Turraea* (1).
Menispermaceae	6/9	*Cissampelos* (1), *Cocculus* (1), *Cyclea* (2), *Stephania* (3), *Tiliacora* (1), *Tinospora* (1).

Contd...

Appendix II Contd...

Moraceae	3/11	*Artocarpus* (1), *Ficus* (9), *Morus* (1).
Moringaceae	1/2	*Moringa* (2).
Musaceae	1/1	*Musa* (1).
Myrtaceae	1/1	*Psidium* ((1).
Myrsinaceae	3/10	*Ardisia* (7), *Embelia* (2), *Maesa* (1).
Nyctaginaceae	2/2	*Boerhaavia* (1), *Mirabilis* (1).
Nymphaeaceae	2/3	*Nelumbo* (1), *Nymphaea* (2).
Ochnaceae	1/3	*Ochna* (3).
Olacaceae	2/2	*Olax* (1), *Ximenia* (1).
Oleaceae	2/2	*Jasminum* (1), *Schrebera* (1).
Onagraceae	2/2	*Jussieua* (1), *Ludwigia* (1).
Orchidaceae	11/11	*Cymbidium* (1), *Dactylorrhiza* (1), *Eulophia* (1), *Geodorum* (1), *Habenaria* (1), *Herminium* (1), *Peristylus* (1), *Rhynchostylis* (1), *Saccolabium* (1), *Satyrium* (1), *Vanda* (1).
Oxalidaceae	2/2	*Averrhoea* (1), *Biophytum* (1).
Pandanaceae	1/2	*Pandanus* (2).
Papaveraceae	2/2	*Argemone* (1), *Meconopsis* (1)
Passifloraceae	1/1	*Passiflora* (1).
Pedaliaceae	2/2	*Martynia* (1), *Sesamum* (1).
Piperaceae	2/4	*Peperomia* (1), *Piper* (3).
Plantaginaceae	1/2	*Plantago* (2).
Plumbaginaceae	1/2	*Plumbago* (2).
Poaceae	10/10	*Agropyron* (1), *Bambusa* (1), *Bromus* (1), *Cynodon* (1), *Desmostachya* (1), *Eleusine* (1), *Heteropogon* (1), *Saccharum* (1), *Thysanolaena* (1), *Vetiveria* (1).
Polygalaceae	1/2	*Polygala* (2).
Polygonaceae	5/17	*Bistorta* (1), *Fagopyrum* (1), *Polygonum* (10), *Rheum* (2), *Rumex* (3). *Actiniopteris* (1), *Adiantum* (2), *Cheilanthes* (2), *Hemionitis* (1),
Polypodiaceae	5/7	*Polypodium* (1)
		Eichornia (1), *Monochoria* (1).
Pontederiaceae	2/2	*Punica* (1).

Contd...

Appendix II Contd...

Punicaceae	1/1	*Monotropa* (1).
Pyrolaceae	1/1	*Aconitum* (11), *Anemone* (3), *Aquilegia* (1), *Caltha* (1), *Cimcifuga*
Ranunculaceae	10/33	(1), *Clematis* (4), *Coptis* (1), *Delphinium* (4), *Paeonia* (1), *Thalictrum* (6).
		Ventilago (2), *Zizyphus* (3).
Rhamnaceae	2/5	*Agrimonia* (1), *Geum* (1), *Gillenia* (1), *Potentilla* (5), *Prinsepia* (1),
Rosaceae	7/12	*Rubus* (2), *Sorbaria* (1).
		Anthocephalus (1), *Canthium* (1), *Catunaregam* (1), *Cephaelis* (1),
Rubiaceae	24/30	*Chasalia* (2), *Cinchona* (1), *Gardenia* (2), *Hedyotis* (2), *Hymenodictyon* (1), *Ixora* (2), *Knoxia* (1), *Meyna* (1), *Mitragyna* (1), *Morinda* (2), *Mussaenda* (2), *Ophiorrhiza* (1), *Paederia* (1), *Pavetta* (1), *Randia* (1), *Rubia* (1), *Spermadictyon* (1), *Tarenna* (1), *Wendlandia* (1), *Xerompsis* (1).
Rutaceae	11/14	*Aegle* (1), *Citrus* (2), *Clausena* (1), *Dictamnus* (1), *Feronia* (1), *Glycosmis* (2), *Murraya* (2), *Naringi* (1), *Paramignya* (1), *Toddalia* (1), *Xanthoxylum* (1),
Santalaceae	1/1	*Osyris* (1).
Salvadoraceae	2/3	*Azima* (1), *Salvadora* (2).
Salviniaceae	1/1	*Salvinia* (1).
Sapindaceae	2/2	*Cardiospermum* (1), *Sapindus* (1).
Sapotaceae	2/2	*Madhuca* (1), *Manilkara* (1).
Saururaceae	1/1	*Houttuynia* (1).
Schizaeaceae	1/2	*Lygodium* (2).
Saxifragaceae	3/5	*Astilbe* (1), *Berginia* (3), *Dichroa* (1).
Scrophulariaceae	4/4	*Artanema* (1), *Picrorrhiza* (1), *Scoparia* (1), *Verbascum* (1).
Simarubaceae	1/1	*Ailanthus* (1).
Solanaceae	6/14	*Atropa* (1), *Capsicum* (1), *Datura* (2), *Physalis* (1), *Solanum* (8), *Withania* (1).
Stemonaceae	1/1	*Stemona* (1).
Sterculiaceae	6/6	*Abroma* (1), *Byttneria* (1), *Helicteres* (1), *Melochia* (1), *Pentapetes* (1), *Sterculia* (1).

Contd...

Appendix II Contd...

Taccaceae	1/1	*Tacca* (1).
Tamaricaceae	1/2	*Tamarix* (2),
Tiliaceae	3/12	*Corchorus* (2), *Grewia* (8), *Triumfetta* (2).
Typhaceae	2/2	*Pentatropis* (1), *Typha* (1).
Ulmaceae	1/1	*Trema* (1).
Urticaceae	7/8	*Boehmeria* (1), *Dendrocnide* (1), *Distemon* (1), *Girardinia* (2), *Oreocnide* (1), *Pouzolzia* (1), *Urtica* (1).
Valerianaceae	2/3	*Nardostachys* (1), *Valeriana* (2).
Verbenaceae	8/17	*Callicarpa* (2), *Clerodendrum* (5), *Gmelina* (2), *Holmskioldia* (1), *Lantana* (2), *Premna* (3), *Symphorema* (1), *Vitex* (1).
Violaeae	2/3	*Hybanthus* (1), *Viola* (2).
Vitaceae	4/7	*Ampelocissus* (2), *Cayratia* (2), *Cissus* (2), *Vitis* (1).
Zingiberaceae	8/15	*Alpinia* (1), *Costus* (1), *Curcuma* (3), *Globba* (1), *Hedychium* (4), *Kaempferia* (1), *Roscoea* (2), *Zingiber* (2).
Zygophyllaceae	2/2	*Balanites* (1), *Tribulus* (1).

□□□

Appendix - III

Predominant Indian Genera of Root Drug Plants

Name of Genus	No. of Species	Name of Genus	No. of Species
Abelmoschus	4	*Blepharis*	2
Abrus	2	*Blumea*	3
Acacia	3	*Bupleurum*	2
Acalypha	2	*Butea*	2
Achyranthes	4	*Caesalpinia*	2
Aconitum	11	*Callicarpa*	2
Adiantum	2	*Calotropis*	2
Aerva	3	*Canavalia*	2
Alocasia	3	*Capparis*	3
Alysicarpus	2	*Carissa*	2
Amaranthus	4	*Casearia*	3
Ampelocissus	2	*Cassia*	8
Andrographis	2	*Cayratia*	2
Anemone	3	*Chasalia*	2
Arctium	2	*Cheilanthes*	2
Ardisia	7	*Chlorophytum*	3
Argyreia	5	*Cichorium*	2
Arisaema	3	*Cirsium*	4
Aristolochia	4	*Cissus*	2
Arnebia	3	*Citrus*	2
Artemisia	3	*Clematis*	4
Asparagus	4	*Cleome*	3
Aster	2	*Clerodendrum*	5
Astragalus	3	*Corchorus*	2
Atylosia	2	*Crotalaria*	5
Barleria	3	*Croton*	2
Bauhinia	3	*Curculigo*	2
Berberis	4	*Curcuma*	3
Bergenia	3	*Cyclea*	2

Contd...

Cyclista	2
Cynoglossum	3
Cyperus	5
Dalbergia	2
Datura	2
Delphinium	4
Desmodium	9
Dioscorea	3
Dipteracanthus	2
Dracaena	3
Echinops	2
Ehretia	2
Elaeagnus	2
Embelia	2
Eranthemum	2
Erythrina	2
Euphorbia	5
Ferula	2
Ficus	9
Flemingia	5
Gardenia	2
Gentiana	2
Geranium	3
Girardinia	2
Glycosmis	2
Gmelina	2
Gossypium	2
Grewia	8
Hedychium	4
Hedyotis	2
Helianthus	2
Heliotropium	2
Heracleum Heterostemma	2
Hibiscus	5
Impatiens	2
Indigofera	5
Inula	3
Iris	3
Ixora	2
Jatropha	3
Justicia	2
Lagerstroemia	2
Lantana	2
Launaea	2
Leea	6
Lepidagathis	3
Lygodium	2
Maerua	2
Mallotus	2
Marsdenia	2
Melothria	3
Mimosa	4
Momordica	2
Morinda	2
Moringa	2
Mucuna	2
Murraya	2
Mussaenda	2
Nymphaea	2
Osbeckia	2
Pandanus	2
Pentanema	2
Phoenix	2
Phyllanthus	4
Piper	3
Plantago	2
Pluchea	2
Plumbago	2
Plumeria	2
Podophyllum	2

Contd...

Pogostemon	3	*Stereospermum*	2
Polygala	2	*Strycnos*	2
Potentilla	5	*Swertia*	4
Premna	3	*Tamarix*	2
Rauvolfia	2	*Tanacetum*	2
Rheum	2	*Tephrosia*	3
Rhynchosia	2	*Thalictrum*	6
Rivea	2	*Thespepsia*	2
Roscoea	2	*Tragopogon*	2
Ru	3	*Trianthema*	3
Rubus	2	*Trichosanthes*	6
Ruellia	3	*Triumfetta*	2
Salacia	4	*Tylophora*	2
Salvadora Selinum	2	*Uraria*	3
Sambucus	3	*Urena*	2
Saussurea	3	*Valeriana*	2
Sida	5	*Ventilago*	2
Smilax	6	*Vernonia*	2
Solanum	8	*Viola*	2
Sonchus	3	*Wrightia*	2
Sonerila	2	*Zingiber*	2
Sphaeranthus	2	*Zizyphus*	3
Stephania	3		

□□□

Appendix - IV

Predominant Indian Root Drug Families with Three or More Species

Families	Total No. of Species	Families	Total No. of Species
Acanthaceae	31	Flacourtiaceae	4
Aizoaceae	6	Gentianaceae	7
Amarantaceae	14	Geraniaceae	3
Amaryllidaceae	4	Hippocrateaceae	4
Apiaceae	16	Iridaceae	4
Apocynaceae	18	Lamiaceae	13
Araceae	15	Leeaceae	6
Arecaceae	7	Liliaceae	26
Aristolochiaceae	7	Lythracea	4
Asclepiadaceae	20	Malvaceae	27
Asteraceae	70	Melastomaceae	6
Berberidaceae	6	Meliaceae	4
Bignoniaceae	4	Menispermaceae	9
Bombacaceae	4	Moraceae	11
Boraginacea	15	Myrsinaceae	10
Brassicaceae	4	Nymphaeaceae	3
Capparidaceae	11	Ochnaceae	3
Caprifoliaceae	4	Orchidaceae	11
Celastraceae	4	Piperaceae	4
Commelinaceae	3	Poaceae	10
Convolvulaceae	12	Polygonaceae	17
Cucurbitaceae	18	Polypodiaceae	7
Cyperaceae	8	Ranunculaceae	33
Dioscoreaceae	3	Rhamnaceae	5
Euphorbiaceae	32	Rosaceae	12
Fabaceae	98	Rubiaceae	30

Contd...

Rutaceae	14	Tiliaceae	12
Salvadoraceae	3	Urticaceae	8
Saxifragaceae	5	Valerianaceae	3
Scrophulariaceae	4	Verbenaceae	17
Solanaceae	14	Violaceae	3
Sterculiaceae	6	Vitaceae	7
		Zingiberaceae	15

□□□

Appendix - V

Index to Botanical Names

Contd...

Contd...

Contd...

Contd...

Contd...

Contd...

Contd...

Contd...

Contd...

Contd...

Contd...

Contd...

Contd...

❑❑❑

Appendix - VI

Index to English Names

Contd...

Contd...

Contd...

Contd...

Contd...

Contd...

Contd...

Contd...

Contd...

Contd...

Contd...

Contd...

□□□

Appendix - VII

Index to Hindi Names

Contd...

Contd...

Contd...

Contd...

Contd...

Contd...

Contd...

Contd...

Contd...

Contd...

Contd...

Contd...

Contd...

Contd...

Contd...

Contd...

Contd...

Contd...

Contd...

Contd...

□□□

Appendix - VIII

Index to Sanskrit Names

Contd...

Contd...

Contd...

Contd...

Contd...

Contd...

Contd...

Contd...

Contd...

Contd...

Contd...

Contd...

Contd...

Contd...

Contd...

Contd...

Contd...

Contd...

Contd...

Contd...

□□□